The Contribution of Mathematics to School STEM Education

Judy Anderson · Katie Makar
Editors

The Contribution of Mathematics to School STEM Education

Current Understandings

Springer

Editors
Judy Anderson
Sydney School of Education
and Social Work
The University of Sydney
Camperdown, NSW, Australia

Katie Makar
School of Education
The University of Queensland
Brisbane, QLD, Australia

ISBN 978-981-97-2727-8 ISBN 978-981-97-2728-5 (eBook)
https://doi.org/10.1007/978-981-97-2728-5

This Springer imprint is published by the registered company Springer Nature Singapore Pte Ltd.
The registered company address is: 152 Beach Road, #21-01/04 Gateway East, Singapore 189721, Singapore

Contents

Editors and Contributors

About the Editors

Dr. Judy Anderson is an honorary associate professor of *Mathematics Education* in the Sydney School of Education and Social Work, at the University of Sydney. She worked at the University of Sydney for more than 18 years as a secondary mathematics education lecturer and researcher. A former secondary mathematics teacher, she lectured in secondary mathematics curriculum across pre-service and postgraduate teacher education programs and was a recipient of a faculty Excellence in Teaching Award. Judy has conducted research into middle years' students' motivation and engagement in mathematics and investigated teachers' implementation of problem solving strategies in mathematics. She was the *director of the STEM Teacher Enrichment Academy* from 2016 to 2020 where she worked with academics from the faculties of Science and Engineering to develop this innovative professional learning program for STEM teachers across the state of New South Wales (NSW), Australia. During her time as the director, the Academy reached over 1200 teachers from 220 primary and secondary school settings in NSW, Australia. By collecting data about teachers' and students' experiences of integrated STEM curriculum, Judy and her colleagues published several journal articles and chapters in books about the impact of the Academy program. She also co-edited a Springer book on *Integrated Approaches to STEM Education: An International Perspective* with Prof. Yeping Li from A&M University, Texas.

Dr. Katie Makar is a professor in *Mathematics Education* at the University of Queensland. She is the current president of the *Mathematics Education Research Group of Australasia* (MERGA) and the co-director of the *International Collaboration for Research in Statistical Reasoning, Thinking and Literacy* (SRTL). Her award-winning research focuses on teachers' learning of innovative pedagogies and children's statistical reasoning. Katie's classroom-based research collaborates with practicing teachers to seek pragmatic solutions to improving teaching and learning. Her recent nationally funded research project collaborated with Dr. Jill Fielding

(University of New England) to investigate how primary teachers initiate, build, and sustain a productive classroom culture and mathematical practices conducive to addressing complex problems that rely on mathematical evidence. Her research also studies teachers' adoption of inquiry-based practices over time, and children's informal statistical reasoning that draws on age-appropriate practices in data science education. Katie has over 100 peer-reviewed publications; the quality and impact of her highly-cited research is evidenced by multiple university and international awards. A former high school mathematics teacher for 15 years, Katie's work with pre-service and practicing primary and secondary teachers continually seeks to have an impact on improving mathematics pedagogies to better prepare students for the future.

Contributors

Jeong Su Ahn Sejong Academy of Science and Arts, Sejong-si, South Korea

Lluís Albarracín Department de Didàctica de la Matemàtica Clències, Universitat Autònoma de Barcelona, Barcelona, Spain

Judy Anderson The University of Sydney, Camperdown, Sydney, NSW, Australia

Jonas Bergman Ärlebäck Department of Mathematics, Linköping University, Linköping, Sweden

Ban Heng Choy National Institute of Education, Nanyang Technological University, Singapore, Singapore

Clodagh Cleary Mary Immaculate College, University of Limerick, Limerick, Ireland

Gabi Cooper Griffith University, Southport, Australia

Anita Crowder Illustrative Mathematics, Richmond, VA, USA

Lyn English Queensland University of Technology, Brisbane, Australia

Michelle Fitzpatrick Mary Immaculate College, University of Limerick, Limerick, Ireland

Rayna Flye Illustrative Mathematics, Richmond, VA, USA

Merrilyn Goos University of the Sunshine Coast, Sippy Downs, Australia

Annette Hilton University of Technology Sydney, Ultimo, NSW, Australia

Geoff Hilton The University of Queensland, Saint Lucia, QLD, Australia

Jihyun Hwang Mathematics Education, Korea National University of Education, Cheongju, South Korea

Nam-Hwa Kang Physics/Science Education, Korea National University of Education, Cheongju, South Korea

Takashi Kawakami Kanazawa Institute of Technology, Utsunomiya University, Utsunomiya, Japan

Mette Als Kristensen University of Southern Denmark, Odense, Denmark

Dorte Moeskær Larsen University of Southern Denmark, Odense, Denmark

Aisling M. Leavy Mary Immaculate College, University of Limerick, Limerick, Ireland

Katja Maass International Centre of STEM Education, University of Education, Freiburg, Germany

Katie Makar The University of Queensland, St Lucia, Brisbane, QLD, Australia

Daniel Clark Orey Universidade Federal de Ouro Preto, Ouro Preto, Brazil

Milton Rosa Universidade Federal de Ouro Preto, Ouro Preto, Brazil

Rob Rouse School of Education, University of Queensland, Brisbane, Australia

Akihiko Saeki Kanazawa Institute of Technology, Utsunomiya University, Utsunomiya, Japan

Lisa Steffensen Western Norway University of Applied Sciences, Bergen, Norway

Oliver Straser International Centre of STEM Education, University of Education, Freiburg, Germany

Connie Svabo University of Southern Denmark, Odense, Denmark

Juan Torralba Department of Teaching and Learning, University of Miami, Florida, USA

Kristin Umland Illustrative Mathematics, Richmond, VA, USA

Part I
Introduction

Chapter 1
The Contribution of Mathematics to an Integrated STEM Curriculum in Schools

Judy Anderson and Katie Makar

Abstract In most education jurisdictions, each school subject has a separate curriculum document presenting knowledge, skills and dispositions. Creating an integrated STEM curriculum requires the design of new, connected learning experiences for students. The challenge in creating a new integrated curriculum is to ensure subject integrity is maintained with students developing both depth and breadth of knowledge in each of the separate subject areas as well as understanding how knowledge and skills can be used across subject boundaries to solve real problems. To date, many efforts to design an integrated STEM curriculum have relegated mathematics to a subservient role in the service of the other STEM subjects. If we are to progress the field and ensure mathematics takes its rightful place as foundational in STEM education, we need to identify how mathematics could be better represented in integrated STEM education.

Keywords STEM subjects · Integrated STEM curriculum · Mathematics · Connected concepts · STEM practices · Integrated curriculum design

Introduction

The school curriculum in most countries is presented as separate disciplines, frequently referred to as a 'siloed' curriculum (World Economic Forum, 2016). Since the mathematics included in a school curriculum document is a subset of the discipline of mathematics, we refer to it, as well as science, technology and engineering, as 'subjects' (Deng, 2013). One criticism against creating an integrated STEM curriculum is the potential to create another 'subject' called 'STEM' in an already overcrowded school curriculum. Why and how school leaders and teachers

J. Anderson (✉)
The University of Sydney, Sydney, Australia
e-mail: Judy.anderson@sydney.edu.au

K. Makar
The University of Queensland, Brisbane, Australia

J. Anderson and K. Makar (eds.), *The Contribution of Mathematics to School STEM Education*, https://doi.org/10.1007/978-981-97-2728-5_1

choose to design and implement an integrated STEM curriculum for their students are important considerations to be addressed in developing a vision for their school (Anderson, 2020; Bybee, 2018; Falloon et al., 2020; Vasquez, 2015).

Subject-specific curriculum implementation is mandatory in most education jurisdictions, with documents typically describing knowledge and skills requirements for students at each grade level, with learning outcomes assessed by teachers (Darling-Hammond, 2012). Taking an integrated approach requires teachers or curriculum developers to design a new curriculum that purposely connects subjects. While there may be some similarities between the STEM subject-specific content and practices, there are differences that raise challenges for those wanting to design an integrated STEM curriculum (Roehrig et al., 2021a). One critical challenge is maintaining individual subject integrity as well as rigour and opportunities for deep learning (English, 2016a; Honey et al., 2014; Roehrig et al., 2021b). Although all STEM subjects are important in the school curriculum, there is evidence mathematics frequently plays a subservient role in integrated curriculum design (Just & Siller, 2022; Maass et al., 2019). We aim to highlight the role that mathematics can play in integrated STEM curriculum and pedagogy.

This first chapter draws on the literature to discuss similarities and differences between the STEM subjects, describes the roles of each of the STEM subjects in an integrated STEM curriculum approach, presents some approaches to integrated curriculum design, highlighting the potential for mathematics to be foregrounded in an integrated STEM curriculum in schools. We also provide background information about the aim of the book and outline the structure and organisation of the chapters to inform our current understandings of the contribution of mathematics to school STEM education.

Differences and Connections Between the STEM Subjects

When considering how the four STEM subjects might be different or how they could be connected, there are several ways this can be approached—through content and skills, through disciplinary practices, or by taking a more holistic view of the role and purpose of a STEM education. For example, in any school education jurisdiction, we could examine the separate curriculum documents such as the science and technology syllabus with the mathematics syllabus for Kindergarten to grade 6 in the Australian state of NSW to align content and skills to be connected through an integrated STEM curriculum approach (see for example Anderson et al., 2019). Alternatively, we could compare the *Next Generation Science Standards* [NGSS] connecting science and engineering practices with the practices from the *Common Core State Standards—Mathematics* [CCSS-Math] from the US to develop a connected STEM curriculum (Honey et al., 2014).

Understanding more about the history and philosophy of the STEM subjects might offer further insights into their similarities and differences (Siu, 2022). There are a range of definitions of the STEM subjects (as in Table 1.1) that indicate mathematics is used in the other three subjects while engineering and technology utilise knowledge from science and mathematics (Honey et al., 2014).

In describing the connections between the STEM subjects, Snyder (2018) drew on the etymological origins of the titles of the STEM subjects arguing science and mathematics provide foundational knowledge, engineering provides "a bridge between foundational knowledge and technological development through thoughtful planning and design" (p. 52) and technology provides the tools, materials and techniques to create artefacts. In addition, Snyder suggested it is through technology and engineering that mathematics and science can be applied to real contexts. Similar views were expressed by the expert panel in the Hallström et al. (2022) study of authenticity in integrated STEM education who also suggested technology and engineering provided a bridge between school education and real-world settings. While Snyder (2018) maintained he did not wish to create a hierarchy between the STEM subjects, his descriptions seem to imply otherwise. His characterisation suggested mathematics provided useful knowledge to be applied in contexts created through the other subjects.

Another representation of the relationships between the STEM subjects in an integrated STEM curriculum was proposed by Lyons (2018) using a Russian nesting doll (matryoshka doll) analogy. Aiming to answer the 'why' question rather than the 'how' question of connecting the subjects, Lyons proposed the dolls, like the STEM subjects, are separate though related and integral to the whole. According to Lyons, mathematics is the smallest doll, followed by science, engineering and technology in increasing size. He argues:

> Technology is often the exterior, public face of STEM, while the contributions of engineering, science and mathematics to technological products are (sometimes intentionally) less obvious. Nevertheless, technological developments are reliant on engineering, which in turn relies on science and mathematics. The contributions of these fields to a particular technology often can only be appreciated by exploring, disassembling or otherwise analysing it. Mathematics is the core of STEM, being foundational to the other fields, with each larger doll contributing an additional layer of applicability. (p. 36)

Table 1.1 Descriptions of the four disciplines

STEM subject	Definition
Science	The study of the natural world
Technology	The modification of the natural world to meet human wants and needs. The entire system of people, knowledge, processes and devices used to create and operate technological artefacts
Engineering	Both a body of knowledge (about the design and creation of human-made products) and a process for solving problems. The process is design under constraint that might include the laws of nature or science, time, money, available materials …
Mathematics	The study of patterns and relationships among quantities, numbers and space. It provides an exact language for technology, science and engineering

Adapted from NRC (2009)

While Lyons addresses some of the limitations of the analogy, particularly the notion that science is merely an application of mathematics, he also argues that mathematics is not merely a 'tool' to be used in science, technology and engineering. Lyons expresses concern his analogy reinforces the notion of a hierarchy and argues the relationship is more of a "reciprocity … with technology and engineering feeding back into science and even pure mathematics" (p. 40). In addressing this limitation, Lyons agrees that the discipline of mathematics provides much more than tools or applications in the service of the other subjects as it is central and fundamental to an integrated STEM curriculum.

These efforts to represent the relationship between the STEM subjects have much in common. They reinforce the notion that mathematics, while foundational to the other subjects, becomes applicable through the contexts presented in science, technology and/or engineering. The view that mathematics becomes contextualised through connections to the other STEM subjects was shared by science teachers in the UK. The researchers, Wong and Dillon (2019), summarised teachers views of the relationship between science and mathematics as "an asymmetry in dependence …[where] science is dependent on mathematics but the reverse is not true" (p. 783). Such views mitigated against teacher collaboration thus making the design and implementation of an integrated STEM curriculum challenging and difficult to sustain, particularly in secondary school settings (Anderson & Tully, 2021; Wong & Dillon, 2019).

Taking a more holistic view of the role and purpose of a STEM education, Siu (2022) argues

> STEM is a comprehensive awareness … [that] embodies a spirit of exploration and a way of thought, combined with mathematical thinking and scientific spirit, to organise theories through experiments and observations in the search for knowledge and further innovation (p. 1644).

This view emphasises the importance of connecting the STEM practices, not just knowledge and skills, suggesting it is through the STEM practices that an integrated STEM curriculum for schools could be designed and implemented. Some refer to a truly integrated STEM curriculum as a 'seamless amalgamation' where subject boundaries are blurred (Zhou et al., 2022) and STEM practices become a focus.

The STEM Practices

In designing an integrated STEM curriculum, some researchers argue disciplinary practices provide a useful way to integrate the subjects (McDonald, 2016; Roehrig et al., 2021b; Tytler, 2020). As one example of how engineering at the professional level is closely connected to mathematics through practices, Alpers et al. (2013) argued solving engineering problems requires a set of mathematical competencies including thinking mathematically; reasoning mathematically; posing and solving mathematical problems; modelling mathematically; representing mathematical entities; handling mathematical symbols and formalism; communicating in, with and about mathematics; and making use of aids and tools.

These connections are also evident in school-based curricula in the US where Honey et al. (2014) argued the core STEM practices were designed to support efforts to make connections across the subjects (see Table 1.2). The core practices in the NGSS link and connect science and engineering practices developed to encourage the design and delivery of more authentic and engaging learning experiences for students (Nadelson et al., 2015). Table 1.2 also contains the practices from the CCSS-Math, that were developed to reflect the work of professional mathematicians and how other STEM professionals use mathematics to conduct research and solve problems. Comparing these sets of practices highlights the more applied nature of solving engineering problems versus the more abstract nature of mathematical reasoning.

While similar language is used in both sets of practices there are differences between them and how they are used—for example 'reasoning' and 'modelling' (Develarki, 2020). Reasoning in school mathematics leads to proof but as noted by McComas and Burgin (2020), there are differences between the STEM subjects when we consider the nature of proof:

> In science, "truth" is impossible—and the philosophy of science acknowledges this. In mathematics, "truth" is self-referential, guided by axioms and similar principles. In engineering and technology, "truth" is what works and/or what works in each situation and/or what works to the level that we can afford it (p. 817).

Similarly, Develarki (2020) concludes that mathematics is concerned with precision, science with evidence-based justification and critical evaluation of evidence, and engineering with balancing multiple constraints. However, acknowledging these differences, she argues an integrated STEM curriculum has its place in the school

Table 1.2 The core STEM practices from the NGSS and the CCSS-Math curriculum

NGSS Science and Engineering	CCSS-Math
1. Asking questions (for science) and defining problems (for engineering)	1. Make sense of problems and persevere in solving them
2. Developing and using **models**	2. Reason abstractly and quantitatively
3. Planning and carrying out investigations	3. Construct viable arguments and critique the reasoning of others
4. Analysing and interpreting data	4. **Model** with mathematics
5. Using mathematics and computational thinking	5. Use appropriate tools strategically
6. Constructing explanations (for science) and designing solutions (for engineering)	6. Attend to precision
7. Engaging in **argument from evidence**	7. Look for and make use of structure
8. Obtaining, evaluating, and communicating information	8. Look for and express regularity in repeated **reasoning**

curriculum where the subjects are used to complement each other as important problems are tackled by students.

To support the implementation of the NGSS core standards and the design of integrated curriculum, a set of seven crosscutting concepts was developed (NRC, 2012) (see Table 1.3). They are also referred to as 'unifying concepts' or 'common themes' to highlight connections across subjects and could be used to design units of work. Mathematics is most evident in the description of the third, 'scale, proportion and quantity' although it could play a role in each of the others. However, integrating mathematics must be purposeful and foregrounded to highlight its contribution (Roehrig et al., 2021a).

Table 1.3 Crosscutting concepts from the framework for K-12 science education (NRC, 2012, p. 84)

Title	Description
1. Patterns	Observed patterns of forms and events guide organization and classification, and they prompt questions about relationships and the factors that influence them
2. Cause and effect: Mechanism and explanation	Events have causes, sometimes simple, sometimes multifaceted. A major activity of science is investigating and explaining causal relationships and the mechanisms by which they are mediated. Such mechanisms can then be tested across given contexts and used to predict and explain events in new contexts
3. Scale, proportion, and quantity	In considering phenomena, it is critical to recognize what is relevant at different measures of size, time, and energy and to recognize how changes in scale, proportion, or quantity affect a system's structure or performance
4. Systems and system models	Defining the system under study—specifying its boundaries and making explicit a model of that system—provides tools for understanding and testing ideas that are applicable throughout science and engineering
5. Energy and matter: Flows, cycles, and conservation	Tracking fluxes of energy and matter into, out of, and within systems helps one understand the systems' possibilities and limitations
6. Structure and function	The way in which an object or living thing is shaped and its substructure determine many of its properties and functions
7. Stability and change	For natural and built systems alike, conditions of stability and determinants of rates of change or evolution of a system are critical elements of study

Developing an Integrated STEM Curriculum for School Education

Recognising that the delivery of school curriculum as separate subjects can lead to poor attitudes and disengagement (e.g., Tytler, 2020), and drawing on Dewey's (1938) goal for students to be engaged in active learning and inquiry, integrated curriculum has been advocated by educators for some time (e.g., Beane, 1995; Toma et al., 2024). With recent heightened interest in STEM education and a focus on increasing students' employment readiness and developing their twenty-first Century skills (e.g., Office of the Chief Scientist, 2016), integrated STEM curriculum in schools has been widely promoted (Bybee, 2013; English, 2016a; Honey et al., 2014). This promotion raises questions about the approach to developing an integrated curriculum, the relationship between the individual subject curriculum outcomes and those created for the new integrated curriculum, and appropriate assessment practices (Anderson, 2020; Honey et al., 2014).

If STEM curriculum for school students has the potential to improve engagement and increase participation, then there remain many challenges in designing and implementing appropriate curricula (Anderson, 2020). One approach is to design learning experiences in which students solve 'real-world problems' requiring the connection and integration of knowledge from STEM subjects (Bybee, 2013; Tytler, 2020; Vasquez, 2015). This raises challenges for teachers when the curriculum is written as separate STEM subjects, recommended STEM pedagogies are unclear, and some teachers may be teaching out-of-field without the requisite deep content knowledge of mathematics or indeed, of any of the STEM subjects. However, given the international interest in STEM education, it is critical we examine the STEM landscape, particularly from the mathematics education perspective, capitalise on current policy advice and research initiatives, and provide practical solutions for implementation in schools (English, 2016b; Fitzallen, 2015; Li & Schoenfeld, 2019; Maass et al., 2019).

Historically, each of the STEM subjects represents a distinct body of knowledge, a unique set of practices, and potentially different pedagogical approaches in schools. How might these differences be reconciled in the design of an integrated school curriculum? Many authors acknowledge there is a place for both the teaching of subject-specific knowledge and skills as well as opportunities for students to experience an integrated curriculum (Bybee, 2013; Roehrig et al., 2021b; Tytler, 2020), so that students can connect knowledge and use their understanding to solve important questions (McComas et al., 2020). Honey et al. (2014) argue for a "strategic approach to implementing integrated STEM curriculum that accounts for the potential trade-offs in cognition and learning" (p. 5). We agree with this position and are not suggesting all the mathematics curriculum should be taught through an integrated curriculum approach. Rather, we are keen to identify the ways in which, and what aspects of the mathematics curriculum, can make a valuable contribution to STEM education and where the learning of mathematics concepts can be enhanced through an integrated curriculum approach.

We begin by describing a set of models that have been used to represent the connections between the STEM subjects—these have been employed in a range of studies to investigate how teachers in schools have designed an integrated STEM curriculum. Given the diversity of approaches, it is clear no single approach to integrated curriculum design will enable students to meet requisite outcomes from all subject-specific curricula.

Models of an Integrated STEM Curriculum

We also understand that not all integrated STEM curriculum efforts provide equal learning opportunities for all four of the STEM subjects. Many definitions of integrated STEM education argue for the purposeful integration of knowledge and understanding from at least two of the subject areas (e.g., Roehrig et al., 2021a). The level of integration of the content and practices in the design of a STEM curriculum could determine whether there is genuine curriculum integration and whether the design is intentional, planned and purposeful. The level of connection between the four subjects has been represented in various ways. For example, Vasquez (2015, p. 13) described an "inclined plane of STEM integration" beginning with disciplinary connections to multidisciplinary, to interdisciplinary to transdisciplinary that represent increasing levels of connection (Fig. 1.1).

To avoid judgement that the most successful integrated STEM curriculum design must be transdisciplinary with all four subjects seamlessly connected, Rennie et al. (2018), proposed six approaches to integration that were successful in their

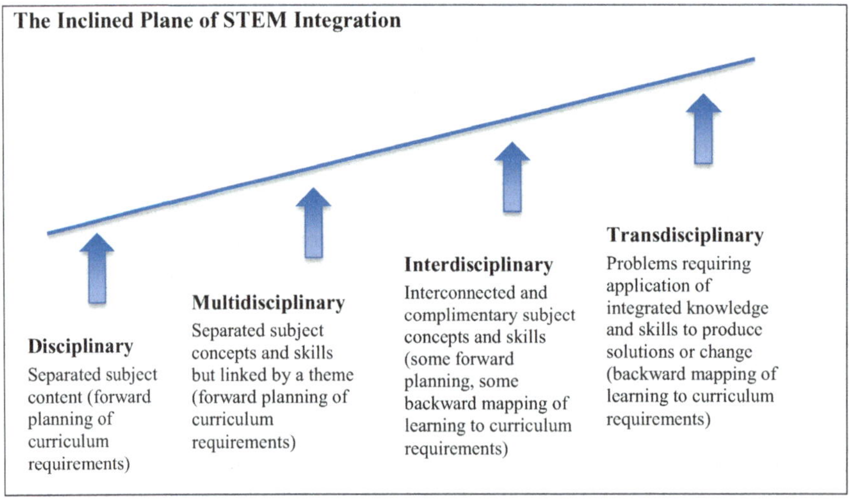

Fig. 1.1 A continuum of STEM integration. Adapted from Vasquez (2015, p. 13)

research project schools. The approaches were synchronised, thematic, project-based, cross-curricular, school-specialised and community focused programs. Interestingly, Rennie et al. did categorise the approaches according to their level of connection by suggesting synchronised, thematic and school-specialised programs tended to be multidisciplinary because teaching still occurred within the separate subjects. Whereas project-based, cross-curricular and community-focused programs were more interdisciplinary with blurring of subject boundaries. According to Rennie et al., the only approach that had the potential to be transdisciplinary was the community-focused program because "meaning-making was focused on the issue, not on the content of the contributing subjects" (p. 97).

Another model enabling the translation of integrated STEM into classroom practice was recently proposed by Toma et al. (2024) involving a two-phase procedure (knowledge development and knowledge application) where students first engage in scientific practices, then engineering practices to apply knowledge to the design of a technological solution for the phenomenon under study, and finally use mathematics and technology through data collection and analysis of results. They argue this didactic model "uses socio-constructivist principles to establish intentional and explicit connections between STEM disciplines via scientific inquiry, engineering design, and computational thinking practices" (p. 1) and that integrated STEM should be considered an educational paradigm "rooted in a long-standing tradition of curricula integration" (p. 6) rather than as a teaching approach. While the model proposed by Toma et al. (2024) presents one approach to integrated curriculum design, it privileges science and relegates mathematics to a 'tool', namely for the collection and interpretation of data during scientific practices.

Teacher-Designed Integrated STEM Curriculum

Creating the right culture for collaborative teams in schools to design integrated STEM tasks, lessons and units of work, particularly in secondary settings, requires support and commitment from school leadership (Anderson & Tully, 2021; Geiger et al., 2023). The need to address constraints related to 'coverage' of curriculum, subject expertise, team composition, timetabling, class allocations, and assessment can have major impact on the function of a school and require full support from senior school leadership as well as subject leaders, STEM team leaders, and STEM teachers (Anderson, 2020). Principals and leaders need to have an appropriate vision of STEM learning environments with teaching and learning practices that use realistic problems, connecting the knowledge and practices of the STEM subjects (Falloon et al., 2020).

Designing an integrated STEM curriculum is challenging and different researchers recommend different approaches—starting with a STEM subject or starting with a STEM problem. Based on a review of 91 publications into STEM education,

Falloon et al. (2020) developed a detailed framework for STEM curriculum integration that recognises "different [subject] entry points, curriculum designs and pedagogical strategies for school programs". Starting with one of the STEM subjects is a view shared by other researchers and curriculum designers (e.g., Bybee, 2013; Roehrig et al., 2021a). To support integrated curriculum design, Falloon et al. (2020) established a set of concepts (in italics) and themes that form the basis of the framework:

- *STEM capabilities, skills and dispositions*—cognitive, creative thinking, personal capabilities, dispositions and attitudes;
- *STEM curriculum and pedagogy*—curriculum and learning design, pedagogy, working processes, design thinking;
- *STEM discipline knowledge*—disciplinarity—priority and positioning of discipline knowledge (orientation);
- *STEM literacy*—the goal and purpose of STEM education. (p. 371)

The framework has at its core skills and dispositions that include discipline knowledge, creativity, higher order thinking, risk taking, and a growth mindset. The skills and dispositions relate to knowledge, engagement, endeavour and awareness. Falloon et al. (2020) argue the centrality of skills and dispositions should be the focus of planning and teaching. Surrounding the core are the four STEM subject areas presented in a different order at each orientation of the framework suggesting any subject can be used as an entry point or emphasis for STEM curriculum. In addition, the framework identifies a further range of contexts or starting points for STEM curriculum design including "wants, opportunities, needs or problems" (p. 377). These contexts are underpinned by the work processes of "partnerships, knowledge exchange, networking, stakeholder engagement, teamwork, collaboration" (p. 377). Overall, the framework is designed to be flexible and as such, incorporates a range of curriculum approaches and pedagogies. Importantly, Falloon et al. (2020) note the pedagogical and curriculum approaches included in the framework "will vary according to the nature and stage of STEM projects, and priority given to outcomes such as discipline knowledge and skill and competency development" (p. 377). The inclusion of "design thinking" reflects its importance as judged by its frequency in the reviewed literature informing the design of the framework.

Another approach to designing an integrated STEM curriculum is to begin with a problem, a driving question, or an engineering design challenge (Beane, 1995; Bybee, 2013; English, 2016a; Rennie et al., 2018; Roehrig et al., 2021a). Some advocate a project-based learning approach, particularly if there is a local issue providing a strong community connection that impacts students' lives (Anderson et al., 2019; Hobbs et al., 2018; Rennie et al., 2018). Engineering design challenges can provide a rich context for the integration of the STEM subjects and these challenges have been popular in school-designed integrated STEM curriculum units (e.g., Anderson et al., 2019). According to Roehrig et al. (2021b), an engineering design challenge requiring meeting a client's needs provides "a motivating storyline for students … when explicit connections to the client are included throughout the curriculum" (p. 17). However, the mathematics required to solve local issues can be below grade level expectations or merely a tool to be applied to address a scientific or engineering problem. The issue remains as to approaches to lift the level of engagement with mathematics in integrated STEM curriculum, and to foreground its usefulness.

Situating Mathematics in STEM Education

Designing quality integrated STEM problems can be a real challenge for teachers, particularly problems that allow for individual disciplinary integrity to be retained (Fitzallen, 2015). Mathematics can suffer from pseudo-realistic problems that ask students to suppress their broader knowledge of the world and risk alienating students traditionally under-represented in the discipline (Palm, 2007; Rampal & Makar, 2012). Researchers have observed that many of the 'real-world' problems available may not require the use of deep knowledge from each of the STEM subjects, something that has become increasingly apparent for mathematics (English, 2016b).

Although there is an appearance that science, technology, engineering, and mathematics are equally represented by the STEM acronym, evidence has suggested that most STEM problems focus on concepts in science and/or engineering. The concern that mathematics has become the 'silent M' in STEM or incidental in integrated STEM problems has had increasing attention (Baldinger et al., 2020; Maass et al, 2019; Shaughnessy, 2013); it has been clear that the perception of mathematics remains as only a means of computation or a set of tools for the other disciplines (Li & Schoenfeld, 2019).

Yet mathematicians seek for education to consider mathematics as consisting of argumentation, structures and tools, and creative design of heuristics (Tietze, 1994). Using mathematics as evidence to a claim, underpinned by reasoning, is an example of argumentation in mathematics (Fielding-Wells, 2014; Makar et al., 2015). This broadens the scope of mathematics in integrated STEM as a means of evidence to justify and explain the outcome of STEM-based activities (Maass et al., 2019). Mathematical structures refer to recognisable patterns that support transfer across contexts (Mulligan et al., 2022; Tytler et al., 2021). For example, at a foundational level, one observes and recognises when relationships are additive in nature and when they are proportional, allowing for efficient methods for describing and modelling relationships. Mathematical tools, most commonly recognised in STEM through algorithms and measurement, also include representations and visualisations which enhance interpretation. Finally, the creativity of mathematics for designing heuristics or analogous scenarios promotes exploration into unfamiliar problems to develop and debate hypotheses that capture potential relationships. Researchers have demonstrated ways that mathematics contributes to integrated STEM in rich, authentic ways, as each of the authors in this book illustrate.

Background Information to the Development of the Book

Our Aim

Our aim for this book is to publish recent STEM education research conducted by mathematics education researchers and their collaborators that highlight the important and pivotal role of mathematics in school STEM education. By inviting researchers who have been working in this field for some time, we plan to discuss the opportunities for highlighting mathematics in school STEM education and collect evidence of integrated curriculum approaches to STEM education (where mathematical concepts can be learnt through integrated tasks). These examples can serve to challenge the idea that the role of mathematics is merely a 'servant' to the other STEM subjects. We also want to highlight the contribution that mathematics can make to the understanding and practices of the other STEM subjects. We believe there is a need to problematize the teaching and learning of mathematics if it is to be better connected to the other STEM subjects and to improve the teaching and learning of integrated STEM curriculum in schools.

For consideration by researchers are questions such as: how do we find the balance between teaching deep knowledge in the individual subjects and teaching how disciplines can be connected and integrated, particularly when addressing real-world problems? What would be the role of each of the STEM subjects in an integrated STEM curriculum in schools? In considering what mathematics might contribute to the STEM education agenda, we draw on advice from Maass et al. (2019), who advocated.

> ... how mathematics might be advanced through three interdisciplinary approaches:
>
> 1. twenty-first century skills;
> 2. mathematical modelling; and
> 3. education for responsible citizenship. (p. 869)

Collecting examples from studies in different international contexts of how researchers and teachers have successfully designed classroom learning experiences that connect the STEM subjects and foreground mathematics will provide useful advice to those working in teacher professional learning (Ärlebäck & Albarracín, 2019). We also aim to propose how we might build this field of research by highlighting potential new questions for research.

Defining STEM Education

When reporting research in this field, one challenge involves the interpretation of the term 'STEM education'. We define 'STEM education' as an approach using an **integrated curriculum** (where mathematics is connected to at least one other of the

STEM subjects), allowing students to draw on knowledge from more than one subject to solve problems. That means approaches in schools connect the **knowledge and practices** from the mathematics curriculum to at least one of the science, technology or engineering curriculum areas—acknowledging global differences in the ways curriculum is written and available for teachers. By practices, we include identifying, appreciating and persevering to solve problems; reasoning; critiquing the reasoning of others and constructing viable alternative arguments; modelling; strategically using tools and procedures; and looking for, making sense of, and generalising the use of structure. These practices are often encompassed in a list of twenty-firstcentury skills that variously include communicating, problem solving and reasoning, critical and creative thinking, and collaborative learning (Care et al., 2017).

It is our intention that this book will fill a void in the current research literature on the role of mathematics in STEM education, providing evidence of the possibilities for designing integrated STEM curriculum and highlighting current understandings of the role of mathematics in school STEM education. For researchers, it will identify and elaborate gaps to encourage further exploration in this field. For education systems personnel and policy makers, it will highlight successful approaches and supports required, for leaders and teachers in schools, it will provide implications for practice and advice into approaches to the design and implementation of integrated STEM curriculum.

Structure and Organisation of the Book

Based on our reading of the literature and advice from Maass et al. (2019), the book has been organised into six parts:

1. Introduction
2. The mathematics in integrated STEM tasks
3. Mathematics and the development of twenty-first century STEM skills
4. Mathematical modelling and the development of STEM competencies
5. Mathematics and the development of responsible citizenship
6. A reflection.

The introductory part includes this chapter summarising key aspects of the literature related to the role of the STEM subjects and potential approaches to foregrounding mathematics in teacher-designed integrated STEM curriculum for school students. The second part includes four chapters describing ways to design an 'authentic' integrated STEM curriculum that foregrounds mathematics, approaches to analysing the mathematics in integrated STEM tasks, including examples of mathematically-rich integrated STEM problems and advice for teachers about designing similar problems, tasks and lessons. The third part describes the role of mathematics in developing twenty-first century STEM skills, the challenges and possibilities for mathematics learning within integrated STEM contexts, one example of a purposefully designed problem-based integrated STEM curriculum, and the potential of using a school makerspace to develop students' proportional reasoning capabilities. The fourth part includes advice about mathematical modelling in integrated STEM curriculum, and provides examples of students' classroom experiences using mathematical modelling to solve integrated STEM problems. The fifth part examines a range of approaches for the use of mathematics in an integrated STEM curriculum to develop responsible citizenship in school education including through social justice perspectives and ethnomodelling, and highlights the need to support teachers in designing and implementing an integrated STEM curriculum. The final part includes a reflection of the chapters in this book addressing five themes including interdisciplinary models; students' experiences; teacher knowledge and learning; classrooms and tasks; and policies and leadership.

References

Alpers, B. A., Demlova, M., Fant, C., Gustafsson, T., Lawson, D., Mustoe, L., et al. (2013). *A framework for mathematics curricula in engineering education: A report of the mathematics working group*. European Society for Engineering Education (SEFI).

Anderson, J. (2020). The STEM education phenomenon and its impact on school curriculum. *Curriculum Perspectives, 40*(2), 217–225.

Anderson, J., & Tully, D. (2021). Factors which sustain integrated STEM curriculum approaches in secondary school settings. In D. Anderson, M. Milner-Bolotin, R. Santos, & S. Petrina (Eds.), *Proceedings of the 6th International STEM in Education Conference (STEM 2021)* (pp. 20–26). University of British Columbia. https://doi.org/10.14288/1.0402129

Anderson, J., Wilson, K., Tully, D., & Way, J. (2019). Can we build the wind powered car again? Students' and teachers' responses to a new integrated STEM curriculum. *Journal of Research in STEM Education, 5*(1), 20–39. https://doi.org/10.51355/jstem.2019.61

Ärlebäck, J. B., & Albarracín, L. (2019). The use and potential of Fermi problems in the STEM disciplines to support the development of twenty-first century competencies. *ZDM, 51*, 979–990. https://doi.org/10.1007/s11858-019-01075-3

Baldinger, E. E., Staats, S., Covington Clarkson, L. M., Contreras Gullickson, E., Norman, F., & Akoto, B. (2020). A review of conceptions of secondary mathematics in integrated STEM education; Returning the voice to the silent M. In J. Anderson & Y. Li (Eds.), *Integrated approaches to STEM education: An international perspective* (pp. 67–90). Springer Nature.

Beane, J. A. (1995). Curriculum integration and the disciplines of knowledge. *Phi Delta Kappan, 76*(8), 616–622.

Bybee, R. W. (2013). *The case for STEM education: Challenges and opportunities*. National Science Teachers Association Press.

Bybee, R. W. (2018). *STEM education: Now more than ever*. National Science Teachers Association.

Care, E., Griffin, P., & Wilson, M. (Eds.). (2017). *Assessment and teaching of 21st century skills: Research and applications*. Springer.

Darling-Hammond, L. (2012). Policy frameworks for new assessments. In P. Griffin, B. McGaw, & E. Care (Eds.), *Assessment and teaching of 21st Century skills* (pp. 301–340). Springer. https://doi.org/10.1007/978-94-007-2324-5_6

Deng, Z. (2013). School subjects and academic disciplines: The differences. In A. Luke, A. Woods, & K. Weir (Eds.), *Curriculum, syllabus design, and equity: A primer and model* (pp. 40–53). Routledge.

Develaki, M. (2020). Comparing crosscutting practices in STEM disciplines: Modelling and reasoning in mathematics, science and engineering. *Science & Education, 29*, 949–979. https://doi.org/10.1007/s11191-020-00147-1
Dewey, J. (1938). *Experience and education.* Collier Macmillan.
English, L. D. (2016a). STEM education K-12: Perspectives on integration. *International Journal of STEM Education, 3*(3), 1–8.
English, L. D. (2016b). Advancing mathematics education research within a STEM environment. In K. Makar, S. Dole, J. Visnovska, M. Goos, A. Bennison, & K. Fry (Eds.), *Research in mathematics education in Australasia 2012–2015* (pp. 353–371). Springer Nature.
Falloon, G. W., Hatzigianni, M., Bower, M., Forbes, A., & Stevenson, M. (2020). Understanding K-12 STEM education: A Framework for developing STEM literacy. *Journal of Science Education and Technology, 29*, 369–385. https://doi.org/10.1007/s10956-020-09823-x
Fielding-Wells, J. (2014). *Developing argumentation in mathematics: The role of evidence and context.* Unpublished doctoral dissertation. The University of Queensland.
Fitzallen, N. (2015). STEM education: What does mathematics have to offer? In M. Marshman, V. Geiger, & A. Bennison (Eds.), *Mathematics education in the Margins.* Proceedings of the 38th annual conference of the Mathematics Education Research Group of Australasia, Sunshine Coast, June 28-July 2 (pp. 237–244). MERGA. Retrieved from http://www.merga.net.au/documents/RP2015-22.pdf
Geiger, V., Beswick, K., Fraser, S., & Holland-Twining, B. (2023). A model for principals' STEM leadership capability. *British Educational Research Journal, 49*(5), 900–924. https://doi.org/10.1002/berj.3873
Hallström, J., Norström, P., & Schönborn, K. J. (2022). International views of authenticity in integrated STEM education. In D. Gill, J. Tuff, T. Kennedy, S. Pendergast & S. Jamil (Eds.), *PATT on the edge: Technology, innovation and education. Conference proceedings of Pupils' Attitudes Towards Technology* (pp. 394–399). Memorial University of Newfoundland.
Hobbs, L., Clark, J. C., & Plant, B. (2018). Successful students—STEM program: Teacher learning through a multifaceted vision for STEM education. In R. Jorgensen & K. Larkin (Eds.), *STEM education in the junior secondary: The state of play* (pp. 133–168). Springer Nature.
Honey, M., Pearson, G., & Schweingruber, H. (2014). *STEM integration in K-12 education: Status, prospects and an agenda for research.* National Academies Press.
Just, J., & Siller, H. S. (2022). The role of mathematics in STEM secondary classrooms: A systematic literature review. *Education Sciences, 12*(629), 1–18. https://doi.org/10.3390/educsci12090629
Li, Y., & Schoenfeld, A. H. (2019). Problematizing teaching and learning mathematics as "given" in STEM education. *International Journal of STEM Education, 6*, 44. https://doi.org/10.1186/s40594-019-0197-9
Lyons, T. (2018). Helping students make sense of STEM. *Teaching Science, 64*(3), 37–43. https://doi.org/10.3316/informit.841550904446404
Maass, K., Geiger, V., Romero Ariza, M., & Goos, M. (2019). The role of mathematics in interdisciplinary STEM education. *ZDM, 51*(6), 869–884.
Makar, K., Bakker, A., & Ben-Zvi, D. (2015). Scaffolding norms of argumentation-based inquiry in a primary mathematics classroom. *ZDM, 47*(7), 1107–1120. https://doi.org/10.1007/s11858-015-0732-1
McComas, W. F., & Burgin, S. R. (2020). A critique of "STEM" education: Revolution-in-the-making, passing fad, or instructional imperative? *Science & Education, 29*, 805–829. https://doi.org/10.1007/s11191-020-00138-2
McDonald, C. V. (2016). STEM education: A review of the contribution of the disciplines of science, technology, engineering and mathematics. *Science Education International, 27*(4), 530–569.
Mulligan, J., Tytler, R., Prain, V., White, P., Xu, L., & Kirk, M. (2022). The role of mathematics learning in the interdisciplinary mathematics and science (IMS) project. In N. Fitzallen, C. Murphy, V. Hatisaru, & N. Maher (Eds.), *Mathematical confluences and journeys* (Proceedings of the 44th Annual Conference of the Mathematics Education Research Group of Australasia, July 3–7) (pp. 410–417). MERGA.

Nadelson, L. S., Seifert, A. L., & Hendricks, J. K. (2015). *Are we preparting the next generation? K-12 teacher knowledge and engagement in teaching core STEM practices.* Paper presented at the 112nd American Society for Engineering Education Annual conference and exposition, June, Seattle, WA.

National Research Council (NRC). (2009). *Learning science in informal environments: People, places, and pursuits.* National Academies Press.

National Research Council. (2012). *A framework for K-12 science education: Practices, crosscutting concepts, and core ideas.* The National Academies Press. https://doi.org/10.17226/13165

Office of the Chief Scientist. (2016). *Australia's STEM workforce: Science, technology, engineering and mathematics.* Commonwealth of Australia.

Palm, T. (2007). Features and impact of the authenticity of applied mathematical school tasks. In W. Blum, P. L. Galbraith, H.-W. Henn, & M. Niss (Eds.), *Modelling and applications in mathematics education: The 14th ICMI Study* (pp. 201–208). Springer. https://doi.org/10.1007/978-0-387-29822-1_20

Rampal, A., & Makar, K. (2012). *Embedding authenticity and cultural relevance in the primary mathematics curriculum.* Paper presented at the 12th International Congress on Mathematics Education (ICME-12), Seoul, Korea. ICME.

Rennie, L., Venville, G., & Wallace, J. (2018). Making STEM curriculum useful, relevant, and motivating for students. In R. Jorgensen & K. Larkin (Eds.), *STEM education in the junior secondary: The state of play* (pp. 91–110). Springer.

Roehrig, G. H., Dare, E. A., Ellis, J. A., & Ring-Whalen, E. (2021a). Beyond the basics: A detailed conceptual framework of integrated STEM. *Disciplinary and Interdisciplinary Science Education Research, 3*(1), 1–18. https://doi.org/10.1186/s43031-021-00041-y

Roehrig, G. H., Dare, E. A., Ring-Whalen, E., & Wieselmann, J. R. (2021b). Understanding coherence and integration in integrated STEM curriculum. *International Journal of STEM Education, 8*(2), 1–21. https://doi.org/10.1186/s40594-020-00259-8

Shaughnessy, M. (2013). Mathematics in a STEM context. *Mathematics Teaching in the Middle School, 18*(6), 324.

Siu, M. K. (2022). The role of M (mathematical worlds) in HPM (history and pedagogy of mathematics) and in STEM (science, technology, engineering, mathematics). *ZDM—Mathematics Education, 54*, 1643–1655. https://doi.org/10.1007/s11858-022-01375-1

Snyder, M. (2018). A century of perspectives that influenced the consideration of technology as a critical component of STEM education in the United States. *The Journal of Technology Studies, 44*(2), 42–57.

Tietze, U.-P. (1994). Mathematical curricula and the underlying goals. In Biehler et al. (Eds.), *Didactics of mathematics as a scientific discipline* (pp. 41–53). Kluwer.

Toma, R. B., Yánez-Pérez, I., & Meneses-Villagrá, J. A. (2024). Towards a socio-constructivist didactic model for integrated STEM education. *Interchange.* https://doi.org/10.1007/s10780-024-09513-2

Tytler, R. (2020). STEM education for the 21st century. In J. Anderson & Y. Li (Eds.), *Integrated approaches to STEM education: An international perspective* (pp. 21–44). Springer Nature.

Tytler, R., Mulligan, J., Prain, V., White, P., Xu, L., & Kirk, M. (2021). An interdisciplinary approach to primary school mathematics and science learning. *International Journal of Science Education, 43*(12), 1926–1949. https://doi.org/10.1080/09500693.2021.1946727

Vasquez, J. (2015). STEM: Beyond the acronym. *Educational Leadership, 72*(4), 10–15.

Wong, V., & Dillon, J. (2019). 'Voodoo maths', asymmetric dependency and maths blame: Why collaboration between school science and mathematics teachers is rare. *International Journal of Science Education, 41*(6), 782–802. https://doi.org/10.1080/09500693.2019.1579945

World Economic Forum. (2016). *The future of jobs: Employment, skills and workforce strategy for fourth industrial revolution.* Executive summary (pp. 1–8). Retrieved from www3.weforum.org/docs/WEF.

Zhou, D., Gomez, R., Wright, N., Rittenbruch, M., & Davis, J. (2022). A design-led conceptual framework for developing school integrated STEM programs: The Australian context. *International Journal of Technology and Design Education, 32*, 383–411. https://doi.org/10.1007/s10798-020-09619-5

Part II
The Mathematics in Integrated STEM Tasks

Chapter 2
Approaches to Analysing the Mathematics in Integrated STEM Tasks

Judy Anderson

Abstract Integrated curriculum in schools is not new but more recently, an integrated science, technology, engineering and mathematics (STEM) curriculum has been promoted, particularly because of the need for more STEM workforce participation. However, for students to be engaged and interested enough to pursue STEM study and careers, they need authentic STEM learning opportunities throughout their schooling experience. Authentic STEM learning requires the use of real-world scenarios where knowledge, skills and practices from the STEM subjects can be used to solve engaging and challenging problems. Since mathematics is foundational to all STEM subjects, the use of mathematics to solve such problems should be evident. One issue involved in designing an authentic integrated STEM curriculum involves using mathematics that challenges students at the appropriate grade level requiring more than basic counting or measuring skills. Mathematics should be more than just a tool, with learning opportunities for students that demonstrate the power of mathematics to model, reason, generalise and predict.

Keywords Authentic STEM integration · Problem solving · Interdisciplinary tasks · Teachers' perceptions · STEM practices

Current Understandings of Mathematics in Integrated Curriculum

Designing an integrated curriculum requires purposefully connecting school subjects to provide students with a more engaging and holistic learning experience (Beane, 1997). Given the promotion of integrated STEM curriculum for school students, the design task for teachers is to connect knowledge, skills and practices from science, technology, engineering and mathematics. Thus, teachers need knowledge of each of the subject areas and an understanding of how best to design integrated STEM

J. Anderson (✉)
The University of Sydney, Sydney, Australia
e-mail: Judy.anderson@sydney.edu.au

J. Anderson and K. Makar (eds.), *The Contribution of Mathematics to School STEM Education*, https://doi.org/10.1007/978-981-97-2728-5_2

tasks that allow students opportunities to select and use their STEM knowledge as appropriate (Anderson & Tully, 2024; Reinholz et al., 2018). Although some argue integrated STEM curriculum should include tasks that use all four of the STEM subjects, many researchers accept that connecting at least two subjects is sufficient (Delahunty et al., 2021; Nadelsen & Seifert, 2017). However, because mathematics is considered foundational to the other STEM subjects (Honey et al., 2014; Lyons, 2018; Martín-Páez et al., 2019; Snyder, 2018), I use the term 'authentic' integrated STEM curriculum if mathematics is evident and foregrounded in STEM tasks, lessons and units of work for students (Roehrig et al., 2021).

The evidence to date suggests examples of an integrated STEM curriculum where mathematics is foregrounded are rare. Since several studies have found mathematics has been under-represented in integrated STEM tasks and frequently used as a 'tool' (Baldinger et al, 2020; English, 2016; Fitzallen, 2015; Goos et al., 2023; Maass et al., 2019; McLure et al., 2022; Stohlmann, 2018), the editors of this book wanted to include a section presenting recent research investigating this issue. Hence, we aimed to identify studies that examined integrated STEM curriculum that included knowledge, skills and practices from mathematics as one of the subjects and the challenges associated with including mathematics as an integral component of an integrated STEM curriculum (Goos et al., 2023; Maass et al., 2019). Given the foundational nature of mathematics, we would have anticipated mathematics would play a more prominent role in integrated STEM tasks, lessons and units of work (Anderson et al., 2020).

This chapter discusses some of the challenges associated with designing and implementing authentic integrated STEM curriculum. Research from several studies that investigated the extent of mathematics integration in integrated STEM tasks is presented, and an overview of the three additional chapters in this section of the book are outlined.

Challenges to Designing and Implementing an Authentic Integrated STEM Curriculum

Several challenges have been identified that are potential barriers to authentic integrated STEM curriculum design and implementation. This section considers the challenges of establishing authenticity, defining and creating quality interdisciplinary STEM tasks, and addressing teachers' and students' perceptions of STEM and mathematics.

Establishing Authenticity in Integrated STEM Curriculum

Authenticity is a well-used term in education that is frequently ill defined. Some have advocated authentic education requires using real-life applications (e.g., Rule, 2006). Hallström et al., (2023, p. 3) suggested authenticity involves "students' participation in practices and activities of [STEM] professionals" although they point out that students may not view the activities as authentic. To address the students' perspective, Shaffer and Resnick (1999) argued authentic learning should be personally meaningful to students. Using a Delphi study with 33 STEM experts, Hallström et al. (2023) identified significant elements of an authentic STEM education including problem solving through modelling or engineering design, technology and engineering underpinning activities, and collaborative teaching to maintain subject integrity. Study participants suggested the definition of authentic STEM education may vary depending on the education level of the students although details about the alternatives at different grade levels were not presented.

While the Delphi study provided advice about the use of real-world scenarios, given the importance of mathematics in the school curriculum and the need for engineering and technology to rely on mathematics, I argue for the prominence of mathematics in authentic integrated STEM curriculum. Having a similar perspective, Baldinger et al. (2020) discuss STEM task authenticity where real-world tasks are used to "resolve mathematical questions beyond merely being an opportunity to practice mathematical routines" (p. 84). However, like Siemon et al. (2019) and Way et al. (2024), I agree that not all mathematics concepts should necessarily be taught through an integrated STEM curriculum approach. The challenge is to identify aspects of the mathematics curriculum that would benefit from solving problems in real-world contexts and that also require the application of knowledge, skills and practices from the other STEM subjects (Baldinger et al., 2020). However, Goos et al. (2023) propose we are still unclear about how to design 'good' interdisciplinary STEM problems or tasks, particularly those that feature mathematics.

Defining and Creating Quality Interdisciplinary STEM Tasks

This commentary about 'good' interdisciplinary tasks and STEM task authenticity suggests a key challenge to progress our understanding of an interdisciplinary STEM curriculum is to determine the salient features of a good interdisciplinary task. In their article, Reinholz et al. (2018) explored the similarities and differences between good disciplinary and good interdisciplinary tasks. They determined good disciplinary tasks were easier to design than good interdisciplinary tasks, even when cross-disciplinary teams worked collaboratively on the design. They concluded good problems needed to have real-world connections, reinforcement of conceptual understanding, a low floor and high ceiling, multiple solution paths, and opportunities to build dispositions of professionals in the discipline. According to Reinholz et al.

(2018), within a *discipline*, a good problem helps students learn about concepts, practices, procedures and dispositions. A good *interdisciplinary* problem helps students learn these four dimensions across multiple disciplines, typically using "a contextual intersection that is interesting and relevant to multiple disciplines" (p. 48) and generate knowledge in multiple disciplines—this can be achieved through crosscutting concepts that are big conceptual ideas such as patterns, systems and conservation. In the Reinholz et al. (2018) study, participants suggested two possible interdisciplinary tasks involving climate modelling and losing fat during dieting. The two questions proposed for students were:

1. What effect would a 10% decrease in the amount of carbon dioxide in the Earth's atmosphere have on the oceans and their ecosystem? (p. 48)
2. If you were to lose five pounds of fat, where does it go? (p. 49)

Answering both questions requires substantial knowledge from science but also mathematical modelling through computer programming. However, use of such problems does require understanding of multiple disciplines if teachers are to support their students. At the very least, they require teams of teachers working collaboratively on task and curriculum design. That level of collaboration can be a challenge in secondary school contexts (Anderson, 2020).

Many real-world challenges aimed at managing and finding solutions to climate change, biosecurity and pandemics, require significant input from mathematicians and those working in the mathematical sciences (AMSI, 2013; Burkhardt et al., 2024; Dhersin et al., 2022). The 17 sustainable development goals (SDGs) designed by the United Nations (UN, 2015) provide a useful platform to create potential integrated STEM tasks for school students. See for example, Oxfam (2019) where suggestions are provided for embedding the SDGs across the curriculum including for science, technology and mathematics as well as other subject areas. However, early research examining teachers' competencies for embedding SDGs in mathematics lessons suggests more professional learning support is required to develop teachers' knowledge and pedagogies (Su et al., 2023). Finding tasks suitable for school students that address global environmental issues and are accessible for the students' age and stage of learning can be challenging, particularly tasks that require the use of mathematics beyond counting, tallying, measuring, or the use of basic statistics. As Baldinger et al. (2020) argue,

> … if mathematics learning outcomes are identified as central alongside learning outcomes in other STEM disciplines, then mathematical practices of proof and justification, creating and analysing representations, and generalisation may be visible in integrated STEM curricula (p. 71).

It appears from several studies that designing integrated STEM curriculum in schools is frequently led by teachers invested in creating richer learning opportunities in science and/or technology/engineering education rather than mathematics education (Anderson et al., 2023; Baldinger et al., 2020; Dare et al., 2021; Just & Siller, 2022; Tytler et al., 2023). Several studies have noted science or engineering design are frequently the starting point of integrated curriculum design (Anderson et al., 2023; Dare et al., 2021; McLure et al., 2022; Wang et al., 2011). While Falloon et al. (2020) advocated any STEM subject could be a starting point for integrated STEM curriculum design, this appears to be rarely the case for mathematics.

Addressing Teachers' and Students' Perceptions of STEM and Mathematics

Further challenges to designing and implementing an authentic integrated STEM curriculum in schools have been identified in several studies. Margot and Kettler (2019) reviewed research into teachers' perceptions of STEM integration with teachers reporting the barriers included pedagogical challenges, curriculum challenges, structural challenges, concerns about students, concerns about assessments, and lack of teacher support. The teachers in the Anderson and Tully (2024) study reported similar concerns and advocated for more resourcing including increased professional learning opportunities and sharing of inspirational STEM project ideas. Both secondary and primary school teachers requested greater access to a range of technologies with support to build their knowledge and skills. In addition, the secondary teachers wanted to learn more about the 'other' STEM subject areas they were not qualified to teach, particularly mathematics.

Teachers perceptions of the purposes of an integrated STEM curriculum vary based on their level of knowledge of the individual STEM subjects. Wang et al. (2011) found teachers with different STEM subject matter expertise had distinctive perceptions about STEM integration that led to dissimilar emphases in their STEM lessons. In their study, Wang and colleagues reported the science teacher thought problem solving was the main purpose of STEM integration and built her lessons around problem-solving processes while the mathematics teacher thought the main purpose of STEM was to bring a real-world context to learning mathematics and that mathematics was a tool for solving a STEM problem. The mathematics teacher was unconvinced he could meet mathematics content requirements through STEM integrated lessons. The engineering teacher in the study focused on students developing skills such as independence and team work through an engineering design project. Given these differences, all three teachers were positive about STEM integration believing it increased students' confidence, willingness to engage in unfamiliar tasks, and interest in STEM careers. Although the three teachers in this study were from the same school and worked collaboratively to design the STEM lessons, they each delivered the lessons separately. This raises one of the issues with designing and implementing an integrated STEM curriculum in secondary settings about who designs and who delivers the curriculum (Baldinger et al., 2020).

If mathematics is to have a more prominent role in integrated STEM curriculum, addressing teachers' and students' perceptions about mathematics will be important (Anderson et al., 2020; Dare et al., 2021; Goos et al., 2023; Little, 2019). Mathematics is frequently perceived to be:

- more abstract than other subjects (Swanson & Williams, 2014);
- less useful than other subjects (Watt et al., 2017);
- a set of rules and procedures to be applied when appropriate (Nardi & Steward, 2003);
- a set of tools and practices (Dare et al., 2021);
- more challenging to learn than other subjects hence creating a barrier for some students (Dare et al., 2021);
- more advantageous for boys (along with the other STEM subjects) (Wieselmann et al., 2020); and
- difficult to apply to problem-solving contexts unless students already have relevant content knowledge (Nadelsen & Seifert, 2017; Wang et al., 2011).

Using suitable interdisciplinary problems that foreground mathematics and are accessible to students is necessary to address such perceptions. The next section considers several studies into the role of mathematics in integrated STEM tasks.

Examining the Mathematics in Integrated STEM Tasks

Several studies have investigated the focus of integrated STEM projects revealing a predominance of science and engineering compared to technology or mathematics (e.g., McLure et al., 2022). However, others have sought to identify studies where mathematics was more evident (Baldinger et al., 2020; Just & Siller, 2022; Stohlmann, 2018, 2020a, 2020b). In early years' education settings, some researchers recommend the use of a different set of practices that help to focus learning experiences on mathematics (e.g., Lowrie et al., 2017).

Of the 35 research articles meeting their inclusion criteria, McLure et al. (2022) reported science content and skills, particularly physics, were evident in 33 of the studies and 31 also had an engineering focus. Although 20 of these articles reported mathematics was integrated into the project, it was less evident than the other subjects. This may have been because almost half of the projects were implemented in either a science or an engineering class rather than in a mathematics or a multidisciplinary class. Twenty-three used an engineering design challenge as the main project although several were teacher guided with limited opportunities for students to independently and creatively engage in engineering design. Twelve of the design challenges involved a design brief, some of which considered the students' contexts and interests. Only 11 of the projects involved teachers as designers of the curriculum and only two gave students agency in choosing the problem to be studied. It appears from these articles that most were designed and lead by the researchers suggesting limited opportunities for teachers to become curriculum designers engaged in creating tasks suitable for their students and that take students' life experiences and interests into account. Even fewer articles reported giving students a choice about solving a problem of interest to them suggesting the students may not have considered the STEM experiences authentic.

One literature review investigating the focus of integrated STEM curriculum sought to identify "research accounts of mathematically rich, secondary level STEM integration" (Baldinger et al., 2020, p. 71). Of the original 4072 papers identified in the search, only 32 articles were determined to include tasks that were 'mathematically rich'. Of these, 19 studies occurred in classroom settings with most occurring in engineering or technology spaces. Seven occurred in science classrooms and only two in mathematics classrooms. Twenty articles blended mathematics with other subjects with only three positioning mathematics as the primary subject. A wide range of mathematical content was required to solve the problems from algebra, number, geometry, and probability and statistics, and a range of mathematical practices were evident—25 addressed communication, 19 representation, 12 generalisation, and 10 proof/justification. Encouragingly 11 studies attended to conceptual understanding as well as procedural fluency and mathematical practices. Finally, the Baldinger et al. study (2020) identified four common themes in the reviewed articles—frequent attention to communication, task authenticity, inquiry, and learning in informal spaces. Along with well-designed tasks, each of these themes provided opportunities for rich mathematics learning in an integrated STEM curriculum. The authors argued mathematics teachers must be involved in designing and implementing tasks in secondary school contexts to ensure better representation of mathematics in integrated STEM curriculum.

Another researcher, Stohlmann (2018, 2020a, 2020b), has published several articles about the mathematics evident in integrated STEM curriculum. In his review of integrated steM approaches between 2008 and 2018 where mathematics was the focus of studies, Stohlmann (2018) categorised 21 papers into stages of schooling—elementary, middle school, and high school—resulting in the proposal of an s.t.e.m. (support, topics, emphasis, and mathematical content) model summarising his findings (see Table 2.1). The 'support' component identifies positive student outcomes from an integrated STEM curriculum. These outcomes include the development of 21st competencies and mathematical modelling as recommended by Maass et al. (2019). Typical 'topics' in integrated STEM curriculum involved investigating global and environmental issues as recommended by Bybee (2018) and Maass et al. (2019) through their notion of developing responsible citizenship (see Part 4 of this book for more details). The 'emphasis' for integrated curriculum implementation typically involved engineering design challenges, mathematical modelling, or game based or open-ended mathematics integrated with technology.

Regarding 'mathematical content', most studies focused on measurement, data analysis, geometry and linear and quadratic equations. Stohlman argued the connections between the STEM subjects must be focused, meaningful and relevant to students and should build on natural connections. Further, he proposed "for integrated STEM, it is essential that the focus is on mathematics through content integration while being supported and enhanced by science, technology, and/or engineering" (Stohlman, 2018, p. 311). A proposal I have referred to as 'authentic' integrated STEM curriculum. In offering advice to secondary mathematics teachers who are tasked with implementing integrated steM, Stohlman (2020a) advocated for using

Table 2.1 The s.t.e.m model for integrated steM research and future work

<table>
<tr><td colspan="3">Support</td></tr>
<tr><td colspan="3">• Develops mathematical understanding through different representations
• Develops mathematical modelling abilities
• Develops twenty-first century competencies including collaboration and communication
• Increases students' confidence, effort and engagement
• Increases students' interests in careers in STEM fields</td></tr>
<tr><td colspan="3">Topics</td></tr>
<tr><td colspan="3">• Global issues
• Environmental issues
• Engineering design</td></tr>
<tr><td colspan="3">Emphasis</td></tr>
<tr><td colspan="3">• Engineering design challenges
• Mathematical modelling
• Game based or open-ended mathematics integrated with technology</td></tr>
<tr><td colspan="3">Mathematical content</td></tr>
<tr><td>Elementary grades</td><td>Middle school</td><td>High school</td></tr>
<tr><td>• Organising, structuring and representing
• Informal inference
• Aggregating quantities
• Informal measures</td><td>• Proportional reasoning
• Area and perimeter
• Linear equations
• Geometry
• Data collection and analysis</td><td>• Measurement
• Data analysis
• Quadratics
• Complex numbers
• Linear functions
• Probability</td></tr>
</table>

Adapted from Stohlmann (2018, p. 315)

open-ended problems using the three emphases in his model in Table 2.1—engineering design challenges, mathematical modelling and mathematics integrated with technology. He also highlighted the importance of using grade level mathematics, problems that have a real-world or social relevance based on students' interests, and developing suitable integrated assessment strategies.

Building on his 2018 review, Stohlman (2020b) conducted a further review of integrated STEM curriculum in kindergarten (K) to fifth grade. For K-2, there were 24 articles and for 3rd to 5th grade, 61 articles published between 2010 and 2020 that met selection criteria. His detailed analyses indicated the use of a variety of contexts with force and motion a popular science topic. In most cases, the curriculum included the engineering design process with students using a variety of materials. Regarding mathematics, data collection and analysis, and measurement were the most popular topics, typically used as an application of knowledge and not to develop new knowledge.

Using a different perspective to Stohlman (2018), Larkin, Lowrie and colleagues (2017, 2018, 2022, 2023) advocated for a focus on practices rather than content, particularly in early years' education. For example, Lowrie et al. (2017) identified a set of STEM practices including: (a) spatial reasoning, (b) location and arrangement, (c) patterns and relationships, (d) problem solving and inquiry, (e) design and making, and (f) understanding change. One could argue the mathematics in these practices is far more evident than in the more general disciplinary practices for science, engineering or technology. The practices are considered an application of an idea, value or method, that relate to a real-world context and are enacted through participation and engagement. Lowrie et al. (2018) proposed their *STEM Practices Framework* "represents a qualitative shift in purpose from the content bound traditions of science, technology, engineering and mathematics education towards developing a greater capacity to use practices in diverse contexts" (p. 1).

While there have been other reviews of STEM task design and focus, the research noted in this section demonstrates the diversity in views about the integration of mathematics in integrated STEM curriculum. As noted earlier, many suggest the extent of the integration of mathematics into STEM is inadequate if all reviews are considered although there have been examples of where mathematics has been foregrounded. I now turn to the chapters in this volume to seek further input into approaches to integrate mathematics into STEM curriculum.

New Contributions to Our Understanding of Mathematics in Integrated STEM Tasks

In this section of the book, the editors have included three chapters by authors who have recently investigated the mathematics in integrated STEM tasks and programs. Each has taken a slightly different approach but together they offer useful advice to both researchers and practitioners. While all three of these chapters do explore the mathematics in integrated STEM tasks, two chapters (Hwang et al., and Larsen et al.) could have been included in other sections of the book. Hwang et al. also examined twenty-first Century skills and/or mathematical modelling and Larsen et al. also investigated productive citizenship. However, the editors have decided to include them in this section as they offer advice about authentic integrated STEM task design to foreground mathematics.

Using a framework of examining the 'centricity' of STEM tasks, the types of connections made between mathematics and the other STEM subjects, and the 'faces' of mathematics presented, Choy and Cooper (2024) reviewed 12 studies that satisfied their selection criteria. This is a new approach to reviewing the quality of mathematics integration offering additional insights into authentic integrated curriculum task design. By examining the tasks described in each of the 12 studies, the authors determined:

- mathematics was the least well integrated of the subject areas (only 5 of 12 had mathematics as a focus;
- most tasks were problem-centric, focused on a science or engineering context;
- most tasks used mathematics as a tool with few where mathematics was used as a 'creative medium' or a 'way of knowing' (Devlin, 2000); and

- future task design should consider technology/mathematics and engineering/ mathematics connections.

Based on the promotion of integrated STEAM curriculum in South Korea, Hwang et al. (2024) examined 81 teacher-designed programs to determine the design principles and practical implications of the programs that were framed using mathematical modelling and creativity. Their analysis framework investigated mathematics content, subject connections, cultural competence, digital literacy, collaboration and creativity, in each of the tasks designed for students at three levels of schooling—elementary, middle school, and high school. At the elementary level, geometry and at the high school level, calculus were favoured mathematics content areas, with statistics and probability evident across all grade levels. At all grade levels, art was the main subject connected to mathematics followed by science and engineering. A focus on communication was more evident at the high school level and creativity was more evident at the other levels. Of note was that few programs addressed local or global issues.

In their study, Hwang et al. (2024) noted only about 10% of the STEAM programs were classified as mathematics centric (referred to as M-STEAM) even though many mathematics teachers and educators were keen to develop further M-STEAM programs. They were surprised by the limited emphasis on digital literacy, and the limited evidence of the promotion of creativity and mathematical modelling during the design of programs. They suggested the teachers from subject areas other than mathematics who were engaged in the design of programs were inclined to view mathematics as a tool. Aiming for a more holistic education, the promotion of STEAM education in South Korea has provided opportunities to enhance the learning and teaching of mathematics through contexts, and through developing students' twenty-first century skills.

The final chapter in this section involves examining the roles of mathematics in 26 STEM activities, also developed and designed by teachers for primary, middle school and lower secondary students in Denmark. Larsen et al. (2024) also explored how the mathematics in STEM tasks could be used for students to understand the role of mathematics in society and develop responsible citizenship. Based on designing activities that allowed students to have autonomy and develop critical thinking skills, mathematics and science teachers participated in 'laboratories' where they collaboratively worked to design tasks. Using a framework on the role of mathematics in STEM (Kristensen et al., 2024) that proposes the main role of mathematics is either a tool or a goal or potentially both, the authors analysed the 26 tasks to identify the mathematics used, links to other STEM subjects, and opportunities to develop responsible citizenship. The analysis suggested:

- teachers were successful in embedding mathematics in the STEM activities with the role of mathematics varying within any one task from a tool to a goal;
- teachers could use an appropriate context (e.g., sustainability) to enable student opportunities to critically explore important social issues;

- tasks within the activities tended to separate the use of mathematics and the engagement with social issues suggesting students may not have connected the two; and
- teachers needed support and guidance from researchers and education leaders to ensure activities met required principles and guidelines.

The three chapters highlight the scarcity of STEM tasks that focus on mathematics unless programs are established that aim to support teachers in the design and delivery of such tasks. For the tasks that did have a mathematics focus, they offer useful approaches to analysing the mathematics in STEM programs, strategies that could be used in targeted professional learning programs. Although they identified problem-solving tasks involving a context that required knowledge and understanding from another STEM subject or the arts, they highlighted the prevalence of using mathematics as a tool. If the professional learning programs include teachers of the other STEM subjects, it seems necessary to have an important discussion about the ways mathematics can be foregrounded and the types of tasks that might afford such opportunities. The missing voice in these chapters is that of the student. While Larsen et al. (2024) noted the students may not have seen the connections in the tasks they investigated, this appears to be an important component for future research.

References

Anderson, J. (2020). The STEM education phenomenon and its impact on school curriculum. *Curriculum Perspectives, 40*(2), 217–223.

Anderson, J., English, L., Fitzallen, N., & Symons, D. (2020). The contribution of mathematics education researchers to the current STEM education agenda. In J. Way, J. Anderson, C. Attard, J. Bobis, K. Cartwright, & H. McMaster (Eds.), *Research in mathematics education in Australasia, 2015–2019* (pp. 27–57). Springer.

Anderson, J., & Tully, D. (2024). Creating new STEM opportunities in schools: Teachers as curriculum designers. In Y. Li, Z. Zeng & N. Song (Eds.), *Disciplinary and interdisciplinary education in STEM: Changes and innovations* (pp. 265–294). Springer.

Anderson, J., Tytler, R., & Williams, G. (2023). Supporting curriculum innovation in integrated STEM for secondary teachers in Australia. *Research in Integrated STEM Education, 1*(1), 30–59. https://doi.org/10.1163/27726673-00101001

Australian Mathematical Sciences Institute. (2013). *In review: Maths of planet earth Australia.* AMSI.

Baldinger, E. E., Staats, S., Covington Clarkson, L. M., Contreras Gullickson, E., Norman, F., & Akoto, B. (2020). A review of conceptions of secondary mathematics in integrated STEM education; Returning the voice to the silent M. In J. Anderson & Y. Li (Eds.), *Integrated approaches to STEM education: An international perspective* (pp. 67–90). Springer Nature.

Beane, J. (1997). *Curriculum integration. Designing the core of democratic education.* Teachers' College Press.

Burkhardt, H., Pead, D., & Stacey, K. (2024). *Learning and teaching for mathematical literacy: Making mathematics useful for everyone.* Routledge.

Bybee, R. W. (2018). *STEM education: Now more than ever.* National Science Teachers Association.

Choy, B. H., & Cooper, G. (2024). Unpacking the 'M' in integrated STEM tasks: A systematic review. In J. Anderson & K. Makar (Eds.), *The contribution of mathematics to school STEM education: Current understandings*. Springer.

Dare, E. A., Keratithamkul, K., Hiwatig, B. M., & Li, F. (2021). Beyond content: The role of STEM disciplines, real-world problems, 21st century skills, and STEM careers within science teachers' conceptions of integrated STEM education. *Education Sciences, 11*(737), 1–22. https://doi.org/10.3390/educsci11110737

Delahunty, T., Prendergast, M., & Ní, R. M. (2021). Teachers' perspectives on achieving an integrated curricular model of primary STEM education in Ireland: Authentic or utopian ideology? *Frontiers in Education, 6*(666608), 1–13. https://doi.org/10.3389/feduc.2021.666608

Devlin, K. (2000). The four faces of mathematics. In M. J. Burke & F. R. Curcio (Eds.), *Learning mathematics for a new century: 2000 yearbook of the National Council of Teachers of Mathematics* (pp. 16–27). NCTM.

Dhersin, J. S., Kaper, H., Ddifon, W., Roberts, F., Rousseau, C., & Zielger, G. M. (Eds.). (2022). *Mathematics for action: Supporting science-based decision-making*. UNESCO.

English, L. D. (2016). STEM education K-12: Perspectives on integration. *International Journal of STEM Education, 3*(1), 1–8. https://doi.org/10.1186/s40594-016-0036-1

Falloon, G. W., Hatzigianni, M., Bower, M., Forbes, A., & Stevenson, M. (2020). Understanding K-12 STEM education: A framework for developing STEM literacy. *Journal of Science Education and Technology, 29*, 369–385. https://doi.org/10.1007/s10956-020-09823-x

Fitzallen, N. (2015). STEM education: What does mathematics have to offer? In M. Marshman, V. Geiger, & A. Bennison (Eds.), *Mathematics Education in the Margins. Proceedings of the 38th annual conference of the Mathematics Education Research Group of Australasia, Sunshine Coast, June 28–July 2* (pp. 237–244). MERGA. Retrieved from http://www.merga.net.au/documents/RP2015-22.pdf

Goos, M., Carreira, S., & Namukasa, I. K. (2023). Mathematics and interdisciplinary STEM education: Recent developments and future directions. *ZDM—Mathematics Education, 55*, 1199–1217. https://doi.org/10.1007/s11858-023-01533-z

Hallström, J., Norström, P., & Schönborn, K. J. (2023). International views of authenticity in integrated STEM education. In D. Gill, J. Tuff, T. Kennedy, S. Pendergast & S. Jamil (Eds.), *PATT on the edge: Technology, innovation and education. Conference proceedings of Pupils' Attitudes Towards Technology* (pp. 394–399). Memorial University of Newfoundland.

Honey, M., Pearson, G., & Schweingruber, H. (2014). *STEM integration in K-12 education: Status, prospects and an agenda for research*. National Academies Press.

Hwang, J., Ahn, J. S., & Kang, N. H. (2024). Analysis of teacher designed integrated STEAM programs with a focus on mathematics. In J. Anderson & K. Makar (Eds.), *The contribution of mathematics to school STEM education: Current understandings*. Springer.

Just, J., & Siller, H. S. (2022). The role of mathematics in STEM secondary classrooms: A systematic literature review. *Education Sciences, 12*(629), 1–18. https://doi.org/10.3390/educsci12090629

Kristensen, M. A., Larsen, D. M., Seidelin, L., & Svabo, C. (2024). The role of mathematics in STEM activities: Syntheses and framework from a literature review. *International Journal of Education in Mathematics, Science and Technology, 12*(2), 418–431. https://doi.org/10.46328/ijemst.3357

Larkin, K. & Lowrie, T. (2022). *STEM Education in the early years: Thinking about tomorrow*. Springer Monographs.

Larkin, K., & Lowrie, T. (2023). Teaching approaches for STEM integration in pre- and primary school: A systematic qualitative literature review. *International Journal of Science and Mathematics Education, 21*(Suppl 1), 511–539. https://doi.org/10.1007/s10763-023-10362-1

Larsen, D. M., Kristensen, M. A., & Svabo, C. (2024). Expanding the horizons of mathematics in STEM: Diverse mathematical roles in STEM activities. In J. Anderson & K. Makar (Eds.), *The contribution of mathematics to school STEM education: Current understandings*. Springer.

Little, J. (2019). Connecting mathematics with science to enhance student achievement—A position paper. In G. Hine, S. Blackley, & A. Cooke (Eds.), *Proceedings of the 42nd annual conference of the Mathematics Education Research Group of Australasia* (pp. 452–459). MERGA.

Lowrie, T., Leonard, S., & Fitzgerald, R. (2018). STEM practices: A translational framework for large-scale STEM education design. *EDeR, Educational Design Research, 2*(1), 1–20. https://doi.org/10.15460/eder.2.1.1243

Lowrie, T., Logan, T., & Larkin, K. (2017). The "math" in STEM practices: The role of spatial reasoning in the early years. In A. Downton, S. Livy, & J. Hall (Eds.), *Proceedings of the 40th annual conference of the Mathematics Education Research Group of Australasia* (pp. 625–628). MERGA.

Lyons, T. (2018). Helping students make sense of STEM. *Teaching Science, 64*(3), 37–43. https://doi.org/10.3316/informit.841550904446404

Maass, K., Geiger, V., Romero Ariza, M., & Goos, M. (2019). The role of mathematics in interdisciplinary STEM education. *ZDM, 51*(6), 869–884.

Margot, K. C., & Kettler, T. (2019). Teachers' perceptions of STEM integration and education: A systematic literature review. *International Journal of STEM Education, 6*(2), 1–16. https://doi.org/10.1186/s40594-018-0151-2

Martín-Páez, T., Aguilera, D., Perales-Palacios, F. J., & Vílchez-González, J. M. (2019). What are we talking about when we talk about STEM education? A Review of Literature. *Science Education, 103*(4), 799–822.

McLure, F. I., Tang, K., & Williams, P. J. (2022). What do integrated STEM projects look like in middle school and high school classrooms? A systematic literature review of empirical studies of iSTEM projects. *International Journal of STEM Education, 9*(73), 1–15. https://doi.org/10.1186/s40594-022-00390-8

Nadelson, L. S., & Seifert, A. L. (2017). Integrated STEM defined: Contexts, challenges and the future. *The Journal of Educational Research, 110*(3), 221–223. https://doi.org/10.1080/00220671.2017.1289775

Nardi, E., & Steward, S. (2003). Is mathematics T.I.R.E.D? A profile of quiet disaffection in the secondary mathematics classroom. *British Educational Research Journal, 29*(3), 345–367. https://doi.org/10.1080/01411920301000155999

Oxfam. (2019). *The sustainable development goals: A guide for teachers*. Oxfam.

Reinholz, D., Slominski, T., French, T. A., Pazicni, S., Rasmussen, C., & McCoy, B. (2018). Good problems within and across disciplines. *Journal of Research in STEM Education, 4*(1), 37–53. https://doi.org/10.51355/jstem.2018.34

Roehrig, G. H., Dare, E. A., Ellis, J. A., & Ring-Whalen, E. (2021). Beyond the basics: A detailed conceptual framework of integrated STEM. *Disciplinary and Interdisciplinary Science Education Research, 3*(1), 1–18. https://doi.org/10.1186/s43031-021-00041-y

Rule, A. C. (2006). Editorial: The components of authentic learning. *Journal of Authentic Learning, 3*(1), 1–10.

Shaffer, D. W., & Resnick, M. (1999). 'Thick' authenticity: New media and authentic learning. *Journal of Interactive Learning Research, 10*(2), 195–215.

Siemon, D., Banks, N., & Prasad, S. (2019). Multiplicative thinking: A necessary STEM foundation. In T. Barkatsas, N. Carr, & G. Cooper (Eds.), *STEM education: An emerging field of inquiry* (pp. 74–100). Koninklijke Brill.

Snyder, M. (2018). A century of perspectives that influenced the consideration of technology as a critical component of STEM education in the United States. *The Journal of Technology Studies, 44*(2), 42–57.

Stohlmann, M. (2018). A vision for future work to focus on the "M" in integrated STEM. *School Science and Mathematics, 118*, 310–319. https://doi.org/10.1111/ssm.12301

Stohlmann, M. (2020a). STEM integration for high school mathematics teachers. *Journal of Research in STEM Education, 6*(1), 52–63. https://doi.org/10.51355/jstem.2020.71

Stohlmann, M. (2020b). Analysing K-5th grade integrated STEM curriculum implemented since 2010. *International Journal of Education Humanities and Social Science, 3*(4), 137–166.

Su, C. S., Díaz-Levicoy, D., Vásquez, C., & Hsu, C. C. (2023). Sustainable development education for training and service teachers teaching mathematics: A systematic review. *Sustainability, 15*, 8435. https://doi.org/10.3390/su15108435

Swanson, D., & Williams, J. (2014). Making abstract mathematics concrete in and out of school. *Educational Studies in Mathematics, 86*, 193–209. https://doi.org/10.1007/s10649-014-9536-4

Tytler, R., Anderson, J., & Williams, G. (2023). Exploring a framework for integrated STEM: Challenges and benefits for promoting engagement in learning mathematics. *ZDM—Mathematics Education, 55*, 1299–1313. https://doi.org/10.1007/s11858-023-01519-x

United Nations. (2015). *Transforming our world: The 2030 agenda for sustainable development.* UN

Wang, H., Moore, T. J., Roehrig, G. H., & Park, M. S. (2011). STEM integration: Teacher perceptions and practice. *Journal of Pre-College Engineering Education Research, 1*(2), 1–13. https://doi.org/10.5703/1288284314636

Watt, H. M. G., Hyde, J. S., Petersen, J., Morris, Z. A., Rozek, C. S., & Harackiewicz, J. M. (2017). Mathematics—a critical filter for STEM-related career choices? A longitudinal examination among Australian and US adolescents. *Sex Roles, 77*(3), 254–271. https://doi.org/10.1007/s11199-016-0711-1.

Way, J., Cartwright, K., Bobis, J., Anderson, J., McMaster, H., & Attard, C. (2024). Yesterday, today and tomorrow in mathematics education research. In C. Mesiti, W.T. Seah, B. Kaur, C. Pearn, A. Jones, S. Cameron, E. Every & K. Copping (Eds.), *Research in mathematics education in Australasia 2020–2023*. Springer.

Wieselmann, J. R., Dare, E. A., Righ-Whalen, E. A., & Roehrig, G. H. (2020). "I just do what the boys tell me": Exploring small group student interactions in an integrated STEM unit. *Journal of Research into Science Teaching, 57*, 112–144. https://doi.org/10.1002/tea.21587

Chapter 3
Unpacking the 'M' in Integrated STEM Tasks: A Systematic Review

Ban Heng Choy and Gabi Cooper

Abstract Despite the promise of integrated STEM for authentic learning, it remains unclear whether such integrated approaches can result in significant learning in STEM disciplines. Mathematics is often seen as an "accessory" discipline to Science in many integrated STEM tasks. In this systematic review paper, we investigate the issue of integrating mathematics by unpacking the centricity of STEM tasks, analysing the connections between mathematics and other disciplines, and highlighting the different faces of mathematics presented in these tasks. Our findings suggest that although current STEM tasks anchored in mathematics are problem-centric, they have relatively weak inter-disciplinary connections. Analyses also revealed less emphases on mathematics as a way of knowing. We discuss possible implications and suggestions for strengthening the M in STEM from this review.

Keywords Mathematics · Integrated STEM · Centricity · Connections

The Promise of Integrated STEM

There have been calls for Science, Technology, Engineering, and Mathematics (STEM) educators to move towards curriculum practices built around authentic integrated STEM tasks—those that involve two or more disciplines (Li et al., 2020; Stohlmann, 2020a). Doing this creates opportunities for students to "use knowledge and skills from multiple disciplines" on tasks that involve "complex phenomena or situations" (Honey et al., 2014, p. 52). Despite the promise of integrated STEM tasks for authentic learning, the notion of integration is problematic (Bybee, 2013; English, 2016a; Honey et al., 2014; Li et al., 2020), and it remains unclear whether such integrated approaches can result in significant learning in the STEM disciplines.

B. H. Choy (✉)
National Institute of Education, Nanyang Technological University, Singapore, Singapore
e-mail: banheng.choy@nie.edu.sg

G. Cooper
Griffith University, Southport, Australia

J. Anderson and K. Makar (eds.), *The Contribution of Mathematics to School STEM Education*, https://doi.org/10.1007/978-981-97-2728-5_3

Mathematics is often under-represented in these integrated STEM tasks (English, 2016a; Fitzallen, 2015; Maass et al., 2019), and is seen as an accessory to science and other STEM disciplines, focusing mainly on calculations (Just & Siller, 2022). For instance, in their systematic literature review, Just and Siller (2022) have found that only 14 documents (less than 1% of their initial shortlist using STEM * AND (math * or mathematics *) AND (school or classroom) as search terms) suggest some "situational evidence" for the role of mathematics in STEM classrooms. Furthermore, mathematics is viewed as a "minor matter and a means to an end" in these documents, where "mainly operative work is attributed to mathematics" with the "problem-solving part assigned to the other disciplines" (p. 16). Although there is consensus that the role of mathematics should be strengthened in integrated STEM tasks (Stohlmann, 2020a), it is not clear how such STEM tasks could be redesigned with a stronger focus on mathematics. In this review paper, we aim to advance this conversation by investigating the centricities of integrated STEM (Teo et al., 2021), the connections made within mathematics and between different STEM disciplines (Tan et al., 2019), and the different faces of mathematics (Devlin, 2000) as presented in the tasks found in the extant STEM education literature. Our findings suggest possible loci of action for mathematics educators to pursue for strengthening the mathematics in STEM.

Integrated STEM Tasks: Perspectives, Connections, and Centricities

There are different conceptions of what the acronym STEM means (Li et al., 2020). First, as Bybee (2013) has highlighted, STEM is often perceived as "equal to" science or mathematics (p. 74). This is a mono-disciplinary view of seeing the four disciplines of science, technology, engineering, and mathematics as independent subjects that often "involve traditional disciplinary coursework" (Fitzallen, 2015, p. 238). This perspective of STEM is not unusual in school curricula as most curricula are still organised along disciplinary lines. However, the agenda for STEM integration highlights the need for educators to make connections within and across the disciplines so that these connections can be harnessed to enhance the learning experiences of our students (Honey et al., 2014).

Enhancing our students' STEM learning experiences calls for a more integrated view of STEM beyond the traditional mono-disciplinary teaching approaches. Since the last decade, many researchers (Honey et al., 2014; Li et al., 2020; McLure et al., 2022; Tan et al., 2019) have begun to conceive STEM as an instructional approach that "removes the traditional barriers separating the four disciplines" and "*integrates* [emphasis ours] them into real-world, rigorous, relevant learning experiences for students" (Vasquez, 2015, p. 11). Yet, the degree of integration can range between a spectrum of disciplinary, multi-disciplinary, inter-disciplinary, or trans-disciplinary levels (Vasquez, 2015). For instance, as Bybee (2013) has pointed out, integrated STEM is sometimes seen as "both science and mathematics"; as "science and incorporates technology, engineering, or mathematics" (p. 75); as a "coordination", "combination" (p. 77), "complementary overlapping" of the four disciplines; or as a "transdisciplinary course" (p. 78). This myriad of conceptions about integrated STEM makes it challenging for teachers to understand what integration means and to design tasks that stimulate enriching integrated STEM learning experiences.

With the aim of supporting teachers to design and implement integrated STEM learning experiences, it is important to focus on how the four disciplines are connected. Honey et al. (2014) suggest that teachers should be supported in noticing how epistemic practices and affective outcomes—in addition to content knowledge—are connected across the four disciplines. These linkages between disciplines are referred to as inter connections. Figure 3.1 illustrates some of these possible interdisciplinary connections in terms of the nature of activities students may engage in. Focusing on these inter connections across the disciplines could provide opportunities for students to bring together knowledge and skills from each discipline as they work on integrated STEM tasks.

There is also a need to pay attention to the development of deep conceptual understanding and epistemic practices unique to each of the disciplines, or simply the connections within the discipline (Tan et al., 2019). Examples of such intra connections include the "learning of the concepts of photosynthesis in science" (Tan et al., 2019, p. 9) or the learning of epistemic criteria for "determining the soundness of a mathematical proof" (p. 10). To strengthen these connections, the idea is to go beyond mere application of a mathematical formula to obtain answers without learning about specific mathematical concepts or engaging with the key epistemic processes, such as reasoning.

In terms of the strength of inter connections between two STEM disciplines, Tan et al., (2019, p. 10) suggest that a strong connection "is one that have all the three applications used" (see Fig. 3.1) when students are working on the STEM tasks. Following that, a moderate connection between two disciplines suggests an engagement in two of the three activities and a weak connection, only one of the three applications. As integrated STEM tasks may involve more than two disciplines, it is useful to consider the strength of connections between the lead discipline and other STEM disciplines, in terms of the conceptual knowledge, epistemic practices and skills needed for the STEM tasks. This leads to the different possibilities as listed in Fig. 3.2. To be clear, we are *not* saying that a good STEM task should involve all four disciplines. Instead, we argue that a sharpened awareness of the strength of the inter and intra disciplinary connections can provide teachers a means to examine and re-design tasks for the purpose of strengthening STEM integration, when appropriate.

But how can researchers and educators manage the tensions and affordances of making connections between and within the STEM disciplines? To that end, Tan et al. (2019) proposed the use of problems or problem solving as a possible integrative mechanism to connect the different STEM disciplines. As highlighted by Tan et al. (2019), problem solving affords a "dual emphasis on disciplinary-specific ways of solving problems and the need to offer multi-faceted solutions to complex problems" (p. 6). Such complex problems are not only "cognitively challenging", but they help foster learners' development of content and processes central to the

Connections between disciplines	Descriptions
S–T	Application of appropriate technological tools to seek causality or to investigate relationships between different variables. Application of technological tools to automate or increase accuracy of evidence collected. Application of technology to create simulations or create models to represent scientific phenomena.
S–E	Application of conceptual knowledge and models of science to seek understanding of the functions of design. Application of scientific knowledge to search for better outcomes. Application of evidence from scientific inquiry to inform and interpret success of solutions.
S–M	Application of mathematical models to share and evaluate scientific knowledge and models. Using mathematical formulation to seek understanding of scientific phenomena. Application of mathematical tools to represent data collected during scientific investigations.
T–E	Construction of models or prototypes using technologies for a specific purpose or product. Application of engineering design to create or enhance technological solutions. Application of technological tools to evaluate and inform engineering solutions.
T–M	Application of mathematical models to create technological algorithms. Application of technological tools to create mathematical models using large data. Application of technological tools to interpret complex mathematical issues to transform them from raw to refined, simple to complex and ill-defined to well-defined.
E–M	Application of precision from mathematical computation in the design of solutions. Application of patterns and regularity in reasoning to define and justify designs. Application of mathematical tools to collect data to search for refinement and improvement of design.

Fig. 3.1 Horizontal connections between disciplines (Tan et al., 2019, p. 10)

discipline by providing "multiple entry points" to engage in "reasoning and high-level thinking" (English, 2016b, p. 359). In addition, these tasks usually require student collaboration, which affords opportunities for learners to think about the same problem from different disciplinary perspectives (Honey et al., 2014). Drawing on these fundamental ideas, Tan et al. (2019) developed the S-T-E-M Quartet as

Levels of integration	Number of disciplines with strong connections	Description
STEM	4	Conceptual knowledge, epistemic practices and skills of all four disciplines are required to solve the problem. Any missing discipline would render the problem unsolvable.
STEm, STeM, StEM, sTEM	3	Conceptual knowledge, epistemic practices and skills of three disciplines are required to solve the problem. Knowledge and skills of the last discipline is either not required or used merely as a tool in the problem-solving process.
STem, StEm, SteM, sTEm, sTeM, stEM	2	Conceptual knowledge, epistemic practices and skills of two disciplines are required to solve the problem. Knowledge and skills of the two other disciplines are either not required or used merely as a tool in the problem-solving process.
Stem, sTem, stEm, steM	1	Conceptual knowledge, epistemic practices and skills of only one discipline is required to solve the problem. Knowledge and skills of the other discipline are either not required or used merely as a tool in the problem-solving process.
stem	0	All the disciplines are used but the problem is not anchored in one specific discipline. A superficial form of integration of the disciplines takes place. Similar to the situation described above.

Fig. 3.2 Levels of STEM integration (Tan et al., 2019)

an instructional framework, via a problem-centric approach, to guide teachers in designing and evaluating STEM tasks that make explicit both horizontal and vertical connections to the learners (See Fig. 3.3).

As explained by Tan et al. (2019), the framework uses an authentic and complex problem, which requires knowledge and skills from the different disciplines, as a starting point. The idea is to anchor integrated STEM activities in the problem-solving processes, similar to those articulated by Pólya (1945)—understanding the problem, formulating a solution, implementing the solution, and looking back. As highlighted by Lester and Kehle (2003), successful problem solving involves

> … coordinating previous experiences, knowledge, familiar representations and patterns of inference, and intuition in an effort to generate new representations and related patterns of inference that resolve some tension or ambiguity (i.e., lack of meaningful representations and supporting inferential moves) that prompted the original problem-solving activity (p. 510).

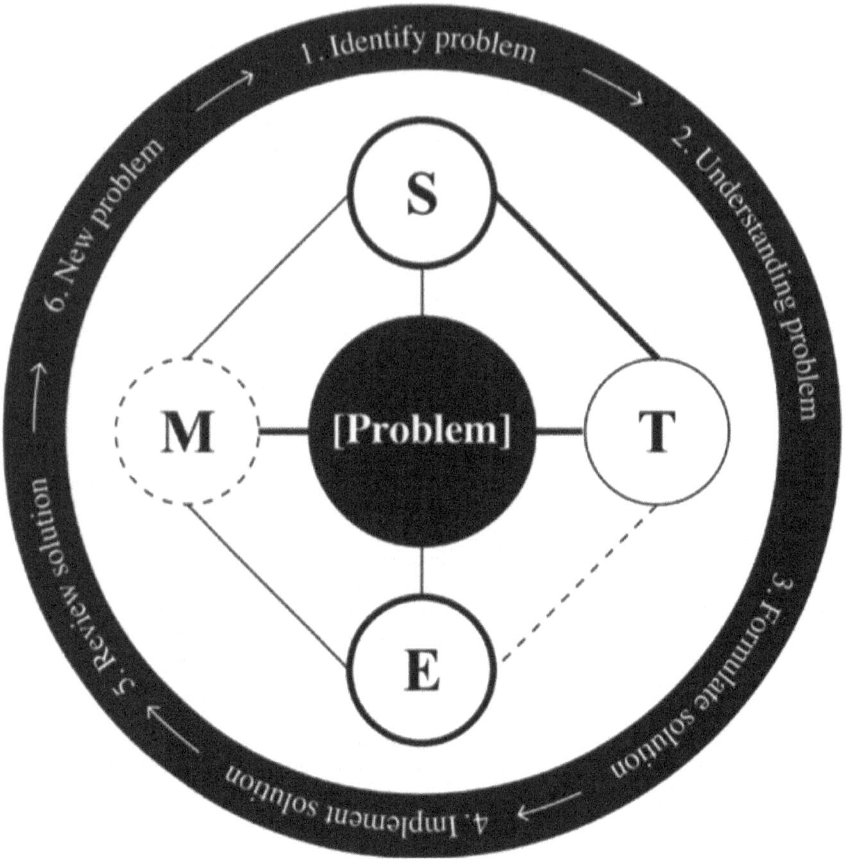

Fig. 3.3 The S-T-E-M Quartet (Tan et al., 2019) illustrating the strength of STEM discipline connections for the insulin delivery problem

In the context of integrated STEM problem tasks, this coordination of experiences and knowledge requires learners to draw on what each of four STEM disciplines has to offer to solve the problem.

Tan et al. (2019) provided an illustrative example of a STEM task designed around the problem of insulin delivery with Biology as the lead discipline (See Fig. 3.4). As argued by the authors, the task is an example of a StEM task with strong connections between S–E and S–M (thick lines), moderate connections between E–M, E–T, and S–T (regular lines), and a weak connection between M–T (dashed lines).

Referring to Fig. 3.4, we see how an unpacking of a STEM task using the S-T-E-M quartet provides opportunities for teachers to conceptualise and review integrated STEM tasks. Doing this empowers teachers to design the task in ways that enables

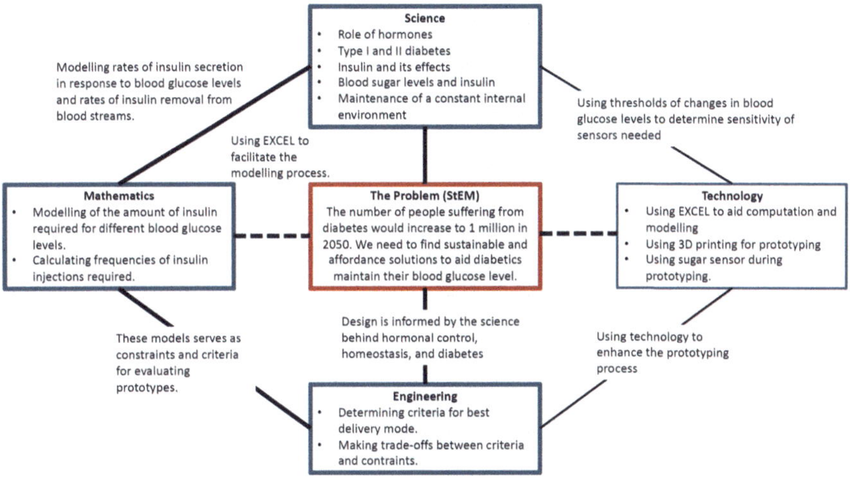

Fig. 3.4 Use of STEM Quartet to unpack a STEM task (Tan et al., 2019, p. 13)

learners to see how different disciplines may be used to view the same problem differently—making use of the (inter) connections between different disciplines to understand the problem, and tapping into the specific knowledge base of each discipline (intra connections) to develop an integrated solution.

However, as Teo et al. (2021) have highlighted, a problem-centric STEM instructional model may not be the only way to think about STEM integration. From their experiences of using the S-T-E-M Quartet with teachers, the problem-centric approach can be challenging for teachers to implement even though it offers possibilities for a coherent integrated curriculum. Building on the ideas in the problem-centric STEM instructional framework, Teo et al. (2021) developed two complementary frameworks—a solution-centric and a user-centric one—which focus on existing solutions and user requirements respectively as starting points to integrate STEM. As described by Teo et al. (2021), a solution-centric approach starts "with existing solutions [to a problem] and engages in the iterative processes of evaluating current solutions, reflecting on the 'gaps' of current solutions, generating improved solutions, and testing new solutions" (p. 144).

From Fig. 3.5, the inter and intra connections between and within STEM disciplines are harnessed by learners as they iteratively improve on the existing solution (V1 solution to V2 solution to V3 solution and so on) to the problem at hand. Similar to how problems and the problem-solving process are used as the integrative mechanisms for integrated STEM, the solution-centric approach engages learners more "productively and extensively in the 'looking back' phase of problem solving (Pólya, 1945), which is often neglected in many problem-solving activities" (Teo et al., 2021, p. 144).

The user-centric approach, on the other hand, starts with the user "to drive the iterative processes of understanding the user and context, ideating solutions for the

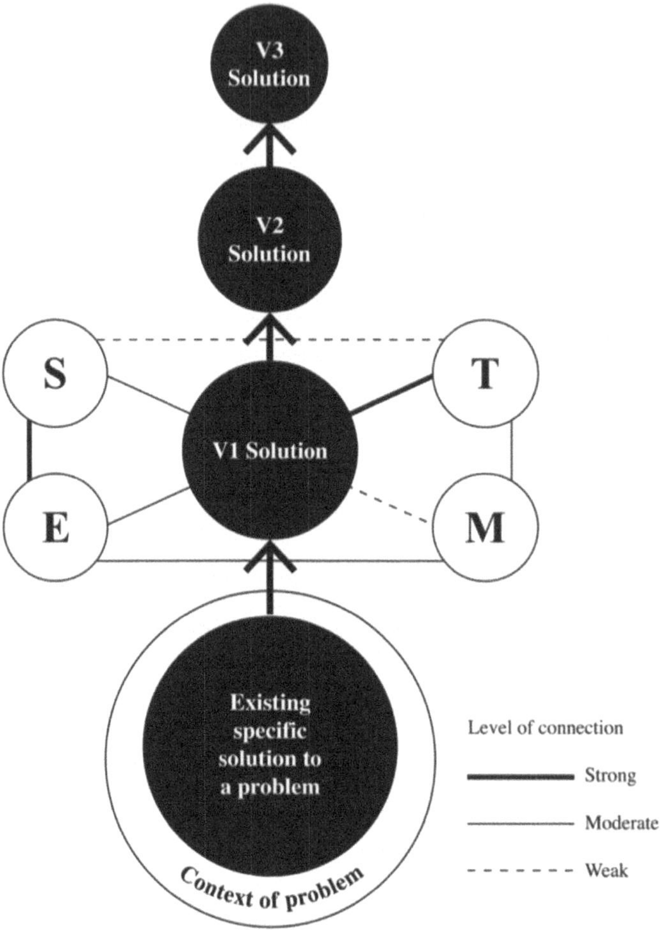

Fig. 3.5 A solution-centric STEM approach (Teo et al., 2021)

user, creating solutions for the user, and evaluating solutions with the user" (p. 145). Here, as shown in Fig. 3.6, the focus is on the user—what the user perceives to be the problem, what the user wants to solve, and what constitutes a solution for the user. The approach is multi-staged and engages learners in learning about the problems faced by the specified users in a particular context, validating and evaluating their proposed design solutions with the users so that the needs identified are satisfied (Teo et al., 2021).

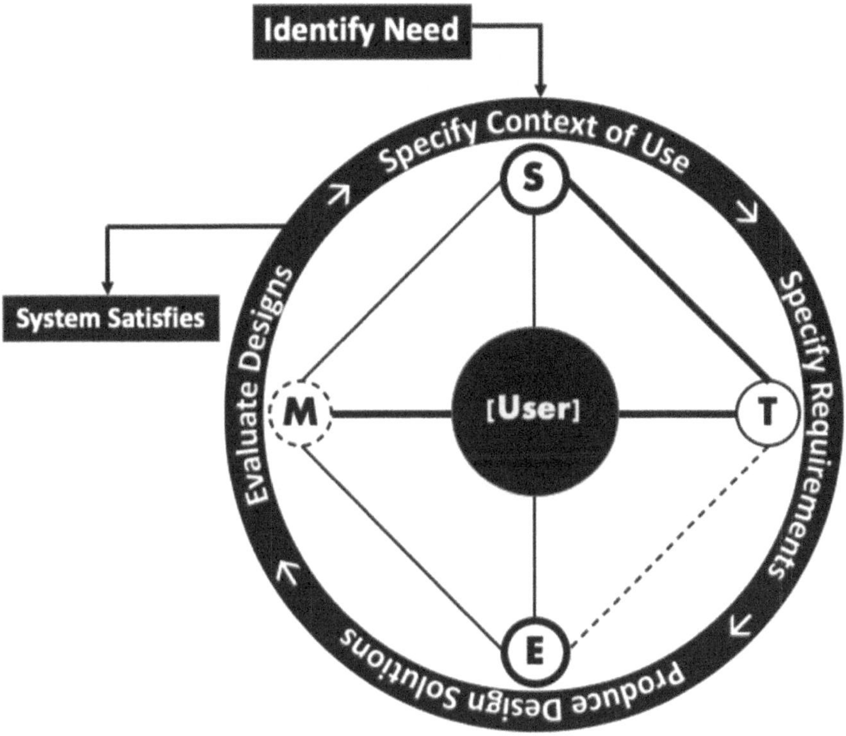

Fig. 3.6 A user-centric STEM approach (Teo et al., 2021)

The 'M' in Integrated STEM Tasks

While the centricity of integrated STEM tasks provided insights into how the four disciplines are connected and integrated, they did not explicitly offer ideas of how students engaged with mathematics when working with these tasks. As Just and Siller (2022) highlighted, learners often engaged with mathematics only as a computation tool, which is only one of the many aspects of the discipline. If mathematics learners were to thrive in an increasingly complex world, they would also need to develop their capacities to think like mathematicians, beyond the proficiency to perform computations. This would involve developing capabilities in understanding the purpose of disciplinary expertise, its inquiry methods, and its forms of communication in addition to an essential base of knowledge (Mansilla & Gardner, 2008). This multi-dimensional view of mathematics was also echoed by Devlin (2000) when he argued that students should experience all the four faces of mathematics—mathematics as "computation, formal reasoning, and problem solving"; "a way of knowing"; "a creative medium", and "applications" (p. 16)—to develop the knowledge and competencies necessary for living "full and active lives" in an ever-changing world (p. 17).

Amongst these four faces of mathematics, most teachers and learners are only familiar with "mathematics as computation, formal reasoning, and problem solving" and to a lesser extent, "mathematics as applications" (Devlin, 2000, p. 18). We agree with Devlin (2000) and Mansilla and Gardner (2008) that it is also crucial for students to see mathematics as a "way of knowing" so that they could learn to see the world through the lens of mathematics. Furthermore, while the other STEM disciplines often provide opportunities for learner to create something tangible (such as a design prototype or device), mathematics as a "creative medium" is often neglected. Here, we go beyond using mathematics to create art pieces (e.g., Escher) to think about how mathematics could be used to create representations and models that make the "invisible visible" (Devlin, 2000, p. 21).

While it is clear that students may have opportunities to see mathematics as computation (Just & Siller, 2022) and mathematics as applications (English & King, 2018; English et al., 2016) in many integrated STEM tasks, we wonder about the other "faces of mathematics" experienced by students as they work with integrated STEM tasks. Experiencing these other faces of mathematics can provide students additional opportunities to engage with mathematics beyond using mathematics for computation, which may promote a better appreciation of mathematics as a discipline (Devlin, 2000), thereby strengthening the 'M' in STEM. By examining the faces of mathematics presented in integrated STEM tasks, together with the centricities of these tasks and the connections afforded by them, we can gain insights into how integrated STEM tasks could be redesigned to strengthen the 'M' in STEM.

In light of our discussion on the centricities, the affordances for making connections, and the faces of mathematics in integrated STEM tasks, this systematic review sought to answer the following overarching research question:

What opportunities are available for students to engage with mathematics in integrated STEM tasks?

There are three sub-questions which framed the systematic review:

1. What is the centricity of such tasks?
2. What are the connections within and between the disciplines as afforded by these tasks?
3. What faces of mathematics are presented in these tasks?

Methods

We followed the *Preferred Reporting Items for Systematic Reviews and Meta-Analyses* (PRISMA) guidelines and suggestions by Alexander (2020) for this review. To ensure the contexts of the literature were specific to integrated STEM rather than singular disciplinary, we examined the databases of journals that explicitly stated "STEM". As such, initial screening revealed seven suitable journals, exclusively dedicated to research on STEM education. These journals included the Journal for STEM Education Research (Springer), International Journal of STEM Education

Research (Springer), Journal of STEM Teacher Education (Illinois State University), Frontiers in Education: STEM Education (Frontiers), Journal of STEM Education: Innovation & Research (Institute for STEM Education & Research), Journal of Research in STEM Education, and Southeast Asian Journal of STEM Education (SEAMEO STEM-Ed). Of these seven, only six were viable, as the search function for the Journal of STEM Education: Innovation & Research (Institute for STEM Education & Research) was not accessible.

Our initial search, which included the search terms "mathematics" and "integrated STEM", produced a total of 252 articles. We included all studies regardless of date of publication or context of study, including studies at all levels of instruction (e.g., level of education, mathematics topic, school sector), The search was however limited to English language publications. The abstracts of these articles were screened for the potential inclusion of a STEM task. For the context of this review, a task is defined as a "prompt" or set of instructions for learners' work (Sullivan et al., 2013, p. 13) to initiate activity by learners (Mason & Johnston-Wilder, 2006). Following screening for inclusion of a task in the literature search, we added "task" for specificity, which yielded a reduced result of 188 articles. Some studies were still considered fit for inclusion even when not yielded by the search term "task". This indicated need for thorough full-text screening to assess eligibility. The full-text versions of 36 studies were examined from initial screening. Inclusion criteria (IC) and exclusion criteria (EC) were formulated and used to assess the viability of these papers, which is summarised in Table 3.1. As per the eligibility criteria established, 19 studies were identified for inclusion and the search was finalised on 6 March 2023.

Our search excluded any literature that were solely theoretical discussions or conceptual frameworks that did not have any explicit description of authentic STEM tasks. Some papers were excluded due to lack of details in the task descriptions. For example, there were papers that did not embed a specific task, but rather only discussed opportunities for a type of task or did not acknowledge the implications of this task for each STEM discipline. Inclusion of these tasks would have required assumptions to be drawn on the nature of the tasks, thereby hindering our analysis. To facilitate our task analyses, we examined the articles that provide sufficient details

Table 3.1 Literature screen inclusion and exclusion criteria

Inclusion criteria	Exclusion criteria
IC1: Studies indexed within the specified databases/journals—document type is journal article IC2: Studies at all levels of education IC3: Studies focusing on integrated STEM OR a broadly STEM context IC4: Studies including an authentic STEM task that embeds mathematical elements and is detailed enough for analysis	EC1: Studies not indexed within the specified databases/journals EC2: Studies not published in English EC3: Studies focused on a single discipline without mention of mathematics/ other STEM disciplines EC4: Studies based around conceptual frameworks/ theory, without an authentic task present EC5: Studies containing a task that is not explicit or detailed enough for analysis

for others to carry out the task (e.g., Stohlmann 2020b), which might include the task instructions, materials needed, the grade levels intended, and the mathematics topics targeted in the task. From the pool of the shortlisted 19 articles, further screening resulted in the elimination of seven more papers, generating a final collection of 12 articles.

We acknowledge a possible delimitation and bias in our literature identification, considering the varied details of task descriptions within each paper included in this review, and how some literature might not be as explicit in their STEM integration as others. There may also be limitations in terms of the approaches to the literature search, as some literature might be overlooked because of the search terms utilised. Another important limitation that we want to acknowledge is the exclusion of useful and important articles about integrated STEM tasks from science education journals and other journals such as the International Journal of Education in Mathematics, Science and Technology (IJMEST) and the International Journal of Science and Mathematics Education (IJSME). Nevertheless, our final sample of 12 articles contained tasks that are sufficiently mathematics-focused as a first step to explore the issue of the 'M' in STEM. The process undertaken for this literature identification, screening and inclusion is summarised in Fig. 3.7.

The tasks of the 12 studies included in the review were examined according to a few categories. Firstly, the task information, including description, level of task and mathematics topic involved was noted. The tasks were then analysed for their centricities, STEM connections, and faces of mathematics presented. In our analyses, we were aware of the different possible definitions of subjects like engineering and technology, which are usually not part of the school curricula. Hence, we took on a more inclusive definition when possible. For instance, we would have interpreted the construction of a device or the use of design thinking as possible indicators of engineering practices. We shared, examined, and discussed each other's analyses and interpretations, before we reached a consensus in our coding of the tasks in terms of their centricities, connections, and faces of mathematics.

Findings

The findings from our systematic review are presented in three parts. First, the integrated STEM tasks of the 12 selected papers are summarised in Table 3.2. From our initial analyses, it was evident that most of the tasks involved quite foundational mathematics topics, such as measurement, which is the most common topic in the 12 tasks surveyed. Given the contexts of the tasks, which mainly focused on designing something (e.g., Tasks 1, 5, 8, 9, 12), it is not surprising that there would be opportunities for measurement as students worked on these tasks. We also noted that several tasks (Tasks 2, 3, 7, 9, 10, 12) provided opportunities for students to learn and apply proportional reasoning. Furthermore, these tasks were designed for a wide range of levels from Grade 3 to Grade 12.

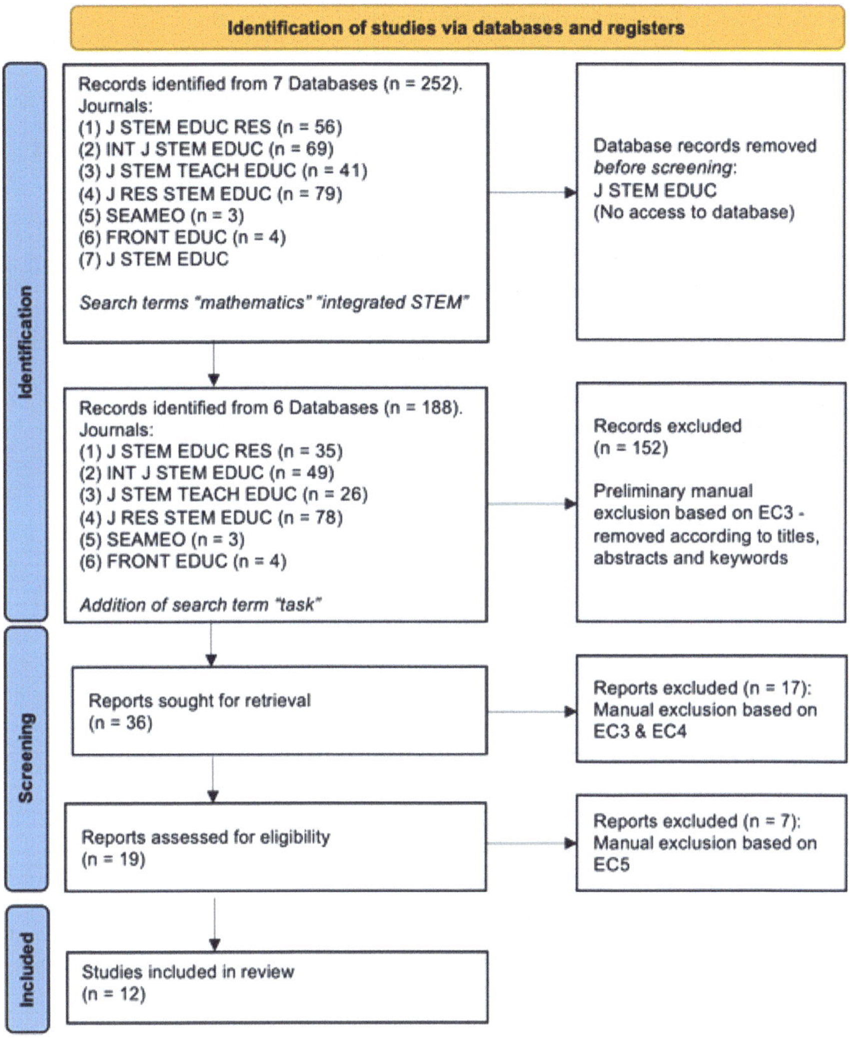

Fig. 3.7 Flowchart of our systematic review

Centricity and STEM Connections

Next, we present the centricity and STEM connections. Of the 12 articles, all the tasks (n = 12, 100%) were designed to embody a predominantly problem-centric approach. These tasks, although developed within varying contexts, are focused on solving a central extended and complex problem. For example, in Task 4 (Roehrig et al., 2021), students were tasked with designing and creating a laser security system using light from a single laser to protect the artifacts in a traveling museum exhibit. This task

Table 3.2 Summary of selected integrated STEM tasks

No.	Authors	Task title/description	Level/mathematics topic
1	Beachner et al. (2015)	***The covered wagon greenhouse*** Design and construct a mini greenhouse for use during the duration of the school year to grow small plants, vegetables, or flowers	3rd–5th Grade Measurement (volume, length, area, perimeter), Geometry, Number (operations)
2	Zhuang et al. (2022)	***Programming ER using mathematics*** Develop a coding sequence for an ER to travel around areas using repetition structures	4th Grade Number (Decimals, Fractions), Proportional reasoning
3	English (2022)	***Modelling volcano lava flow & viscosity*** Model the rate of model flow and use it to predict evacuation times for villages	5th Grade Measurement (length, volume, surface area), Rates (time/ speed, velocity), Scaling
4	Roehrig et al. (2021)	***Museum security*** Design and create a laser security system using light from a single laser	6th Grade Geometry (angles)
5	Stohlmann (2020b)	***The rocket MEA*** Design a rocket that will go the highest in the context that this rocket could be used to deliver supplies to the international space station	6th–8th Grade Trigonometry, Measurement, Geometry (angles)
6	English (2016a)	***Rebuilding a damaged bridge*** Create a model for comparing the different bridge types and determine an appropriate design for a new bridge that meets specific parameters	6th Grade Measurement (length, surface area), Proportional reasoning, Number (cost)
7	Alfieri et al. (2015)	***Programming robots*** Use Interactive simulator game (Expedition Atlantis) to program a robot to navigate through a 3D aquatic environment	6th–8th Grade Measurement, Rates, Proportional reasoning

(continued)

Table 3.2 (continued)

No.	Authors	Task title/description	Level/mathematics topic
8	Aydeniza and Bilican (2018)	***Balloons: testing pressure and volume*** Design a system of two balloons with differing pressure rise at different rates, based on the relationship between pressure and volume	7th Grade Number (operations), Measurement
9	Baker and Galanti (2017)	***Survivor MEA*** Design a shelter for people to survive while stranded on an island, given a list of structural requirements and materials	8th Grade Measurement, Ratios, Scaling, Proportional reasoning, Number operations
10	Leung (2020)	***A day as a palaeontologist*** Model the relationship between the size of a type of fossil bone (thigh bone) and the weight of the dinosaur	9th Grade Measurement (length, circumference), Scaling, Geometrical reasoning, Estimation, Extrapolation
11	Ponnaiyan et al. (2022)	***Routes—investigations for designing routes*** Determine the total time taken by the smart car using two different routes and develop programs to traverse each route to infer which route should be taken	9th to 12th Grade Statistics (average, standard deviation), Rates (speed, distance, time),
12	Mentzer (2011)	***Design project: electric car*** Design and construct a 1/10 scale model prototype maglev car with propeller propulsion and race it in the shortest amount of time	11th Grade Measurement, Proportional reasoning, Rates, Ratios

was considered problem centric as problem-solving is the overarching process, where students were required to connect the interdisciplinary STEM knowledge to create a solution that is unique to the given context (Teo et al., 2021). Although the leading centricity for all the 12 tasks was problem-centric, there were elements of other centricities present.

An example of a task that embeds elements of a solution-centric approach is Task 10, "A Day as a Palaeontologist" (Leung, 2020). This task required students to model the relationship between the size of a type of fossil bone and the weight of the dinosaur. Students were also asked to evaluate the validity of an existing method for modelling this relationship between size of a fossil and the dinosaur weight. This, in turn, reflects a solution-centric approach since students worked with a specific existing solution and evaluated it. Students then modified the existing method and explored the feasibility and important features of this new solution (Teo et al., 2021). In addition, elements of a solution-centric approach were also clearly embodied in several user-centric tasks.

Task 11 (Ponnaiyan et al., 2022) is an example of one such task that embeds elements of user-centric approach to drive the project by focusing on how autonomous vehicles affect differently abled people of society. For this task, students needed to develop a deep understanding of the demographics of differently abled people and how they interact with autonomous vehicles to program a suitable route for the car. This is consistent with the pattern of a user-centric approach because students had to consider the users and produce an appropriate solution output fit for purpose (Teo et al., 2021).

The 12 tasks had varying strengths of disciplinary connections of STEM. To represent the strength of STEM alignment of tasks, each task was analysed and coded with each discipline as being either capitalised or lowercase. For example, a task with very weak alignment could be represented by "Stem", whereby the only discipline emphasised in the task would be science. Additionally, each task was analysed for presence of vertical connections within mathematics, measured to be either "weak", "moderate" or "strong" based on the depth and complexity of the mathematics content embedded within the task. Following that, the inter connections were also assessed, with a focus on the other disciplinary connections to mathematics and identifying which of these were most evident in each task. The analysis of these connections, STEM alignment and centricity within the tasks is summarised in Table 3.3.

Following the guidelines on the strength of disciplinary connections by Tan et al. (2019), only one task ($n = 1, 8\%$) was coded as a "STEM" task with all the disciplines

Table 3.3 Centricity, connections and alignment within the tasks

Task no.	Centricity	Vertical connections within mathematics	Horizontal connections with mathematics	STEM alignment
1	Problem	Weak	S-M	StEm
2	Problem	Moderate	T-M	sTeM
3	Problem	Moderate	S-M, E-M	StEM
4	Problem	Weak	S-M, E-M	StEm
5	Problem	Weak	S-M	STEm
6	Problem	Weak	S-M, E-M	StEm
7	Problem	Weak/moderate	T-M	sTeM
8	Problem	Weak	E-M	StEm
9	Problem	Weak	E-M	stEm
10	Problem/ solution-elements	Moderate	S-M	STeM
11	Problem/ user-elements	Moderate/strong	T-M, E-M	STEM
12	Problem	Weak	S-M	STEm

emphasised. This was Task 11 in Ponnaiyan et al. (2022), which had significant inter connections between each discipline, particularly T-M and E-M. In this task, the application of interdisciplinary knowledge is crucial for students to devise a solution for designing routes for autonomous vehicles. Each part of the task embeds varying connections between the STEM disciplines, and thus designates the task has comparatively more integration than other tasks analysed. For example, in one section of the task students are required to use statistical analysis tools (e.g., formulas for calculating average and standard deviation) to find estimated time of arrival (ETA) to make informed decision on optimal route and then proceed to design an algorithm for this route (Ponnaiyan et al., 2022). This reflects both strong intra connections within mathematics and inter connections between mathematics and the other disciplines.

Four tasks (n = 4, 33%), comprising of Task 3 (English, 2022), Task 10 (Leung, 2020), Task 5 (Stohlmann, 2020b), and Task 12 (Mentzer, 2011) emphasised three disciplines, producing combinations of "StEM", "STeM" or "STEm". With Task 3 (English, 2022), which is denoted as a "StEM" task, students explore how volume, viscosity and slope are factors that affect the surface area that lava covers and then apply this knowledge to take on the roles of geochemical engineers and volcanologists to predict evacuation times for villages. The "S-M" interdisciplinary connections embedded in this task are very clear, as the task seeks to link students' knowledge of the viscosity of liquids with flow/time rate to represent this flow rate. There is also presence of "E-M", where students design of volcano model for modelling of varying lava viscosities and apply mathematics in the design stages to create dimensions for the volcano and pre-emptive rates of flow.

The remaining tasks (n = 7, 58%) embedded weaker inter-disciplinary connections, placing emphasis only on an individual or a pair of disciplines. An example is with Task 1 (Beachner et al., 2015) where students are challenged with designing and constructing their own mini greenhouse for use during the duration of the school year to grow small plants, vegetables, or flowers. This task reflected a weakly integrated approach, as the only disciplines emphasised and connected were science and engineering. The role of mathematics in this task was to be used simply as a tool of computation, aiding the design and physical construction process. Overall, this was a common trend in the analysis of these tasks, where certain STEM disciplines appeared to be more commonly and easily integrated than others.

Besides science, engineering also featured strongly as a primary discipline because many of the tasks were design-based and required students to apply background knowledge to construct a physical prototype. As expected, mathematics was the least integrated of all the disciplines, with five tasks (n = 5, 42%) being coded as a "M" anchored STEM task. Interestingly, mathematics was typically paired with technology (T-M), in which these tasks require some coding or the use of mathematics to develop a digital algorithm, such as in Zhuang et al. (2022) [Task 2] and Alfieri et al. (2015) [Task 7] where students programmed educational robots. This is demonstrated in Fig. 3.8 below, which compares the number of tasks according to the strength of integration of interdisciplinary connections and depicts how many tasks emphasised mathematics as a discipline within each of these categories.

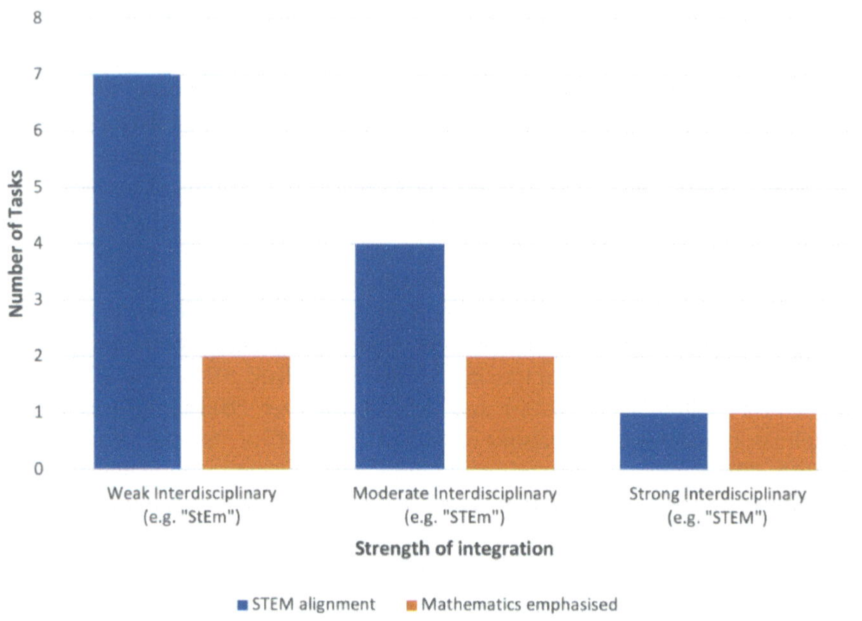

Fig. 3.8 Strength of integration of interdisciplinary connections and mathematics as singular discipline in selected integrated STEM tasks

Overall, our findings reflect relatively weaker inter mathematical connections in the 12 integrated STEM tasks reviewed. When considering the intra connections and inter connections as afforded by these 12 tasks, we noted that Tasks 2, 3, 7, 10, and 11, which had a strong mathematics focus, also had stronger intra and inter connections. This suggests that designing more strongly integrated STEM tasks may promote a better integration of mathematics with the other disciplines. In addition, these tasks suggest that educators could tap into the "T-M" and "E-M" interdisciplinary connections to strengthen STEM integration.

The Faces of Mathematics Presented in the Tasks

In terms of the faces of mathematics presented in these tasks, we see that all the tasks (n = 12, 100%) saw mathematics as "Computation, Formal Reasoning, and Problem Solving (C, FR &PS)" and "Applications" (See Fig. 3.9). For example, Task 5 (Stohlmann, 2020b), the face of "C, FR & PS" is evidenced through students having to apply mathematical problem-solving processes in the context of designing a rocket to provide supplies to the international space station. The face of "Applications" involved students applying knowledge of trigonometric ratios and related mathematical concepts to design a rocket flight path that meets the conditions of the

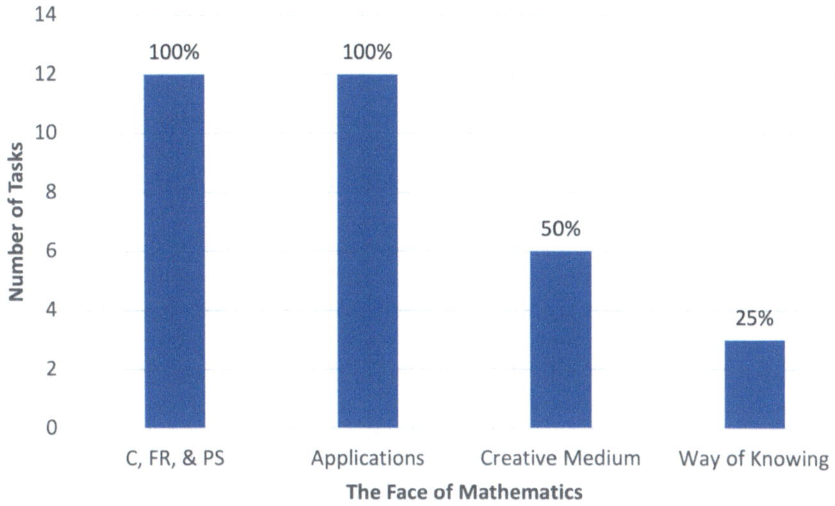

Fig. 3.9 Faces of mathematics presented in selected integrated STEM tasks

task. This finding of all the tasks embodying these two faces reflects the common but limited utilitarian view of mathematics in STEM (Just & Siller, 2022). Six tasks (n = 6, 50%) had some elements of mathematics as a "Creative Medium" and three tasks (n = 3, 25%) presented mathematics as a "way of knowing". To different extents, these six tasks involved students in using mathematics as a representational or explanatory tool, often to represent or explain a real-world scenario. For example, in English (2022) [Task 3], students created a mathematical model for the rate of lava flow based on viscosity. This provided students opportunities to understand how lava flow rate can be examined through a mathematical lens (way of knowing). Similarly, in Ponnaiyan et al. (2022) [Task 11], students had to create a smart car algorithm to program routes for autonomous vehicles (creative medium). Incidentally, these two tasks are ranked highly in terms of their disciplinary connections. This suggests that embedding a more comprehensive view of mathematics in the tasks may prompt the use of mathematics beyond computation.

Discussion

Our analyses revealed insights into the current centricities, levels of connectedness between and within disciplines, and the faces of mathematics presented in 12 selected tasks. Addressing the first sub-question framing the review, we found that the leading approach observed in all the tasks was problem centric. It was apparent that many of the problem-centric tasks were designed around a context that was science or engineering based. However, there were two tasks (Tasks 10 and 11) that had solution-

and user-centric elements, which afforded learners to see the world through the lens of mathematics in different ways other than computation. Specifically, we see the use of statistics to see the "patterns of chance" or patterns of data in these tasks (Devlin, 2000, p. 19). This suggests that different centricities of tasks may offer different affordances for mathematics content featured in these tasks. For example, a user-centric STEM task may offer more opportunities for using statistics to collect and analyse data about the specified users in the tasks. Likewise, in a solution-centric task, there might be more opportunities for learners to evaluate existing solutions and test new solutions through computational simulation. As highlighted by Teo et al. (2021), centricities of tasks provide teachers a way to make explicit the disciplinary connections within and across the discipline. Here, we make it clearer for mathematics teachers by highlighting the need to consider how specific mathematical topics, such as measurement and statistics, can be integrated into tasks by also considering their connections with other disciplines. This suggests that employing tasks with different centricities can provide new possibilities for better integration of mathematics in STEM going beyond performing calculations.

As we have presented, our findings reflect relatively weaker inter connections across disciplines even though there are evidence for a rather strong intra connection within mathematics. Referring to Fig. 3.2, all 12 tasks featured more than one mathematical topic, which highlighted possible opportunities for learners to engage with different topics simultaneously as they work with integrated STEM tasks. However, as noted in our review, only five tasks had moderately strong connections to other disciplines. Although this finding is not surprising (English, 2016b), we gained insights to how certain interdisciplinary connections may offer a more natural way to integrate mathematics. For instance, we see that strong T-M connections (e.g., Tasks 2 and 7) may provide opportunities for students to develop both their computational thinking and mathematical thinking through coding or design of algorithms. Likewise, the affordances of design in tasks with stronger E-M connections suggest measurement as a natural fit to integrate mathematical competencies. Perhaps, what is needed to promote integration is to think about how the topic of measurement could be conceived beyond measuring to thinking about the measures themselves. For example, when evaluating a proposed solution to a design problem, students should have opportunities to consider the kind of measures that can be used for this purpose. We see a clear interaction between how mathematics is embedded in a task and the depth of interdisciplinary STEM connections.

This brings us to the faces of mathematics that were presented in integrated STEM tasks. For the tasks (e.g., Tasks 3, 10, and 11) that engaged students in seeing mathematics as a "Creative Medium" and "Way of Knowing", there are opportunities to use mathematics as a representational or explanatory tool beyond its utility as a computational tool. Our findings suggest a possible locus of action to integrate mathematics in ways that promote more integrated STEM learning experiences: consider how the four different faces of mathematics (Devlin, 2000) can be incorporated into STEM tasks. Mathematics as computation or problem solving is a low hanging fruit in integrated STEM. What we can do is to think of ways to engage learners in the other faces of mathematics beyond mathematics as computation. Inquiry-based learning approaches (Artigue & Blomhøj, 2013) to learning mathematics seem to be a promising way to bring other faces of mathematics (e.g., way of knowing) into integrated STEM tasks. How this can be done in relation to the centricities and inter-disciplinary connections will be a valuable area for future research.

What can we say about 'M' in integrated STEM tasks? More work is needed to strengthen the 'M' in STEM. While the problem-centric approach to integrated STEM is helpful for thinking about the design of integrated STEM tasks, our findings suggest teachers need to consider a solution-centric or a user-centric approach when designing integrated STEM activities. It is also possible to look at the same problem from different perspectives, providing students opportunities to choose whether they want to start their inquiry with the problem, the existing solutions, or to pose new problems and propose new solutions based on the "users" in the problem context. Instructional frameworks such as the ones proposed by Tan et al. (2019) and Teo et al. (2021) are good starting points for mathematics teachers as they design these tasks so that they can begin to explore other centricities beyond the use of a complex real-world problem.

To encourage students to make connections within mathematics and across different STEM disciplines, teachers should provide opportunities for them to see "mathematics from different vantage points" (Choy & Toh, 2021, p. 7). This suggests an urgency for teachers to hone their horizon content knowledge, which refers to the "awareness of how mathematical topics are related over the span of mathematics included in the curriculum", and this domain of knowledge may include "vision useful in seeing connections to much later mathematical ideas" (Ball et al., 2008, p. 403). However, it may be challenging for mathematics teachers to see connections across other STEM disciplines because they may not always be well-versed in science, engineering, or technology.

Here, the ideas proposed by Teo and Choy (2021) may be useful. Instead of thinking about a

> Da-Vinci type of teacher, who is a universalist—someone who is strong in each of the disciplines—it may be more productive to consider how teachers from different disciplinary backgrounds can collaborate in a STEM classroom to design, implement, and review integrated STEM tasks (p. 56).

Having a multi-disciplinary team of teachers to think, plan, and implement integrated STEM tasks may promote the likelihood of making some of the inter connections more explicit to students as they engage with the tasks. This approach largely reflects how today's wicked problems are more likely to be solved by a team of experts with different disciplinary backgrounds. Clearly, time and space are needed for teachers from different disciplines to collaborate and this demands special attention from the school leadership to consider how scheduling can be re-worked.

Finally, while we do not expect every mathematics teacher in STEM education to be a "Da-Vinci" teacher, it is useful for teachers to adopt a "Da-Vinci" mindset—to be curious, creative, and adopt a willingness to be the change. If we want our students to develop productive dispositions and mindsets, then as teachers, we will need to model these dispositions for our students! We need to walk the talk: teachers need to be curious about exploring different possibilities of making STEM connections; they need to think outside the box to consider how an integrated STEM task can be designed and implemented within the constraints of our curricula; and they need to be the change by adopting a variety of teaching approaches and noticing new affordances of the tasks they are using every day (see Choy and Dindyal 2021) for some examples). The 'M' in STEM is not just found in integrated STEM tasks but opportunities to strengthen the 'M' in STEM abound in everyday mathematics classrooms!

References

Alexander, P. A. (2020). Methodological guidance paper: The art and science of quality systematic reviews. *Review of Educational Research, 90*(1), 6–23. https://doi.org/10.3102/0034654319854352

Alfieri, L., Higashi, R., Shoop, R., & Schunn, C. D. (2015). Case studies of a robot-based game to shape interests and hone proportional reasoning skills. *International Journal of STEM Education, 2*(1), 4. https://doi.org/10.1186/s40594-015-0017-9

Artigue, M., & Blomhøj, M. (2013). Conceptualizing inquiry-based education in mathematics. *ZDM, 45*(6), 797–810. https://doi.org/10.1007/s11858-013-0506-6

Aydeniza, M., & Bilican, K. (2018). The impact of engagement in STEM activities on primary pre-service teachers' conceptualization of STEM and knowledge of STEM pedagogy. *Journal of Research in STEM Education, 4*(2), 213–234.

Baker, C. K., & Galanti, T. M. (2017). Integrating STEM in elementary classrooms using model-eliciting activities: Responsive professional development for mathematics coaches and teachers. *International Journal of STEM Education, 4*(1), 4–10. https://doi.org/10.1186/s40594-017-0066-3

Ball, D. L., Thames, M. H., & Phelps, G. (2008). Content knowledge for teaching: What makes It special? *Journal of Teacher Education, 59*(5), 389–407. https://doi.org/10.1177/0022487108324554

Beachner, M., Daugherty, M. K., Carter, V., Swagerty, L., & Wissehr, C. (2015). Expanding earth science in STEM: A model. *Journal of STEM Teacher Education, 50*(1). https://doi.org/10.30707/JSTE50.1Beachner

Bybee, R. W. (2013). *The case for STEM education: Challenges and opportunities*. National Science Teachers Association Press.

Choy, B. H., & Dindyal, J. (2021). Productive teacher noticing and affordances of typical problems. *ZDM—Mathematics Education, 53*(1), 195–213. https://doi.org/10.1007/s11858-020-01203-4

Choy, B. H., & Toh, T. L. (2021). Mathematics: Connections and beyond. In T. L. Toh & B. H. Choy (Eds.), *Mathematics: Connections and beyond* (pp. 1–8). World Scientific. https://doi.org/10.1142/9789811236983_0001

Devlin, K. (2000). The four faces of mathematics. In M. J. Burke & F. R. Curcio (Eds.), *Learning mathematics for a new century: 2000 yearbook of the national council of teachers of mathematics* (pp. 16–27). NCTM.

English, L. D. (2016a). STEM education K-12: Perspectives on integration. *International Journal of STEM Education, 3*(1). https://doi.org/10.1186/s40594-016-0036-1

English, L. D. (2016b). Advancing mathematics education research within a STEM environment. In K. Makar, S. Dole, J. Visnovska, M. Goos, A. Bennison, & K. Fry (Eds.), *Research in mathematics education in Australasia 2012–2015* (pp. 353–371). Springer. https://doi.org/10.1007/978-981-10-1419-2_17

English, L. D. (2022). Fifth-grade students' quantitative modeling in a STEM investigation. *Journal for STEM Education Research, 5*(2), 134–162. https://doi.org/10.1007/s41979-022-00066-6
English, L. D., & King, D. (2018). STEM integration in sixth grade: Desligning and constructing paper bridges. *International Journal of Science and Mathematics Education, 17*(5), 863–884. https://doi.org/10.1007/s10763-018-9912-0
English, L. D., King, D., & Smeed, J. (2016). Advancing integrated STEM learning through engineering design: Sixth-grade students' design and construction of earthquake resistant buildings. *The Journal of Educational Research, 110*(3), 255–271. https://doi.org/10.1080/00220671.2016.1264053
Fitzallen, N. (2015). STEM Education: What does mathematics have to offer? In M. Marshman, V. Geiger, & A. Bennison (Eds.), *Mathematics education in the margins (Proceedings of the 38th Annual Conference of the Mathematics Education Research Group of Australasia)* (pp. 237–244). MERGA.
Honey, M., Pearson, G., & Schweingruber, H. (2014). *STEM integration in K-12 education: Status, prospects, and an agenda for research.* The National Academies Press.
Just, J., & Siller, H.-S. (2022). The role of mathematics in STEM secondary classrooms: A systematic literature review. *Education Sciences, 12*(9). https://doi.org/10.3390/educsci12090629
Lester, F. K. J., & Kehle, P. (2003). From problem solving to modeling: The evolution of thinking about research on complex mathematical activity. In R. Lesh & H. M. Doerr (Eds.), *Beyond constructivism: Models and modeling perspectives on mathematics problem solving, learning, and teaching* (pp. 501–517). Erlbaum.
Leung, A. (2020). Boundary crossing pedagogy in STEM education. *International Journal of STEM Education, 7*(1), 15. https://doi.org/10.1186/s40594-020-00212-9
Li, Y., Wang, K., Xiao, Y., & Froyd, J. E. (2020). Research and trends in STEM education: A systematic review of journal publications. *International Journal of STEM Education, 7*(1), 11. https://doi.org/10.1186/s40594-020-00207-6
Maass, K., Geiger, V., Ariza, M. R., & Goos, M. (2019). The Role of Mathematics in interdisciplinary STEM education. *ZDM Mathematics Education, 51*(6), 869–884. https://doi.org/10.1007/s11858-019-01100-5
Mansilla, V. B., & Gardner, H. (2008). Disciplining the mind. *Educational Leadership, 65*(5), 14–19.
Mason, J., & Johnston-Wilder, S. (2006). *Designing and using mathematical tasks.* Tarquin Publications.
McLure, F. I., Tang, K.-S., & Williams, P. J. (2022). What do integrated STEM projects look like in middle school and high school classrooms? A systematic literature review of empirical studies of iSTEM projects. *International Journal of STEM Education, 9*(1). https://doi.org/10.1186/s40594-022-00390-8
Mentzer, N. (2011). High school engineering and technology education integration through design challenges. *Journal of STEM Teacher Education, 48*(2). https://doi.org/10.30707/JSTE48.2Mentzer
Pólya, G. (1945). *How to solve it.* Princeton University Press.
Ponnaiyan, T., Czerniak, C. M., Javaid, A., Mathur, A., Majumder, S., Oluoch, J., & Shockey, T. (2022). Applying Project-Based Learning (PBL) to enhance learning in mathematics and Career & Technology Education (CTE) classrooms using robotic smart vehicles. *Southeast Asian Journal of STEM Education, 3*(2), 94–119.
Roehrig, G. H., Dare, E. A., Ring-Whalen, E., & Wieselmann, J. R. (2021). Understanding coherence and integration in integrated STEM curriculum. *International Journal of STEM Education, 8*(1). https://doi.org/10.1186/s40594-020-00259-8
Stohlmann, M. (2020a). Analyzing K-5th grade integrated STEM curriculum implemented since 2010. *International Journal of Education Humanities and Social Science, 3*(4), 137–166.
Stohlmann, M. (2020b). STEM integration for high school mathematics teachers. *Journal of Research in STEM Education, 6*(1), 52–63.

Sullivan, P., Clarke, D., & Clarke, B. (2013). *Teaching with tasks for effective mathematics learning*. Springer.

Tan, A.-L., Teo, T. W., Choy, B. H., & Ong, Y. S. (2019). The S-T-E-M Quartet. *Innovation and Education, 1*(1). https://doi.org/10.1186/s42862-019-0005-x

Teo, T. W., & Choy, B. H. (2021). STEM education in Singapore. In O. S. Tan, E. L. Low, E. G. Tay, & Y. K. Yan (Eds.), *Singapore math and science education innovation: Beyond PISA* (pp. 43–59). Springer. https://doi.org/10.1007/978-981-16-1357-9_3

Teo, T. W., Tan, A. L., Ong, Y. S., & Choy, B. H. (2021). Centricities of STEM curriculum frameworks: Variations of the S-T-E-M Quartet. *STEM Education, 1*(3), 141–156. https://doi.org/10.3934/steme.2021011

Vasquez, J. A. (2015). STEM: Beyond the acronym. *Educational Leadership, 72*(4), 10–15.

Zhuang, Y., Foster, J. K., Conner, A., Crawford, B. A., Foutz, T., & Hill, R. B. (2022). Teaching elementary mathematics with educational robotics. *Journal of STEM Teacher Education, 57*(1). https://doi.org/10.30707/jste57.1.1664998343.900405

Chapter 4
Analysis of Teacher-Designed Integrated STEAM Programs with a Focus on Mathematics

Jihyun Hwang, Jeong Su Ahn, and Nam-Hwa Kang

Abstract In Korea, integrated STEAM education aims to enhance mathematical skills outlined in the national curriculum: problem-solving, reasoning, creativity/integration, communication, information processing, and attitude toward mathematics. Mathematics educators in Korea have designed and implemented mathematics-centred STEAM (M-STEAM) programs to maximize the potential of STEAM education. Through a literature review, we explored design principles and practical implications of M-STEAM programs framed around mathematical modelling and creativity. To understand how STEM might be implemented in the classroom, we also examined M-STEAM programs developed and implemented by teachers in South Korea from 2018 to 2021: 46 programs for elementary schools (grades 1–6), 14 programs for middle schools (grades 7–9), and 21 programs for high schools (grades 10–12). These programs were analysed in terms of mathematics content areas, overarching ideas, and outcomes related to 21st-century skills. The programs provide opportunities to integrate ideas from multiple disciplines centred on mathematical ideas. The findings indicate that Korean teachers identified greater potential of the component of arts in the M-STEAM programs. However, it is important to note that there was a lack of sufficient discussion on 21st-century skills during the development and implementation process of M-STEAM programs.

Keywords Mathematics-centred STEAM · STEAM programs · South Korea · Mathematical competency · Twenty-first century skills

J. Hwang (✉)
Mathematics Education, Korea National University of Education, Cheongju, South Korea
e-mail: Jihyun-hwang@knue.ac.kr

J. S. Ahn
Sejong Academy of Science and Arts, Sejong, South Korea

N.-H. Kang
Physics/Science Education, Korea National University of Education, Cheongju, South Korea

J. Anderson and K. Makar (eds.), *The Contribution of Mathematics to School STEM Education*, https://doi.org/10.1007/978-981-97-2728-5_4

Introduction

STEM education has garnered significant attention in recent years as there is a growing demand for a workforce equipped with technical skills in science, technology, engineering, and mathematics (STEM). However, it is important to note that STEM education can also provide opportunities for fostering higher-order thinking skills such as creativity, critical thinking, and communication skills (Maass et al., 2019). In South Korea (Korea hereafter), the integration of arts with STEM education, known as STEAM education, significantly broadens the spectrum of opportunities available. By blending arts and STEM fields, STEAM education provides an extensive platform for cultivating advanced cognitive abilities and fosters a more inclusive and well-rounded approach to learning.

In various countries, the motivation for promoting STEM or STEAM education differs significantly, with noticeable distinctions evident between the United States (US) and Korea. The US emphasizes STEM education due to a relative shortage of STEM degree holders and workers, as well as students' mediocre performance on international mathematics and science tests like TIMSS and PISA (Chen, 2009; Kuenzi, 2008). The goal is to boost interest in STEM careers and elevate education quality in STEM areas.

Although Korea has achieved high performance in international comparison studies, the Korean government and educators recognize the importance of incorporating diverse subjects including arts and design into STEM education to foster students' creativity, innovation, and critical thinking skills. This led to the promotion of integrated STEAM education, where the "A" signifies the inclusion of all subjects other than STEM, recognizing the pivotal role of arts and humanities in nurturing well-rounded and forward-thinking individuals (Baek et al., 2012). While the wide range of subjects within STEAM might appear to weaken its core focus, capturing the full essence of S, T, E, A, and M is a complex endeavour. With the Korea Foundation for the Advancement of Science and Creativity (KOFAC) requiring either mathematics or science as an essential component, the term 'STEAM' serves as a beacon for educators, emphasizing the rich possibilities of holistic teaching (Kim, 2011).

Over a decade, STEAM education has become a top priority in Korea, with various initiatives and programs being launched to promote it in schools and universities (Kang, 2019; Kim, 2011). The Korean government has also provided funding and resources to support STEAM education, with a focus on interdisciplinary learning, hands-on project-based learning, teacher professional development, and creating STEAM programs for K-12. KOFAC is responsible for organizing these STEAM initiatives.

In terms of school implementation, most STEAM programs are designed to be delivered within standard class hours. While teachers in elementary school are in charge of all subjects, implementing STEAM programs is relatively easy. However, in secondary schools, the approach is more collaborative. Teachers from each discipline included in STEAM programs instruct part of their subject. Notably, core subject teachers, like mathematics, often have a greater portion of their class time devoted to STEAM programs. In the context of South Korea's educational system, variation in course selection among students is minimal, except for the last two years of high school. This uniformity enables the design of STEAM programs applicable to all enrolled students. Additionally, certain initiatives have been designed and implemented through avenues such as STEAM clubs and supplementary extracurricular undertakings.

The recent national curriculum revisions in Korea (Ministry of Education, 2015, 2022) have placed an emphasis on the development of mathematical competencies, including problem-solving, reasoning, creativity, integration, communication, information processing, and attitude toward mathematics. To harness the full potential of STEAM education, researchers and teachers in Korea have designed and implemented mathematics-centred STEAM (M-STEAM) programs. M-STEAM programs were created to provide students with opportunities to apply mathematical concepts to real-world problems and enhance their mathematical competencies through the integration of STEAM subjects. These programs have the potential to improve students' mathematical skills as well as to develop 21st-century skills, such as critical thinking, creativity, and collaboration (Seo, 2014). Thus, the integrated STEAM education in Korea highlights the importance of preparing students for the future, and the specific emphasis on M-STEAM education is to enhance students' mathematical competencies.

The purpose of the study presented in this chapter was to explore the theoretical ideas underlying the design of M-STEAM programs and their implications in Korean literature. Through a review of the literature, the study aimed to identify the meaningful M-STEAM design principles. In addition, the study investigated M-STEAM programs designed and implemented by Korean teachers between 2018 and 2021, encompassing various grade levels from elementary to high school. In relation to 21st-century skills, the programs were analysed in terms of mathematical content, overarching ideas, use of technology, implementation of collaborative tasks, and students' own creative works. In doing so, this chapter provides a comprehensive understanding of M-STEAM education in Korea, including its potential benefits and limitations. The results and discussion will inform future research and implementation of M-STEAM programs in Korea and beyond.

Literature Review

STEAM Education in Korea

The emergence of integrated STEAM education in Korea can be attributed to a government policy agenda introduced in 2011. This agenda aimed to nurture and support the science and technology workforce by transforming science and mathematics education in Korea through the integration of arts into STEM subjects, giving rise to the concept of STEAM (Kang, 2019). The motivation behind this reform was the recognition that STEM fields were becoming increasingly interconnected, necessitating the development of the "competency of convergence" (Kang, 2020, p. 227), which involves collaboration across different disciplines, a crucial skill for the future STEM workforce. In Korea, the terms 'convergence education' and 'STEAM education' are used interchangeably, and the Korean Ministry of Education has played a pivotal role in promoting this integrated approach to education.

In recent years, the Korean Ministry of Education has emphasized the importance of higher-order thinking skills such as problem-solving, critical thinking, and creativity through the *Second Convergence Education Comprehensive Plan 2020–2024* (KOFAC, 2018; Ministry of Education, 2020). This plan aims to transform the learning paradigm by recognizing the importance of incorporating digital technologies into education and considers higher-order thinking skills as foundational for generating new fields and values, enhancing core competencies necessary for future work readiness (OECD, 2023).

STEAM education in Korea aims to cultivate an interdisciplinary and holistic approach to learning, fostering the development of comprehensive problem-solving skills and stimulating innovation across diverse fields. Jun and Park (2015) categorized the arts component of STEAM in Korea into two distinct categories: a narrower definition of arts that exclusively pertains to fine arts and design, and a broader definition that includes liberal arts, language arts, physical arts, and practical arts. The broader definition has gained widespread acceptance in Korea as it aligns with the fundamental aim of STEAM initiatives, facilitating a more comprehensive and well-rounded education.

Given the diverse approaches to STEAM education since 2011, there has been a call for a clear definition of STEAM education to show what STEAM classes look like. Drawn from the STEAM research and practices since 2011, KOFAC, the main government-funded agency for STEAM initiatives, has published specific guidelines for designing STEAM classes for teachers to use (KOFAC, 2018). The document provides teachers with a definition and goals of STEAM:

> STEAM is a term created by combining the initials of science, technology, engineering, humanities and arts, and mathematics. Humanities and arts are added to STEM, a field of science and technology, in consideration of humanistic literacy and artistic sensibility. STEAM education is to increase students' interest and understanding in scientific technology and to grow STEAM literacy based on scientific technology and the ability to solve problems in the real world (KOFAC, 2018, p. 8).

From this statement, there are three clear goals: (1) increasing students' interest and understanding in scientific technology, (2) growing STEAM literacy based on scientific technology, and (3) enhancing the ability to solve problems in the real world.

Given the definition and goals, KOFAC provides STEAM teaching and learning standards that have three key aspects (Fig. 4.1). The standards require STEAM classes to situate lessons in students' everyday experiences, to have students solve real-world problems creatively, and to provide students with an emotionally rich learning environment. The focus on 'Creative Design' and 'Emotional Touch' are drawn from

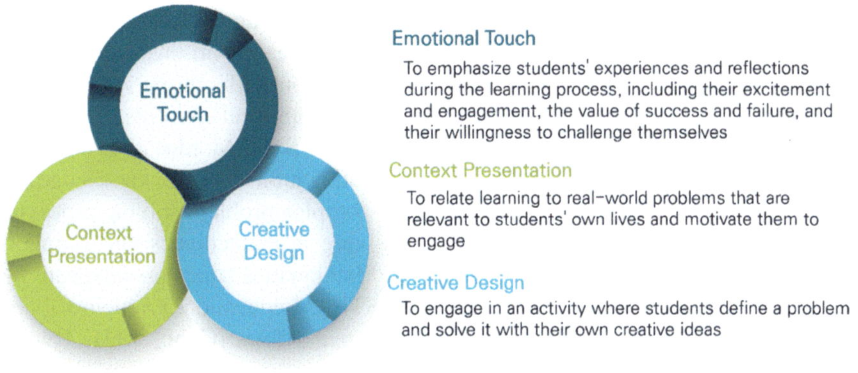

Fig. 4.1 STEAM teaching and learning standards (KOFAC, 2018)

engineering design (Baek et al., 2012). Because engineering is not a school subject, the standards have built-in engineering perspectives. Practical applications like 3D-printed air purifiers, furniture construction based on ratio and proportion concepts, and LEGO wheel-based rangefinder manufacturing are engineering designs integrated with mathematics. These standards, combined with the goals, are presented as criteria for deciding whether a teacher's lesson encompasses STEAM or not (KOFAC, 2018, p. 11).

The Role of Learning Mathematics in STEAM Education

In the context of STEAM education, mathematics is perceived as an epistemological lens through which events and phenomena in both societal and natural realms can be comprehended. Rather than being approached through holistic integration, learning mathematical concepts is deemed a prerequisite for understanding relevant contexts (Cuoso & Simarro, 2020). Students are expected to possess a solid mathematical foundation before fully grasping the interconnections between events and phenomena and their relations with other STEAM disciplines. Within this epistemological framework, mathematics in STEAM assumes the role of a tool for representation and problem-solving within the context of other disciplines (Coad, 2016). This, in turn, poses a challenge when attempting to seamlessly integrate mathematics into STEAM activities in a meaningful way.

To achieve a meaningful integration of mathematics into STEAM programs, the emphasis should shift from mere comprehension of mathematical concepts in isolation to fostering competencies like creativity and complex problem-solving (Seo, 2014). Creativity is a key aspect of STEAM education as it promotes innovative thinking and problem-solving skills by combining ideas from multiple disciplines. In mathematics education, creativity is equally emphasized but defined differently (e.g., Chamberlin & Moon, 2005). In a mathematics learning context, creativity refers to the ability to use mathematical concepts and techniques in new and unconventional ways to solve problems. It involves not just creating original solutions, but also having the flexibility to think about problems from different perspectives and the ability to identify new connections between mathematical concepts.

Mathematical modelling is another key aspect of STEAM education in Korea. Integrated STEAM problem-solving requires the utilization of diverse knowledge and interdisciplinary perspectives, leading to more complex problem-solving processes (Kim & Pang, 2015). Traditional problem-solving in mathematics classrooms is often too simplistic for real-world situations. Mathematics educators advocate for incorporating real-world tasks into integrated STEAM education. Problem-solving in STEAM addresses individual and collaborative efforts (Cobb, 1994; Hwang et al., 2023) while navigating the complexities of real-world situations. Complex real-world tasks play a vital role in stimulating students' mathematical modelling process (Zawojewski, 2010), which encompasses a cyclical progression of exploration, mathematical observation, specification, mathematization, combination, analysis, association, highlighting, interpretation, and examination (Zbiek & Conner, 2006).

Collectively, the role of learning mathematics in STEAM education has predominantly been the acquisition of fundamental concepts as tools for comprehending phenomena, rather than fostering substantial integration with concepts from other disciplines. For meaningful interdisciplinarity within STEAM programs, the creativity for promoting innovative thinking and the application of mathematical modelling to address intricate real-world problems are necessary.

21st Century Skills and STEAM

The Partnership for 21st Century Learning (n.d.) underscores the significance of 21st-century skills as essential proficiencies for students to thrive in contemporary society. These skills encompass critical thinking, problem-solving, communication, collaboration, and others (Trilling & Fadel, 2009). STEAM education is widely acknowledged as an effective approach to nurturing the indispensable 21st-century skills (Hadinugrahaningsih et al., 2017). Among the various 21st-century skills, cultural competency, collaboration, creativity, and digital literacy (ICT literacy) can be developed through STEAM/STEM learning environments (Maass et al., 2019). In particular, the incorporation of non-STEM subjects into the traditional STEM framework encourages innovative thinking and offers room for imaginative expression (Mariano & Chiappe, 2021).

In addition, STEAM learning environments offer abundant opportunities for student collaboration (Quigley et al., 2020). STEAM activities typically involve teamwork, promoting interpersonal communication. In doing so, students learn how to work effectively and respectfully, assuming shared responsibility for collaborative work and valuing the individual contributions made by each team member. In an increasingly digital world, developing ICT literacy—also known as digital literacy—has become a paramount objective of 21st-century education. STEAM education can contribute significantly towards ICT literacy by teaching students how to use digital technology, communications tools, and/or networks to access, manage, integrate, evaluate, create, and communicate information (Park & Hyun, 2020). Within a STEAM education setting, cultural competency emerges as a vital skill that can be fostered (Dell'Erba, 2019). The interdisciplinary nature of STEAM enables students to develop an appreciation for diverse cultures, ideas, and perspectives. By blending various disciplines, STEAM provides a broader context of the world, empowering students to effectively and respectfully interact and engage with individuals and groups from diverse cultural backgrounds (Taylor & Taylor, 2019).

Approaches to Designing Mathematics-Centred STEAM Programs

Mathematical Modelling

Some studies on M-STEAM (e.g., Kim & Han, 2021; Kim & Pang, 2015) have adopted mathematical modelling as their design principle for M-STEAM program development. These studies emphasized problem-solving and creative design processes to maximize students' emotional experiences and enable cyclical learning, in accordance with the principles of integrated curriculum structure suggested by Baek et al. (2012). For example, Kim and Pang (2015) proposed an instructional model for M-STEAM (see Table 4.1), highlighting the significance of mathematical modelling. This process is crucial in M-STEAM, starting from mathematical observation to summarization. Mathematical modelling involves the investigation of the problem context and mathematical observation. Through mathematizing, students can translate the assumptions and conditions of a problem into mathematical properties. To make individual and collective decisions, students must be able to solve mathematical problems and interpret the results of the mathematical modelling process. The summarization stage of the M-STEAM model involves reviewing and validating the results of the mathematical modelling process. In doing so, students can experience authentic mathematical thinking and doing processes.

Creativity to Solve Real-World Problems

Others have emphasized creativity in their mathematics-centred STEAM programs (e.g., Hwang et al., 2017; Jun & Park, 2015). These researchers provided detailed tasks used in the STEAM programs to demonstrate what students have designed. For example, Hwang et al. (2017) described a STEAM education outreach program in

Table 4.1 Mathematics-centred STEAM instructional model (Kim & Pang, 2015, p. 114)

M-STEAM model	Mathematical modelling process
Context presentation ► Complicated context with multiple concepts included ► Context with concrete examples that can stimulate various thoughts ► Context with cases that can be solved using students' mathematical knowledge	
Activity extraction ► Activities necessary for achieving learning objectives ► Presentation of the process of solving problems in real life, but in a condensed form	
Mathematical observation ► Observing the conditions and assumptions of problems from a mathematical perspective	*Investigating, mathematical observation* ► Collecting information about the problem
Connect with mathematics ► Transforming the conditions and assumptions of the problem into mathematical properties and variables	*Specializing, mathematizing* ► Listing the conditions of the problem ► Modelling the conditions and assumptions of the problem into mathematics
Individual problem solving ► Executing the solution plan for the transformed problem	*Combining, analysing, relating, emphasizing, interpreting* ► Combining mathematical properties and variables into reality ► Manipulating activities to find other properties ► Connecting with the real world ► Emphasizing mathematical conclusions ► Connecting mathematical conclusions to real-world conclusions
Collective decision-making ► Reviewing the process of solving problems individually ► Making a decision about the most rational outcome	
Summarizing ► Points learned and felt through the lesson ► Internalizing interest in solving real-life problems ► Sharing emotional experiences	*Review* ► Checking the connection between the real world and the mathematical conclusions
Connection to next lesson ► Connect emotional experiences to next lesson ► Reflect on next lesson plans	

Korea (see Table 4.2) that aimed to enhance students' creative thinking abilities in mathematics by encouraging integrated problem-solving and developing students' creativity. The program entitled, "Design My Idea through STE*M*!", provides various activities tailored to each grade band but commonly asks students to design something that requires mathematical problem-solving in everyday contexts. Thus, the study found that the STEAM outreach program increased STEAM competency and the perception of convergence education for all students.

Given the design principles for meaningful M-STEAM in Korea, in the following section, we present an analysis of recent M-STEAM programs designed and implemented by Korean teachers spanning from elementary to high schools. The analysis focused on several 21st-century skills including mathematical content, overarching ideas, use of technology, implementation of collaborative tasks, and students' own creative works.

Analysis of Teacher-Designed M-STEAM Programs

Data Sources

We collected STEAM programs designed and implemented by teacher communities from the STEAM portal run by KOFAC. Since 2011, KOFAC supports the STEAM research group of teachers in developing STEAM programs and deposited the programs on the STEAM portal (Kang, 2019). As of Spring 2023 (KOFAC, 2023), the STEAM portal has a total of 2,051 STEAM programs. Among these, a total of 938 were developed between 2018 and 2021. All programs on the STEAM portal are intended to align with the Korean STEAM standards (KOFAC, 2018).

When the teacher research groups shared their programs, they reported the core subjects covered in their programs. To identify M-STEAM programs, we specifically focused on programs where the teacher research groups indicated mathematics as the core subject. Our investigation covered M-STEAM programs published on the STEAM portal between 2018 and 2021 for elementary, middle, and high schools. Initially, we identified 24 programs for high schools, 14 for middle schools, and 52 for elementary schools. However, after initial screening, we excluded some programs due to misidentification or missing information. Consequently, the final number of programs analysed was 81 in total (21 for high schools, 14 for middle schools, and 46 for elementary schools).

It is important to note that relating the meaningful M-STEAM design principles discussed in the literature review to the collected M-STEAM programs is challenging. The available programs primarily focus on the practical aspects of the programs and are intended to serve as examples for other schools to apply. Thus, most programs do not provide information about their theoretical background, making it difficult to determine how well the programs align with the theoretical ideas. Therefore, inferences made during the analysis were cross-checked among the researchers.

Table 4.2 Tasks of STEAM outreach programs (Translated; Hwang et al., 2017, p. 396)

Task (grade band)	Descriptions
Designing buildings (Grades 3–4)	The activity discusses architectural design plans, modular buildings, and their connection to science and engineering. Using modules, students complete a mathematics curriculum activity to understand three-dimensional objects viewed from the top, front, and side, and design their own buildings with plans, elevation, and section. They express their creativity by designing buildings in a virtual space called the "research base"
Designing an optical illusion spinning top (Grades 5–6)	This activity focuses on the common phenomenon of optical illusions in everyday life, in which students observe and create various optical illusion shapes. They experience optical illusions due to the refraction of light, and through observing and exploring optical illusions using mathematical and scientific methods, they can actively engage and experience the usefulness of mathematics and science. Additionally, students learn that the principles of optical illusions are applied in various fields such as animation, movies, and architecture, and through this activity, they can understand that mathematics and science form the basis of these applications
Designing a balanced bird (Grades 7–9)	This activity explores the concept and application principles of the centre of mass by examining specific real-life situations where it can be observed. Students are given a few situations where they can directly find the location of the centre of mass. Additionally, they explore the principle of the centre of mass in various situations and use it to find the centre of mass of a shape with non-uniform density. Using this knowledge, students can then design and create their own balancing bird and present their ideas while comparing and contrasting their features
Designing structures (Grades 7–9)	To design a model of a lunar lander that minimizes shock absorption, students engage in an activity that involves selecting and calculating the weight of the cargo to be carried in the lander and selecting the materials needed to create the lander. They then compare the structures of the landers designed by each group and discuss their advantages and disadvantages, allowing them to confirm each group's unique ideas. Additionally, through the process of considering factors to be taken into account during lunar landing, designing for the situation, and testing the designed product, students experience a series of processes that result in creative outputs
Designing a polyhedral lamp using star-shaped polyhedral (Grades 10–12)	This activity involves exploring the meaning of star shapes in the history and culture of humankind, as well as investigating the properties of regular polygons. Students will also examine the structural relationship between regular polyhedra and star-shaped polyhedra, and learn about the application of star-shaped polyhedra in design in daily life. In addition, they will explore the structure of small star-shaped dodecahedron units, assemble them, and create a lamp, which will help them understand the similarities and differences between the structures of convex and concave polyhedra

Analysis Framework

Our study aimed to investigate the concept of "mathematics-centred" across a range of programs claimed by teachers as M-STEAM programs. Programs were analysed using a list of components (Table 4.3)—mathematics content, subjects, cultural competency, digital literacy, collaboration, and creativity. We analysed the mathematics content taught in these programs, as well as the relationships between mathematics and other disciplines, such as science, engineering, technology, and arts. As proposed by KOFAC (2018), each M-STEAM program can include multiple subjects other than mathematics in the program. Thus, the subjects within each M-STEAM program were counted individually. The mathematical content covered in M-STEAM programs was classified into five categories: algebra, functions, geometry, probability and statistics, and calculus. Measurement was included in geometry to align with the Korean national mathematics curriculum.

As we were interested in 21st-century skills development, in our analysis of competency goals and their assessment, we identified whether topics addressed global and local issues such as climate change. While all programs included a specific problem situation in a real-world context, programs that had global and local issues were counted for having cultural competency goals. We evaluated the presence of

Table 4.3 Analytical framework of M-STEAM programs

Component	Description
Mathematics contents	Geometry Algebra Calculus Function Statistics and Probability
Subjects	**S**: Science (Physics, Chemistry, Biology, and Earth Science) **T**: Technology **E**: Engineering **A**: Art (Fine Arts, Music, Korean language, Korean History, Society and Culture, Geography, Moral education)
Culture competency (overarching topics)	**Y**: Setting learning context based on cross-disciplinary topics (e.g., global climate change) **N**: Others
Digital literacy (Use of technology)	**Y**: Using Technology (computers and tablets) **N**: Others
Collaboration (Collaborative tasks)	**Y**: A program includes collaboration tasks **N**: Others
Creativity (program outputs)	**Y**: Program outputs include students' works that they designed and created themselves **N**: Others

collaboration tasks by assessing whether the programs included group activities for students. To assess program outputs, we utilized codes derived from actual student work provided in the program descriptions to determine if students were encouraged to design and create their own projects, rather than merely completing worksheets or solving pre-existing problems. Additionally, we examined if the program provided digital literacy training utilizing technology or computer software.

Using the analytical framework, the authors initially identified each component and then coded the content following the guidelines given by Cohen et al. (2018). The researchers mainly observed the frequency of the component and did not apply any inferential statistical analysis. Therefore, the results were interpreted descriptively. By using a systematic and thorough approach to analysing the data, the study aimed to provide a comprehensive and accurate understanding of the components of each M-STEAM program.

Results

Exemplary M-STEAM Programs

In this section, we share two exemplary programs that show how mathematics can be meaningfully integrated with other subjects and how 21st-century skills can be developed in the classroom. The first exemplary program is entitled, "Life is a Graph"—Answering Life with Mathematics" designed for high school students (Woo et al., 2020). The program addresses the mathematical concepts of the increase, decrease, local maximum, local minimum, and inflection point of functions, and their application in creating graphs. During the program, students are introduced to various social issues such as health and the economy, and select a topic related to these issues. They then form groups and investigate the topic using sources such as national statistics portals, newspaper articles, and books to identify the key variables related to the social issue. Once variables are identified, the students create graphs using technology and analyse their graphs using mathematical concepts such as local maximum and minimum points to discover new interpretations that are overlooked during the initial investigation. Finally, the students summarize their inquiry in posters as creative and attractive as possible and then present them to their peers.

We argue that this program exemplifies a 21st-century skill-based activity for several reasons. Firstly, the process of exploring graph topics exposed students to various social issues and helps to develop their global or social awareness. Secondly, the collaborative nature of the activity helped students to develop their communication and collaboration skills. Thirdly, the use of technology tools to create graphs, while not the primary focus, contributes to students' digital literacy skills. Finally, the students produced creative outputs (see Fig. 4.2) that go beyond the teacher's guidance or intention.

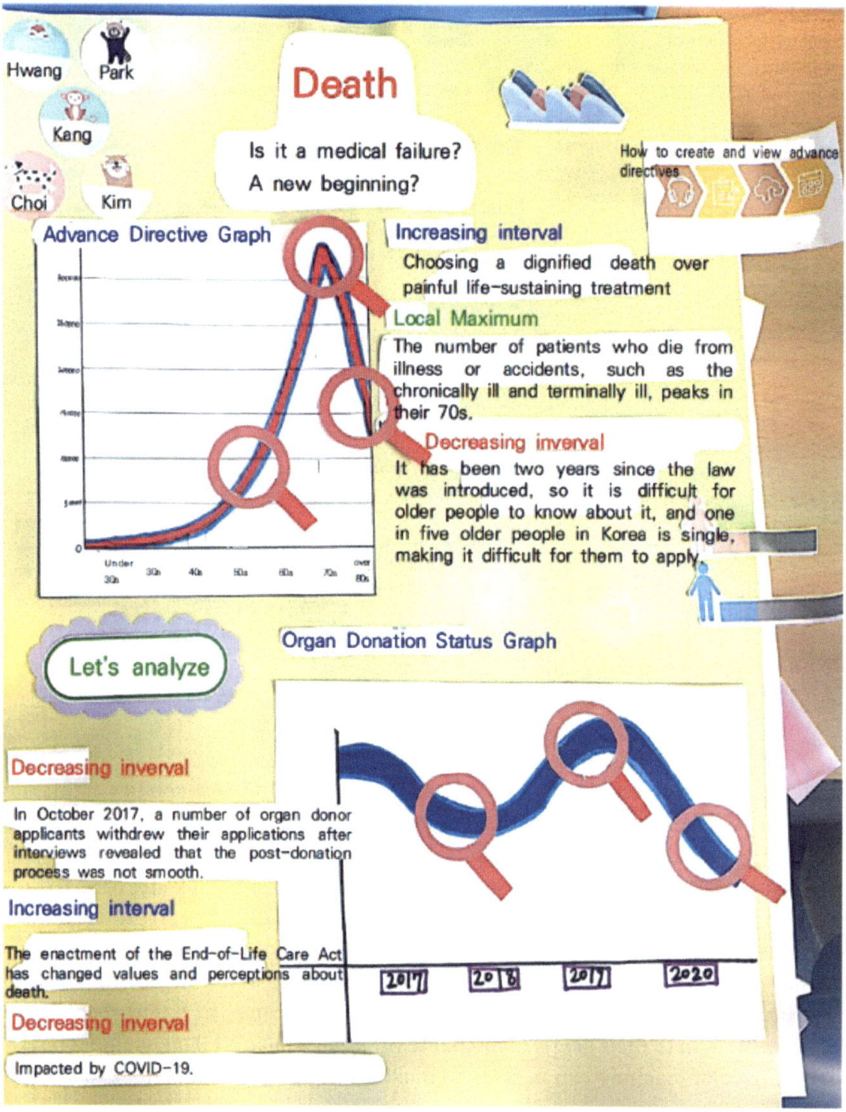

Fig. 4.2 A translated example of Creative outputs of the program "Life is a Graph" (Woo et al., 2020)

The second exemplary program is one for elementary school students titled "Probability and Statistics in Our Proverbs" (Oh et al., 2019; see Fig. 4.3). The program integrates mathematical concepts of bar graphs, pie charts, and ring charts into the context of the popular Korean proverb "When the swallows fly low, it rains." We argue this program is outstanding because it successfully integrates mathematics with

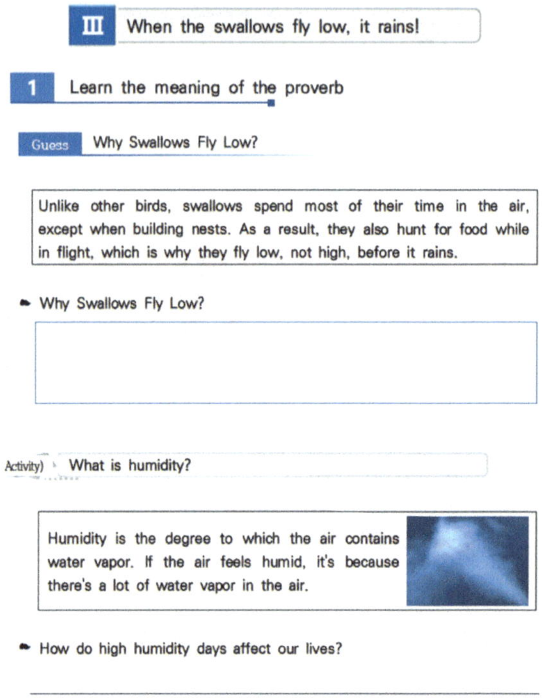

Fig. 4.3 Student activity sheets of the program "probability and statistics in our proverbs"

language arts and science and facilitates students to develop 21st-century competencies. The program encourages students to improve their communication skills by discussing their ideas with peers and presenting their results to others.

The program begins by having students discuss the reasons why sparrows fly low, followed by learning about scientific facts related to humidity, weather forecasts, and pressure. Students acquire data from the National Meteorological Agency Website to study humidity levels in their local area over the past 100 days and represent them in an appropriate graph. Then, students create posters to report their results and present them to their peers. Overall, the program is an excellent example of how cross-curricular teaching can be used to make learning more engaging and relevant for students.

Results of the Analysis

The coverage of mathematical content areas varied across different school levels (Table 4.4). For example, the programs for elementary schools mainly addressed

Fig. 4.3 (continued)

2 Finding scientific principles in proverbs

Guess Discover the science behind weather forecasts

We've always been interested in the weather. Depending on where you live, what you're interested in is different, but the main thing you're interested in is whether it's raining or sunny.

What does the weather forecast show? Let's take a look at what we see in the weather forecast.

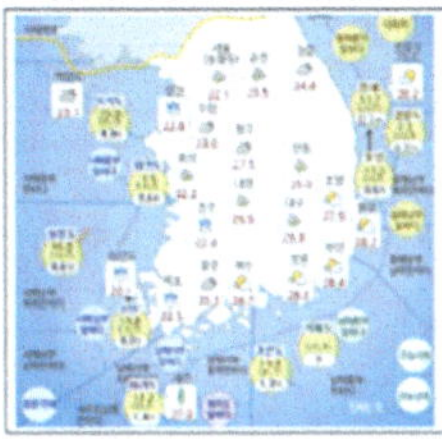

Activity Understanding why air moves

Barometric pressure is the pressure caused by the weight of air. If there is a large amount of air in the same area, it is called high pressure, and if there is less air in the same area than the surrounding area, it is called low pressure. This difference in pressure causes air to move from high pressure to low pressure, which is wind.

Fig. 4.3 (continued)

3 Designing exploration

Guess Find the right graphs

- What is the best graph to show the humidity for 100 days in my area? Let's talk to your friends.

Activity) Find the right graph for your purpose or situation

A pie graph is a graph that shows the ratio of each item to the whole in a circle shape. Pie graphs are easier to compare overall than tables because they are expressed as percentages.

- When is it convenient to plot a pie graph?

Activity) Pie Graphing

· Let's ask 100 of your friends what their favorite animal is.

Animal names						
Student counts						

geometry while high school programs mainly addressed calculus. Obviously, mathematical content levels are reflected in the program just like the national mathematics curriculum. Regardless, some content areas seem to be favoured over others. Since elementary school students benefit from hands-on and interactive learning experiences, geometry may be more suitable than other content areas. In the same vein, programs for high school students tended to focus on upper-level mathematical concepts, with calculus being the most popular area of study.

Across all school levels, function was the least popular content area. These results could be attributed to factors such as adapting to students' proficiency levels, the manner of integration with other subjects, and addressing complex real-world issues. Nonetheless, it is essential to carefully consider exposing students to a wide range of mathematical content areas in M-STEAM programs to prevent an unintended misconception that some content is less relevant to real-world issues.

In the analysis of content areas, the influence of Korea's recent mathematics education policy and mathematics education community was found with 21% of the analysed M-STEAM programs focused on statistics and probability, encompassing all school levels. This is a result of the effort to improve statistics education in Korea (Woo et al., 2020; Ministry of Education, 2015), encouraging students to understand

Table 4.4 Mathematics content areas of the M-STEAM programs

	Elementary schools (G1–G6)	Middle schools (G7–G9)	High schools (G10–G12)	Total
Geometry	28 (60%)	4 (29%)	4 (20%)	36 (45%)
Algebra	9 (20%)	4 (29%)	2 (10%)	15 (19%)
Calculus	–	–	8 (40%)	8 (10%)
Function	3 (7%)	1 (7%)	1 (5%)	5 (6%)
Statistics and probability	6 (13%)	5 (36%)	6 (30%)	17 (21%)

Note The percentages indicate the proportion of M-STEAM programs out of the total number of programs analysed at each school level

and use statistics in real-world situations. Statistics and probability are becoming increasingly important in many fields, including finance, sports, and medicine, and it is essential for students to have a solid foundation in these areas if they are to succeed in these fields. The development of M-STEAM programs in these areas can help to inspire students and deepen their understanding of these subjects.

In terms of disciplines integrated with mathematics, typically two or more subjects were integrated with mathematics, but the subjects categorized as arts were integrated the most (Table 4.5). This is not surprising given the category includes all disciplines other than STEM. Nonetheless, it is noteworthy that many M-STEAM programs have taken advantage of the opportunity to learn mathematics with a connection to non-STEM content. M-STEAM programs provide opportunities to achieve the educational goal of preparing well-rounded citizens with opportunities to be creative.

Other than arts, no clear patterns were found in the preferences for disciplinary areas for integration. However, it is noteworthy that for elementary and high school programs integration with either science or engineering was preferred to integration with technology. One example of integrating mathematics and engineering was a program titled "Separate the Ping-Pong Balls and Golf Balls in the Basket!", designed for elementary school students. The program applies approximation (the area of a circle) over four lessons. Students started by comparing the properties of two types

Table 4.5 STEAM subjects integrated with mathematics in the M-STEAM programs

	Elementary Schools (G1–G6)	Middle schools (G7–G9)	High schools (G10–G12)	Total
Art	37 (80%)	10 (71%)	13 (65%)	60 (75%)
Engineering	17 (37%)	3 (21%)	4 (20%)	24 (30%)
Science	16 (35%)	3 (21%)	11 (55%)	30 (38%)
Technology	10 (22%)	7 (50%)	4 (20%)	21 (26%)
Sum	80	23	32	135

Note The percentages indicate the proportion of M-STEAM programs out of the total number of programs analysed at each school level. The sum is bigger than the number of programs as each program integrates multiple subjects or subject areas of balls (size, weight, etc.) and designed a device to separate them—an opening needs to be created that allows one ball to pass through but not the other. This kind of design activity is commonly found in the programs but few M-STEAM programs for the elementary and high schools used technologies. On the other hand, half of M-STEAM programs for middle school students adopted technology. This might be because the middle school curriculum has a compulsory ICT course but whether the difference is due to the curriculum or teacher perceptions needs further investigation.

Competency goals were analysed through student activities in terms of whether students had opportunities to develop 21st-century skills (see Fig. 4.4). The analysis showed several recognizable patterns. Firstly, across all school levels, digital literacy was less emphasized than other competencies. This finding was surprising given the increasing ubiquity of technology in modern society and the potential benefits of incorporating it into M-STEAM education.

Secondly, elementary M-STEAM programs incorporated collaboration tasks (36%) much less than secondary school M-STEAM programs, indicating that collaboration is increasingly emphasized as students develop better communication skills and cooperative behaviours as they mature. On the other hand, the demand for creativity was far less evident in high school programs than at the other school levels.

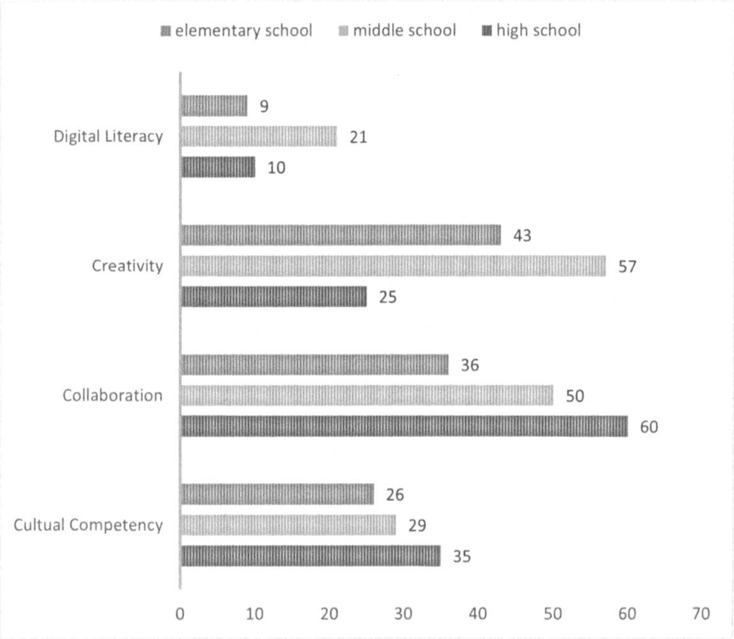

Fig. 4.4 M-STEAM program components related to the 21st skills. *Note* The data indicate percentages for the proportion of M-STEAM programs out of the total number of programs analysed at each school level. The sum is bigger than the number of programs as each program may aim for multiple competency goals

This might be because M-STEAM programs for high school students addressed high-level mathematics that required time and rigor at the cost of time for students' creative thinking. Furthermore, high school teachers felt pressure to cover content to prepare students for the college entrance examination. This might have led them to design M-STEAM programs for content learning rather than competency development. To examine these hypotheses, further analysis of program content and research on teacher perceptions are needed.

Thirdly, not many M-STEAM programs addressed global and local issues that provided opportunities for enhancing cultural competencies and global awareness. Some examples of global and local issues addressed in the M-STEAM programs analysed in this study included COVID-19, global warming, air quality, school safety, and unification of North and South Korea. Other notable examples included using drones to deliver emergency kits in the event of an accident in the city centre at the high school level. By addressing such relevant and pressing issues, students could connect with mathematics and its usefulness.

Discussion and Conclusion

This study analysed 81 teacher-designed M-STEAM programs for mathematics content areas, overarching ideas, use of technology, collaborative tasks, and program outputs in relation to 21st-century skills. In doing so, the findings provided a comprehensive understanding of M-STEAM education in Korea, including its potential benefits and limitations. Based on the analysis, we discuss possibilities for future research and implications for the implementation of M-STEAM programs.

The results showed a prevalent trend of integration of arts in teacher-designed M-STEAM programs. The diverse scope of arts, encompassing physical education, fine arts, language, culture, philosophy, psychology, and history, underpinned the promotion of integrated STEAM education in Korea (Kim, 2011). This movement stemmed from the recognition of the potential of the "A" component in nurturing well-rounded and forward-thinking individuals. Korean in-service teachers embraced this potential, and they integrated subjects other than STEM. By combining STEM competencies with the creative and expressive elements of the arts, STEAM education offered a more comprehensive and holistic approach to learning (Baek et al., 2012). In this way, STEAM education effectively bridges the gap between STEM and the humanities, and can prepare students to become more well-rounded, versatile, and adaptable individuals in both their academic and professional lives.

The results of this study revealed that collaboration tasks were more prevalent in secondary school M-STEAM programs compared to elementary programs. This emphasis on collaboration at the middle and high school levels may be attributed to several factors. One plausible explanation is that the contexts used in the M-STEAM programs at the secondary school levels could be more complex, making collaboration essential for effective problem-solving (Quigley et al., 2020). Additionally, the need for teamwork and communication skills for college admission might contribute to this focus on collaboration in middle and high school programs. Korean high school teachers place significant importance on preparing students for college or university, and collaboration skills are seen as valuable for students' success in higher education. Teachers can also document student behaviours in group tasks, which can be advantageous for students during the college admissions process.

The limited emphasis on digital literacy in M-STEAM programs across all school levels is somewhat surprising, particularly when juxtaposed against the considerable significance accorded to this competency within the Korean context. In Korea, the experience of digital learning during the COVID-19 pandemic and the increasing prominence of Artificial Intelligence (AI) across diverse occupations and everyday life are recurrent topics of examination within the media and policy declarations including educational directives (KOFAC, 2018). Digital literacy has become increasingly vital in modern society, enabling individuals to navigate and effectively utilize technology for personal, academic, and professional purposes. By prioritizing digital literacy within M-STEAM programs, students can enhance their ability to engage with technology-driven innovations and be better prepared for the demands of future society.

Another intriguing finding pertains to a noticeable pattern observed in programs at the elementary school level programs, wherein a recurring emphasis was observed on geometry concepts, in contrast to programs at the high school level which more frequently explored calculus or algebra concepts. This disparity could be attributed to the fact that many elementary school programs aim to teach fundamental concepts like unit, measurement, and ratio of shapes while creating tangible products that are easily applicable in everyday contexts. In contrast, high school programs aim to explore changes derived from scientific facts and mathematically express them using calculus or algebra concepts, aligning with the emphasis on connecting mathematics to various professions and advanced scientific applications.

STEAM education is undeniably valuable for fostering 21st-century skills in students (Hadinugrahaningsih et al., 2017). However, we observed that sufficient discussion and consideration of these skills were lacking during the development and implementation of M-STEAM programs. Thus, we advocate for the need to design programs that not only focus on students' understanding of mathematics concepts but also encompass the development of essential 21st-century skills.

Among the various cases we reviewed, the program "Life is a Graph"—"Answering Life with Mathematics" (Woo et al., 2020) stands out as exemplary. This program was not designed with explicit consideration of 21st-century skills, but students are still able to acquire global awareness, communication, and collaboration skills, digital literacy skills, and creativity through the nature and content of the program. This demonstrates the potential of M-STEAM initiatives to develop vital skills crucial for success in the twenty-first century. By intentionally and explicitly promoting the development of these skills, M-STEAM education can further empower students to thrive in a rapidly evolving global landscape.

In this study, we explored the design principles and practical implications of M-STEAM programs in the literature review. Then, we analysed the M-STEAM programs developed by teacher research communities in consideration of design principles promoted in the literature. We were unable to find evidence that each program analysed was designed based on the design principles that promote creativity and mathematical modelling. This might be attributed to the point stated at the beginning of this chapter that the teacher communities, consisting of teachers of various subjects including mathematics, have an epistemological view of mathematics as a tool for solving problems and fail to meaningfully integrate mathematics with the other subjects.

Regardless, Korean mathematics teachers are actively exploring diverse forms of integrated education, leveraging their expertise in the field, as evidenced by the M-STEAM program analysis. The development of over 20 new M-STEAM programs annually underscores the efforts of Korean mathematics educators to foster successful M-STEAM education. Some programs like "Probability and Statistics in Our Proverbs" (Oh et al., 2019) increase the feasibility of meaningful M-STEAM, showing great balance between mathematics and other subjects by pursuing appropriate overarching topics and well-connected content of each subject.

The findings of this study have highlighted several areas for further research and improvements in M-STEAM education. Notably, a mere 9.6% of STEAM programs spanning the period from 2018 to 2021 manifest as M-STEAM. This is highly problematic given the importance of the role of mathematics and mathematical thinking in contemporary society. One possible explanation is the perception of mathematics as merely a supplementary tool (Coad, 2016), a notion suggested in this chapter's introduction. Conversely, challenges in creating effective M-STEAM programs could be the culprit. As many STEAM initiatives replace traditional subject-specific curricula, designing M-STEAM programs that align developmentally is a significant challenge. Delving deeper into the causes of this M-STEAM underrepresentation will aid in better integrating mathematics into STEAM initiatives.

Currently, there is a lack of concrete information available on the effects of implementing these programs with diverse groups of students. A meta-analysis of the effect of STEAM education with students in Korea has reported a positive effect on student competencies as well as conceptual learning in general (e.g., Kang, 2020), but there is a need for examining how M-STEAM affects students with diverse backgrounds and characteristics. This study examined M-STEAM programs that were developed before and during the COVID-19 pandemic, but it was difficult to obtain specific data on their implementation. In conjunction with the contemporary surge in generative AI, there exists a compelling imperative for the development and execution of M-STEAM programs in the recent timeframe. Research on such progress and educational effects is needed.

Although the case reported in this chapter pertains to the Korean context, where STEAM education receives intentional and vigorous endorsement from the Korean Ministry of Education, comparative research on the design and execution of M-STEAM programs would offer insights into the best practices that can contribute to equipping students for the fulfilment of their potential.

References

Baek, Y., Kim, Y., Noh, S., Park, H., Lee, J., Jeong, J., Choi, Y., Han, H., & Choi, J. (2012). *Basic research for establishing the direction of STEAM education in Korea*. Korea Foundation for the Advancement of Science and Creativity [KOFAC].

Chamberlin, S. A., & Moon, S. M. (2005). Model-eliciting activities as tool to develop and identify creativity gifted mathematicians. *Journal of Secondary Gifted Education, 17*(1), 37–47.

Chen, X. (2009). Students who study science, technology, engineering, and mathematics (STEM) in postsecondary education. Stats in Brief. NCES 2009-161. U.S. Department of Education, National Center for Education Statistics. Retrieved from http://nces.ed.gov/pubs2009/2009161.pdf

Coad, L. T. (2016). The M in STEM: What is it really? *Australian Mathematics Teacher, 72*(2), 3–6. https://search.informit.org/https://doi.org/10.3316/informit.191508122817376

Cobb, P. (1994). Where is the mind? Constructivist and sociocultural perspectives on mathematical development. *Educational Researcher, 23*(7), 13–20. https://doi.org/10.3102/00131_89X02_30070_13

Cohen, L., Manion, L., & Morrison, K. (2018). *Research methods in education* (8th ed.). Routledge.

Couso, D., & Simarro, C. (2020). STEM education through the epistemological lens: Unveiling the challenge of STEM trans disciplinarity. In C. C. Johnson, M. J. Mohr-Schroeder, T. J. Moore, & L. D. English (Eds.), *Handbook of research on STEM education* (pp. 17–28). Routledge. https://doi.org/10.4324/9780429021381-3

Dell'Erba, M. (2019). *Preparing students for learning, work and life through STEAM education: Policy brief*. Education Commission of the States.

Hadinugrahaningsih, T., Rahmawati, Y., & Ridwan, A. (2017). Developing 21st century skills in chemistry classrooms: Opportunities and challenges of STEAM integration. *AIP Conference Proceedings, 1868*(1), 030008. https://doi.org/10.1063/1.4995107

Hwang, J., Choi, K. M., & Hand, B. (2023). Epistemic actions and mathematics achievement. *International Journal of Science and Mathematics Education, 21*(3), 787–809. https://doi.org/10.1007/s10763-022-10278-2

Hwang, S., Kim, N., Son, J., Song, W., Lee, K., Choi, S., & Lew, K. (2017). Development and effectiveness of steam outreach program based on mathematics. *Journal of the Korean Society of Mathematical Education Series e: Communications of Mathematical Education, 31*(4), 367–407.

Jun, M. S., & Park, M. H. (2015). Effects of STEAM program development and application for the 1st grades of elementary school. *Education of Primary School Mathematics, 18*(2), 91–106. https://doi.org/10.7468/JKSMEC.2015.18.2.91

Kang, N. H. (2019). A review of the effect of integrated STEM or STEAM (science, technology, engineering, arts, and mathematics) education in South Korea. *Asia-Pacific Science Education, 5*, 6. https://doi.org/10.1186/s41029-019-0034-y

Kang, N. H. (2020). What can integrated STEAM education achieve? A South Korean case. In J. Anderson & Y. Li (Eds.), Integrated approaches to STEM education: An international perspective (pp. 227–249). Springer. https://doi.org/10.1007/978-3-030-52229-2_13

Kim, J. (2011). A cubic model for STEAM education. *The Korean Journal of Technology Education, 11*(2), 124–139.

Kim, M., & Han, H. (2021). The effects of mathematics-centered STEAM program on mathematical modeling ability of first grade students in middle school. *Communications of Mathematical Education, 35*(3), 295–322. https://doi.org/10.7468/JKSMEE.2021.35.3.295

Kim, Y., & Pang, J. (2015). A study of the potentials of math based convergence instructional model. *Journal of the Korean Society of Mathematical Education Series c: Education of Primary School Mathematics, 18*(2), 107–122. https://doi.org/10.7468/jksmec.2015.18.2.107

Korea Foundation for the Advancement and Creativity [KOFAC]. (2018). STEAM 교육으로 채우는 즐거운 수업: 선생님을 위한 안내서 [Fun lessons filled with STEAM education: A guide for teachers]. KOFAC. https://steam.kofac.re.kr/?p=12632

KOFAC. (2023). *About STEAM*. https://steam.kofac.re.kr/?page_id=11269

Kuenzi, J. J. (2008). *Science, technology, engineering, and mathematics (stem) education: Background, federal policy, and legislative action*. CRS Report for Congress. http://digitalcommons.unl.edu/crsdocs/35/

Mariano, W. K., & Chiappe, A. (2021). 21st-century skills and their relationship to STEAM learning environments: A review. *Revista de Educación a Distancia, 66*(21). https://doi.org/10.6018/red.470461

Maass, K., Geiger, V., Ariza, M. R., & Goos, M. (2019). The role of mathematics in interdisciplinary STEM education. *ZDM Mathematics Education, 51*(6), 869–884. https://doi.org/10.1007/s11858-019-01100-5

Ministry of Education. (2015). *2015 revised mathematics education curriculum*. Ministry of Education. http://ncic.re.kr/mobile.dwn.ogf.inventoryList.do

Ministry of Education. (2020). *Mathematics education master plan* (2020–2024). https://steam.kofac.re.kr/steam-includes/download.php?mode=bbs&n=1064

Ministry of Education. (2022). *2022 Revised Korean mathematics curriculum*. Ministry of Education.

Oh, G., Jung, S., Choi, Y., & Ha, J. (2019). *Probability and statistics in our proverbs* (KOFAC 2019 STEAM Teacher Institute Results Reports). KOFAC. http://steam.kofac.re.kr/?p=15581

OECD. (2023). *OECD learning compass 2030: A series of concept notes*. https://issuu.com/oecd.publishing/docs/e2030-learning_compass_2030-concept_notes?fr=xKAE9_zU1NQ

Park, C. J., & Hyun, J. S. (2020). K-TIHM: Korean technology integration hierarchy model for teaching and learning in STEAM education. *Journal of Advanced Information Technology and Convergence, 10*(2), 111–123. https://doi.org/10.14801/JAITC.2020.10.2.111

Partnership for 21st Century Learning. (n.d.). *Framework for 21st century learning*. Retrieved from https://www.battelleforkids.org/networks/p21

Quigley, C. F., Herro, D., Shekell, C., Cian, H., & Jacques, L. (2020). Connected learning in STEAM classrooms: Opportunities for engaging youth in science and math classrooms. *International Journal of Science and Mathematics Education, 18*(8), 1441–1463. https://doi.org/10.1007/s10763-019-10034-z

Seo, D. Y. (2014). STEAM on the viewpoint of didactics of mathematics. *Journal of Educational Research in Mathematics, 24*(3), 429–442.

Taylor, P. C., & Taylor, E. (2019). Transformative STEAM education for sustainable development. In Y. Rahmawati & P. C. Taylor (Eds.), *Proceedings of the Science and Mathematics International Conference* (SMIC 2018) (pp. 125–131). CRC Press. https://doi.org/10.1201/9780429461903-19

Trilling, B., & Fadel. C. (2009). *21st century skills: Learning for life in our times*. Wiley.

Woo, J. H. (2000). An exploration of the reform direction of teaching statistics. *School Mathematics., 2*(1), 1–27.

Woo, I., Kwon, B., Hwang, B., Lim, S., & Yu, J. (2020). *Life is a graph—Answering life with mathematics* (KOFAC 2020 STEAM Teacher Institute Results Reports). KOFAC. https://steam.kofac.re.kr/?p=16600

Zawojewski, J. (2010). Problem solving versus modeling. In R. Lesh, P. L. Galbraith, C. R. Haines, & A. Hurford (Eds.), *Modeling students' mathematical modeling competencies: ICTMA 13* (pp. 237–244). Springer.

Zbiek, R. M., & Conner, A. (2006). Beyond motivation: Exploring mathematical modeling as a context for deepening students' understandings of curricular mathematics. *Educational Studies in Mathematics, 63*, 89–112. https://doi.org/10.1007/s10649-005-9002-4

Chapter 5
Expanding the Horizons of Mathematics in STEM: Diverse Mathematical Roles in STEM Activities

Dorte Moeskær Larsen, Mette Als Kristensen, and Connie Svabo

Abstract Mathematics plays a vital role in enabling citizens to understand, analyse, and determine solutions to real-world problems. The research presented in this chapter explores the roles of mathematics in 26 STEM activities that were developed and implemented by teachers as part of the LabSTEM research and development project conducted in primary and lower secondary schools in Denmark. The findings reveal the teachers succeeded in developing activities that centred on mathematics and also, to some extent, addressed crucial societal issues. The analyses expose the challenges of developing activities which established a genuine connection between mathematical tasks and understanding the role of mathematics in society. This balance is difficult for teachers to achieve, without support, time, professional development, and guidance from school leaders.

Keywords Mathematics · Citizenship · Integrated STEM

Introduction

In the contemporary global landscape, active and informed citizenship requires more than a rudimentary comprehension of political frameworks and societal concerns; it demands a repertoire of abilities and knowledge that empowers individuals to navigate challenges and make contributions to their communities. Mathematics can be perceived by students as detached from reality, but it has a vital role in equipping citizens to comprehend, analyse, and solve real-world issues (Maass et al., 2019). This chapter considers the significance of mathematics as an important component of developing global citizenship education in schools through an integrated science, technology, engineering, and mathematics (STEM) curriculum approach.

One effective way to integrate mathematics into citizenship education is through interdisciplinary approaches, such as STEM education. A STEM curriculum

D. M. Larsen (✉) · M. A. Kristensen · C. Svabo
University of Southern Denmark, Odense, Denmark
e-mail: dmla@imada.sdu.dk

J. Anderson and K. Makar (eds.), *The Contribution of Mathematics to School STEM Education*, https://doi.org/10.1007/978-981-97-2728-5_5

programme provides a platform for students to engage with authentic, real-world issues by offering opportunities to tackle complex, context-rich problems (Larsen et al., 2022). By working collaboratively across disciplines, students can advance their mathematical skills as well as their capacities for critical thinking, problem-solving, modelling, and teamwork, which are all essential competencies for active global citizenship in the twenty-first century (Bybee, 2018). By implementing an integrated STEM curriculum, teachers can enhance individual subject-specific knowledge and promote the importance of connected knowledge and processes for solving real-world challenges (Larsen et al., 2022). The inclusion of integrated STEM activities in school curricula has therefore emerged as a worthy research focus.

The integration of mathematics into STEM teaching and learning has often been described as a complex endeavour due to the underlying epistemological differences amongst the four STEM disciplines. Mathematics, engineering, and the sciences differ in terms of their areas of study, their methods of validating conjectures, and the potential outcomes of such validation. Mathematics explores abstract representations, while science investigates tangible, real-world phenomena (Baldinger et al., 2020). Honey et al. (2014) acknowledged that integrating the STEM disciplines often leads to one discipline dominating over the others. For instance, engineering tends to dominate the study of robotics and engineering design problems, whereas mathematics is frequently underrepresented in integrated STEM activities (English, 2016; Fitzallen, 2015), even though it is recognised as a foundational discipline in science, technology, and engineering (Martín-Páez et al., 2019).

When mathematics is integrated into STEM tasks and activities, its role is frequently minimised to that of a tool. Given the increasing emphasis on STEM education in schools, it is becoming even more crucial to consider which aspects of mathematics should be integrated into STEM contexts and how to strengthen the role of mathematics (Shaughnessy, 2013; English, 2016). A study by Honey et al. (2014) revealed that STEM approaches seem to have a less positive influence on mathematics outcomes compared to science outcomes. This finding aligns with Hurley's (2001) research indicating mathematical performance was significantly lower than science performance when the two subjects were integrated in a transdisciplinary approach. In addition, a meta study by Becker and Park (2011) examined STEM achievement through integrative approaches, with the results showing that mathematics had the smallest effect size.

There is a potential risk of insufficiently emphasising the interconnections between mathematics and other STEM disciplines, that could undermine the effectiveness of mathematics learning within STEM activities (English, 2016). In addition, Thibaut et al. (2018) argued that "integrated STEM education should also focus on learning goals and standards in the individual STEM subjects, so as not to inadvertently undermine students learning in those subjects" (p. 6). Therefore, the task at hand is to find a balance that allows students to acquire deep knowledge of mathematics while also grasping its importance for solving real-world problems and shaping society (Skovsmose, 2023). Hence educators need to design interdisciplinary STEM learning experiences that genuinely integrate mathematical concepts, methods, and reasoning in problem-solving as well as make connections between mathematics and other STEM disciplines that help students gain a deeper understanding of the role of mathematics in society.

Our research sought to answer the following question:

> How can the balance be achieved between students cultivating mathematical understanding and simultaneously grasping the significance of mathematics when solving real-world problems that impact society?

This chapter includes an argument for advancing the role of mathematics through education for responsible citizenship, an approach to analysing the role of mathematics in teaching activities in a STEM curriculum context; a description of the role of mathematics in 26 teaching activities developed by teachers as part of the research and development project LabSTEM. LabSTEM was conducted in Denmark from 2020 to 2022 with the goal of developing STEM teaching with a mathematical focus.

The chapter also includes an in-depth analysis of one activity from LabSTEM that successfully incorporated mathematics as an objective and addressed social aspects of society. However, because the integration of these two aspects was not clear, and the students did not appear to engage with them simultaneously, the mathematics did not become relevant for the students, raising an issue about the importance of foregrounding mathematics in integrated STEM tasks.

Advancing the Role of Mathematics Through Education for Responsible Citizenship

There are various perspectives on how integrated STEM programmes can incorporate and utilise mathematics in conjunction with the other disciplines as well as how mathematics can be given more attention in STEM approaches. Maass et al. (2019) described three possibilities to advancing the role of mathematics that involve connecting learning in STEM by promoting:

1. The acquisition of twenty-first-century skills;
2. The meaningful inclusion of mathematical modelling in school education; and
3. Education for responsible citizenship (p. 873).

These three categories are interconnected. However, to answer our research question, the chapter focuses on the third category to explore how mathematics can be incorporated into STEM activities to support responsible citizenship. The aim of education for responsible citizenship is to equip young people with the ethical and moral decision-making skills they need to navigate a rapidly evolving world. Initiatives within the natural sciences involve addressing socio-scientific issues to foster discussions of the ethical aspects of scientific solutions (Maass et al., 2019). Critical mathematics approaches emphasise mathematics as a tool for informed decision-making (Skovsmose, 2023). Mathematics education can build responsible citizens through critical reasoning and social critique, but its precise role in the STEM context still requires exploration.

Numeracy and socio-scientific issues represent two potential approaches to supporting responsible citizenship within STEM (Maass et al., 2019). Numeracy focuses on developing individuals' mathematical skills for navigating private and public life, while addressing socio-scientific issues can raise awareness and generate solutions for global challenges, such as sustainability and climate change. Social scientific issues are inherently controversial and engage students in dialogue, discussion, and, ultimately, decision-making. For example, wind power is considered a sustainable energy source, but the installation of wind power plants can lead to undesirable effects, such as risks to wildlife, shade impacts, and infrasound (Maass et al., 2019). Maass et al. (2019) have advocated for the importance of cultivating students' responsible citizenship while developing their subject-specific STEM understandings and mathematical competencies. This perspective resonates with the intent and results of the research and development project LabSTEM, which balanced an orientation towards students' general competence, personal and civic formation, and mathematical understanding.

The LabSTEM Project

LabSTEM was a three-year research and development project that commenced in January 2020. Its primary objective was to create a STEM-based teaching approach tailored to Denmark that would be applicable to teaching and learning practices that support society-oriented, sustainable, and interdisciplinary STEM education. The project established temporary communities of practitioners and researchers known as 'STEM Laboratories', which were event-based gatherings rather than physical laboratories. From 2020 to 2022, 19 such laboratories were formed. They were comprised of approximately 200 participating teachers, including kindergarten, primary, and secondary school teachers.

In the LabSTEM project, the activities were guided by five STEM principles, which were iteratively developed within the LabSTEM research and development work. The principles stated STEM activities should:

1. significantly contribute to the development of personal and civic formation;
2. prioritise participant-centred approaches, such as problem-based or inquiry-based teaching;
3. focus on the real world using a relevant context;
4. support learning objectives that are relevant to the curriculum at a particular level or stage of education; and
5. integrate two or more disciplines within STEM in a meaningful manner—for example, both mathematics and engineering or both science and mathematics (Larsen et al., 2022).

The first principle is the primary focus of this chapter. While the other principles are also important, the first principle involves an emphasis on cultivating students' personal and civic formation while developing their subject-specific STEM understandings. The term 'personal and civic formation' is a translation of the Danish word *almendannelse* and the German word *Bildung*, for which there is no exact English translation. However, "all papers describe Bildung as a positive and fruitful concept that aims at the formation of the learner as a person and contributes to the development of his/her full potential" (Sjöström et al., p. 175). We elaborate on the meaning of 'personal and civic formation' in this section in light of the Danish school system and the LabSTEM project. By nature, personal and civic formation not only encompasses vocational functions and professional life but also concerns persons as individuals (Schnack, 2003). According to Schnack (2003), personal and civic formation has five dimensions. The first two dimensions relate to the individual: first, personal and civic formation is for everyone and accessible to all individuals; second, the development of personal and civic formation concerns all aspects of a person, including their cognitive, emotional, social, physical, communicative, and other competencies. The next three dimensions regard the individual in society: personal and civic formation concerns are significant and relevant to all of us; personal and civic formation involves an understanding of cultural tradition with its accompanying values and norms, which are neither arbitrary nor immutable; and there is a need to understand the tension between the understanding of the tension between individuality and community and the ability and willingness to participate in various communities (Schnack, 2003).

In LabSTEM, the first of the five STEM principles involves the development of a person's independence and autonomy to allow them to make independent decisions, have control over their life, and resist manipulation (Sjøberg, 2007). Additionally, personal and civic formation includes helping individuals develop a wide range of versatile skills and knowledge (Sjøberg, 2007) and takes place in all aspects of life (Oettingen, 2016), but schools and their subjects support the development of students to participate independently, reflectively, and critically in democratic society (Sjöberg, 2007). To this end, natural sciences, mathematics, and technical sciences (i.e. STEM curricula) offer an holistic approach that can help us comprehend the complexity of the world and transdisciplinary issues.

Throughout the LabSTEM project, the teachers convened in small 'laboratories' located across the country. These laboratories consisted of small workshops of approximately 5 to 20 teachers from different institutions. The participants gained inspiration and knowledge about STEM from university professors, and they worked together to develop STEM activities. The teachers had diverse motivations for participating in the project: some expressed a personal interest in STEM and had voluntarily joined the initiative, while others had been encouraged to participate by their school administrators, who sought to promote interdisciplinary collaboration within their institutions.

The purpose of the STEM laboratories was to enable the teachers to design STEM teaching activities which included the five principles and had mathematics as the central focus. These activities were intended to be empirically tested in real classroom settings under the guidance of university college lecturers or researchers. The primary and lower secondary school teachers in the laboratories developed a total of 26 STEM teaching activities. The participating teachers were specialised in mathematics education or science education, and many had expertise in both domains, which might have influenced the role of mathematics within the activities. The STEM activities were collaboratively developed by teams of two to four teachers in different STEM laboratories. The teachers documented each activity, including details such as the title, grade level, theme, duration, prerequisites, alignment with the national curriculum, goals, assessment methods, learning process, and involvement of each of the STEM disciplines. Some of these activities have been published in various publications for STEM teachers in preschools and primary and secondary schools (Christensen et al., 2023; Mariegaard et al., 2023; Larsen et al., 2023; Svendsen et al., 2023).

In the next section, we consider the 26 STEM activities that were developed for primary and lower secondary schools in the LabSTEM project to investigate the extent to which the STEM teachers successfully centred mathematics as the focal point of the activities and the role of mathematics within those activities. We were keen to explore the ways in which mathematics can be used in STEM activities to develop the citizenship and personal and civic formation of students. For this investigation, we first employ a framework developed by Kristensen et al. (2024). Then, we apply the argumentation of Maass et al. (2019) regarding education for responsible citizenship in STEM education.

The Role of Mathematics in STEM Activities

The framework was developed by the LabSTEM research group on the basis of a comprehensive literature review and analytical process reported by Kristensen et al. (2024). In this framework, mathematics is classified either as a primary goal of the activity or as a tool. When mathematics acts as a tool, it can be used to improve students' understanding of science or technology, to qualify an engineering design process, or to solve problems. When it is the primary objective of an activity, the goals concern mathematical content, skills, and procedures as well as mathematical competencies, as defined and described by Niss and Højgaard (2011). When mathematical goals are the focus, any of the other three disciplines, or a combination of them, can serve as the context for achieving the mathematical goals.

Notably, the framework should not be viewed as an 'either/or' classification tool. In some STEM activities, mathematics is both a tool and the primary goal. For example, mathematical modelling can be regarded as both a crucial tool to acquire a deep understanding (e.g. of science, as in Michelsen, 2016) and the primary objective of teaching (Niss & Højgaard, 2011). The framework can be used for planning and discussion with the aim of highlighting and nuancing the various roles of mathematics in STEM. Kristensen et al. (2024) have recommended adopting this framework to achieve a balanced focus amongst the disciplines. This approach is effective for increasing attention to mathematics and underscoring the importance of prioritising mathematics as the primary objective when designing STEM activities.

The 26 STEM Activities Developed in LabSTEM

The teachers in LabSTEM worked on achieving a balance between cultivating a deep mathematical understanding in students and developing their personal and civic formation, which reflects the first principle of the project. To assess how mathematics was integrated into STEM activities, this section analyses the 26 activities developed by the teachers according to the framework in Fig. 5.1. The 26 activities are presented in three tables (Tables 5.1, 5.2 and 5.3) corresponding to primary school (grades 1–3), middle school (grades 4–6), and lower secondary school (grades 7–9) activities, respectively. Each table provides a brief description of the activities in the first column. The second column specifies whether mathematics is an explicit objective of the activity, and which mathematical content is targeted by the activity. The third, fourth, fifth, and sixth columns correspond respectively to the four distinct categories of 'mathematics as a tool'. It is important to note that our analysis of the 26 activities is based solely on the course descriptions provided by the teachers and not on any empirical observation of the activities.

All of the activities identify mathematics as an objective. The goals range from learning to use an Excel spreadsheet to applying statistical concepts to analyse and utilise data. They also include objectives related to drawings and scale factors in mathematics. Many of the activities use mathematics as a tool to qualify the engineering process (16 activities), and/or to improve scientific understandings (10 activities) and/or to improve understandings of technology (2 activities). However, some activities use mathematics as more of a problem-solving tool for mathematical problems which do not connect to the other disciplines in the same way (3 activities). In one of the activities, mathematics is not directly related to the overall problem; the focus is on the history of pi and not on designing jewellery, which is the real-world context. The results presented here align with the trend observed in Kristensen et al.'s (forthcoming) review, which shows that mathematics is commonly used in engineering and secondarily in science but not frequently observed in problem-based approaches.

When examining the mathematics-specific content targeted by the 26 activities, there is a broad spectrum of content. However, it is difficult to determine whether the content aims to review already-acquired mathematical knowledge and skills or to introduce new mathematical concepts, processes, or understandings. Our analysis suggests that the mathematical aims of the activities predominantly emphasise the application of familiar mathematical concepts, which is understandable since it can be challenging for students to learn new content and procedures while simultaneously solving more complex STEM problems.

In relation to Maass et al.'s (2019) argument that STEM activities should be grounded in real-world contexts, we observe in the first column that the objectives and intentions of nearly all STEM activities align with the aim of fostering

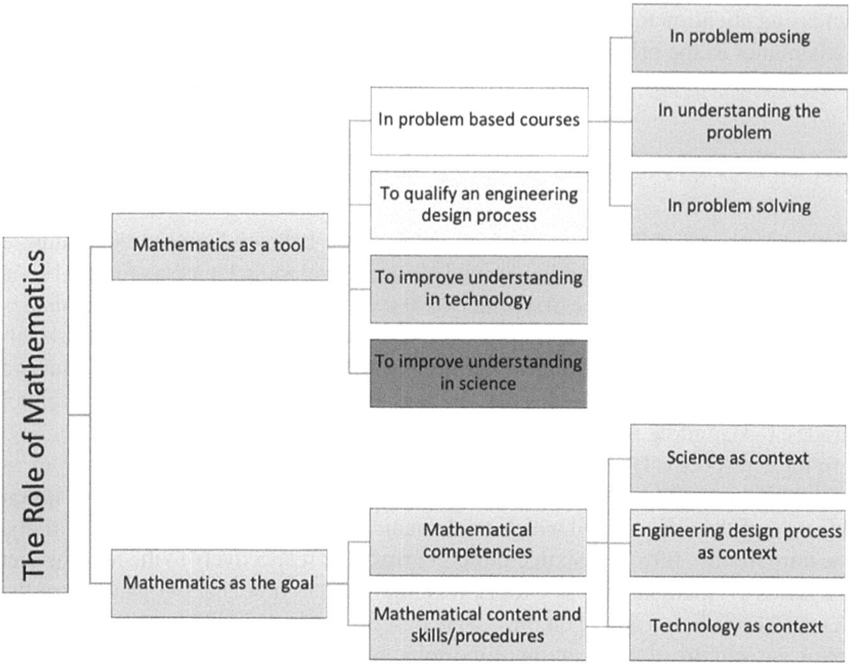

Fig. 5.1 Framework on the Role of mathematics in STEM (adapted from Kristensen et al., 2024, p. 423)

students' personal and civic formation, which is highlighted in the first principle of the LabSTEM project. Many of the topics explored in the activities are universal and facilitate discussion and the cultivation of students' decision-making skills. The STEM approach helps to ensure that decisions are based on evidence. For instance, in the first primary school activity in Table 5.1, first-graders learn about litter in nature, recyclable materials, and environmental degradation timelines. Similarly, two fourth-grade activities involve the reduction of plastic and cigarette litter in the local environment. In these activities, students begin by investigating whether the issue genuinely constitutes a problem, often relying on specific observations or statistical data. Subsequently, they are tasked with formulating potential solutions and, in some cases, acting on the proposed measures.

Several activities encourage collaboration and discussion, particularly in engineering and design processes where students must reach a consensus about the product's specifications. For instance, in one activity from Table 5.2, fourth-grade students work in groups to design a hedgehog habitat that improves the living conditions of the animals. This activity not only promotes active participation in a community but also addresses the relation between the individual and the community, as students may be willing to reject their own ideas in favour of other students' proposals. Some activities are explicitly centred on sustainability, such as one activity for lower secondary students in Table 5.3, which focuses on constructing a sustainable house

Table 5.1 Overview of the role of mathematics in the 6 activities developed for primary school

Primary school (Grades 1–3)	Mathematics as a goal	Mathematics as a tool			
Brief description of the LabSTEM teaching activity	Is M explicit as a goal? M content	M for science	M for tech	M in engine. process	M in problem-solving
Focus on litter in nature and what can be recycled as well as the longevity of degradation in the environment	Yes—Statistics on the found litter	✓			
Teamwork and heart rate in playing Counter-Strike	Yes—Statistics on the class heart rate	✓			
Birds in the woods and in gardens (statistics)	Yes—Statistics on the different birds	✓			
Energy for heating bath water	Yes—Statistics on the temperature	✓			
Germination of cress seeds and sunflower seeds under different conditions	Yes—Measurements and statistics on the sprouts	✓			
Designing a bicycle with the best wheels	Yes—Statistics			✓	

or bicycle shed using environmentally friendly materials. Another activity for lower secondary school revolves around wolves in Denmark. This activity initiates ethical and moral discussions; the presence of wolves in Denmark is a relatively recent phenomenon, and opinions about their existence in the country vary widely, with arguments for and against their eradication.

In summary, it is evident that teachers in the LabSTEM project succeeded in creating STEM activities which approach mathematics as a distinct goal while simultaneously contributing to the development of other STEM areas in which mathematics serves as a valuable tool. All of these activities are firmly grounded in real-world contexts, and many are designed to nurture students' personal and civic formation. Thus, the activities allow students to experience first-hand how mathematics can assist and enhance engineering processes, deepen their comprehension of science and technology, and contribute to problem-based activities, while acquiring essential general competences.

A more in-depth examination of the 26 activities in Tables 5.1, 5.2, and 5.3 reveals that they not only have distinct mathematical objectives but also differ in how they focus on responsible citizenship and personal and civic formation. However, from the tables, it is difficult to ascertain whether there is any relationship between the

Table 5.2 Overview of the role of mathematics in the 9 activities developed for middle school

Middle school (Grades 4–6)	Mathematics as a goal	Mathematics as a tool			
Brief description of the LabSTEM teaching activity	Is M explicit as a goal? M content	M for science	M for tech	M in engine. process	M in problem solving
Developing a dream house with sustainable materials and insulation	Yes—Mathematical floorplan drawing and geometric shapes			✓	
Designing, placing, and developing sustainable planters and gardens at school	Yes—Creating working drawings and sketches as well as optimising space			✓	
Designing a hedgehog habitat to enhance their living conditions	Yes—Creating working drawings and sketches as well as optimising space			✓	
Designing and constructing an insulated container based on the insulation of environmentally friendly houses	Yes—Slopes and intersections of functions in a coordinate system	✓		✓	
How to build houses that can withstand earthquakes	Yes—Creating working drawings and statistics			✓	
How to reduce plastic waste in our city	Yes—Statistics			✓	
How to reduce cigarette litter in our city	Yes—Statistics			✓	
Designing a place to live in nature (solo in the wilderness)	Yes—Creating working mathematical drawings			✓	
Designing a house with a focus on natural light and sun as the primary energy source	Yes—Creating working drawings			✓	

role of mathematics and the activity's approach to incorporating responsible citizenship formation. Regrettably, COVID-19 restrictions prevented us from accessing the classrooms during the project, so only one out of the 25 activities was recorded on video—the last activity in Table 5.3—which we analyse in greater depth to gather more empirical data. This activity was video-recorded on a handheld camera by a research assistant who followed a group of students throughout the entire activity.

Table 5.3 Overview of the Role of Mathematics in the 11 Activities Developed for Lower Secondary School

Lower secondary school (Grades 7–9)	Mathematics as a goal	Mathematics as a tool			
Brief description of the LabSTEM teaching activity	Is M explicit as a goal? M content	M for science	M for technology	M in engine. process	M in problem solving
Designing a mini-golf course	Yes—Creating working drawings and sketches as well as optimising space			✓	
Developing a game that helps with understanding family finances	Yes—Excel and various financial concepts		✓		✓
Optimising students' high jump in physical education through technology	Yes—Optimisation of the parabola based on angles	✓			✓
Incorporating pi and sustainability into jewellery design for the royal family	Yes—Understanding and using pi			✓	
Adaptability of a bird to its environment—significance of conserving the environment for regeneration of its food	Yes—Statistics		✓		
Designing and constructing a bicycle shelter to encourage more cycling to school	Yes—Creating working mathematical drawings and making the economics			✓	✓
Investigate 1 square metre of nature for different plants and animals (solo in the wilderness)	Yes—Statistics	✓			
The energy-efficient house: what does it look like, and why?	Yes—Analysing data from sustainable materials and designs			✓	
The distribution of wolves in Europe and Denmark	Yes—Working with data about the distribution of wolves in Europe and Denmark	✓			
Designing a throwing machine that can throw as far as possible (solo in the wilderness)	Yes—The appearance of a parabola			✓	
Designing the best hydropower plant	Yes—The volume and velocity of water	✓		✓	

Developing and Designing a Hydroelectric Power Plant

The last activity in Table 5.3 tasks seventh-grade students with studying hydroelectric power and energy in a full-day workshop. The activity is structured with three initial exercises leading up to the fourth and final exercise, consisting of building a model of a hydroelectric power plant (Fig. 5.2).

During the observed activity, the first exercise prompted students to express their opinions about whether more artificial water channels should be excavated to increase water flow to the hydropower plant, which would allow the plants to generate more green energy but would also impact the biodiversity of the area. Students were required to position themselves under the statement they found most favourable and provide the best possible argument to support their position. The second exercise consisted of a photo safari in which students had to capture images of the various components of the power plant and explain them. These components included turbines, generators, water supply systems, and eel ladders, amongst others. The aim of this exercise was to familiarise the students with the different components of the plant.

The third exercise was divided into four workshops. These workshops aimed to facilitate a detailed examination of the individual parts of the plant and guide students in utilising mathematics to explain and enhance the efficiency of the hydroelectric plant. The question for the first workshop was, 'How much water can you transport

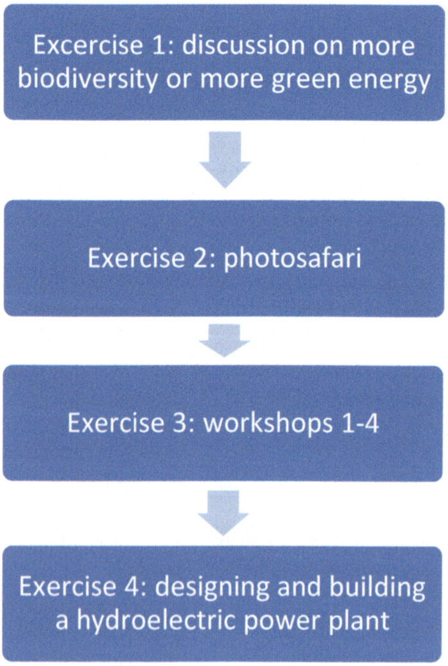

Fig. 5.2 The sequence of exercises in the seventh-grade field day

from the top of the reservoir to the bottom in 10 min, and at what speed?' In this workshop, each student was given a bucket to fill with water and then instructed to run downhill with it. At the bottom of the hill, there was a tub in the shape of a truncated cone, which the students had to fill. The mathematical task was to calculate the amount of water they managed to transport down the hill and the speed of the water. To calculate the volume of the water, they measured the height of the water and the diameter of the bucket and used a formula provided by the teacher. In the second workshop, the task was to measure the velocity of the water stream exiting the hydroelectric plant. The students were required to experiment by placing various objects into the water, such as cream, leaves, and sticks, and measuring the time it took for them to move a certain distance. They were then asked to calculate the average velocity of their measurements. The third workshop required students to investigate the question, 'How much water can the water supply pipe hold?' For this, the students had to measure the circumference and length of the pipe and calculate its volume. Finally, in the fourth workshop, the goal was to find the volume of the box-shaped water reservoir by measuring the length and width of the box and determining the depth by using a string with a stone attached to it.

The fourth and last exercise was linked specifically to the question, 'How can we build the most efficient hydroelectric plant by using different materials (different pipes, jugs of water, small turbines, and voltmeters)?' This part could be considered an engineering project and an evaluation of the day. The intention was for the students to apply all the knowledge they gained throughout the day to build their own mini hydroelectric plant.

Overall, this activity exemplifies the first principle of LabSTEM as well as Maass et al.'s (2019) concept of responsible citizenship. The first exercise necessitates students' active involvement in grappling with an intricate problem, which nurtures their self-reliance, aptitude for constructing arguments, reflective capabilities, and critical thinking. The addressed problem holds universal significance, and the aim is to engage students with the concept of responsible citizenship in relation to the feasibility of developing the hydroelectric power plant. The experiences of the students while discussing this issue can be extrapolated to more diverse and expansive scenarios and predicaments, thus prompting them to consider whether it is advisable to construct additional water channels that may impact the area's biodiversity or if it is of greater significance to preserve ecological diversity. This aspect could be perceived as a socio-scientific issue characterised by a heightened ethical consideration of such matters.

The subsequent exercises similarly centre on socio-scientific issues but delve more specifically into the operational mechanics of a hydroelectric power plant—an approach by which students come to understand that energy can be generated through the movement of water and without perpetual reliance on the combustion of fossil fuels. Such comprehension is essential for students to acquire to effectively grasp the challenges prevalent in society regarding sustainability and the environment.

By examining this sequence through the lens of Kristensen et al.'s (2024) framework, we can observe that the role of mathematics varies significantly across the exercises, and mathematics is an explicit objective primarily in the four workshops.

The intention of these workshops is to have students utilise mathematics to enhance their comprehension of hydroelectric power plant operations. The students calculate the volume of both the reservoir and the pipes not necessarily to improve their understanding of volume but rather to practically apply volume concepts in a real-world context. Additionally, the workshops help them recognise the significance of having a substantial reservoir and the role of water velocity in generating electricity.

In the fourth exercise, where the students develop a hydroelectric power plant, the explicit involvement of mathematics is somewhat limited. Nonetheless, the overarching intention is for students to draw on the knowledge they acquired during the preceding workshops and effectively apply it. In this context, their objective is to devise and construct a highly efficient hydroelectric power plant characterised by the presence of high-velocity water flow and a substantial volume of water within the pipes. Within the framework in Fig. 5.1, this activity places mathematics in two roles: one supporting the comprehension of science, specifically of enhanced flow capacity, and the other as an integral component of the engineering aspect.

Balancing Mathematical Understanding with the Role of Mathematics in Society

The activity reviewed here is a STEM-oriented activity that yields numerous benefits and contributions to students' STEM education. The challenge of using such STEM activities is that, while teachers successfully incorporated both mathematical content and broader societal dimensions, these elements tend to remain somewhat segregated. Consequently, students tend to either immerse themselves in the mathematics by applying formulas, or they engage in a more comprehensive exploration of critically related inquiries connected to society.

On the one hand, developing mathematical understanding is vital to STEM disciplines and obtaining the analytical tools for comprehending and solving complex problems across STEM domains. A solid foundation in mathematical concepts equips students with, for example, quantitative skills for engaging with intricate scientific models, devising engineering solutions, and analysing data-driven outcomes. On the other hand, an emphasis on mathematical concepts alone may risk overshadowing the broader context in which mathematics operates in society. Therefore, we find it important to complement the development of mathematical concepts with an exploration of the pragmatic relevance of mathematics beyond academic settings.

It has been argued by Fitzmaurice et al. (2021), among others, that many students believe that mathematics will be neither necessary nor useful for them in school or in their future studies and careers. Mathematics is viewed as an isolated school subject that lacks connections to life and other subject areas. Recognising the relevance of school lessons is important for both the intrinsic and extrinsic motivation of students (Mujtaba et al., 2018). In the activity reviewed here, we observed that this integration did not occur. Because of the different and somewhat segregated exercises, students may not experience this connection and fail to realise the broader societal implications of mathematics, which would hinder their ability to comprehend the significance of mathematics for informed citizenship. Fostering an understanding of the societal relevance of mathematics can support students' intrinsic motivation to engage with the subject once they acknowledge its practical applications and meaningful connections to the real world.

Several other activities from the LabSTEM project reveal a similar pattern. For example, in the activity about constructing a bicycle shelter for their school, students focus on creating sketches and construction drawings to guide the actual building process of the shelter, but the broader societal issues, such as material selection, remain somewhat disconnected from the mathematics, which lacks distinct integration. In essence, we find it important for STEM activities to include not only one aspect and instead try to incorporate objectives for both developing mathematical concepts and understanding of the role of mathematics in society. Such a balanced approach can ensure that students develop mathematical competence while cultivating an holistic understanding of the contribution of mathematics to an informed and responsible citizenry in our increasingly quantitative world.

Implications

There are several noteworthy implications of this analysis. The first significant implication is the importance of the model in Fig. 5.1. This model is especially useful for planning STEM activities because it provides teachers with a reflective tool to think about the role of mathematics in their own teaching. Thus, it can empower educators in a practical way to synergistically integrate mathematics into STEM activities and focus on cultivating students' mathematical understanding alongside issues such as socio-scientific matters and citizenship. In this way, the implication is the need for a stronger focus on the connections between mathematics and other disciplines in the planning and development of STEM activities. In this regard, it is crucial to begin the planning by explicitly defining the mathematical goals of the activity in terms of mathematical content, skills, procedures, and competencies. However, teachers must also carefully consider the contexts in which such mathematical content, procedures, and competencies naturally emerge within integrated STEM programmes. As the model illustrates, this can occur in conjunction with one of the other disciplines or in programmes that incorporate all the disciplines.

One mathematical process or competency that encapsulates mathematics in the real world is mathematical modelling. As Maass et al. (2019) have underscored, it can be beneficial to enhance the role of mathematics in STEM programmes while concurrently focusing on responsible citizenship by incorporating mathematical modelling when addressing socio-scientific issues. Furthermore, it is pertinent to critically consider the model's validity and simplifications in addition to the integration of mathematics into problem-based scenarios where mathematics is essential to understanding and problem-solving. In contrast to the Harteværket example, where mathematical modelling should have perhaps been incorporated into the initial stage but was entirely absent, the role of mathematics should be integrated and justified across all phases of the programme.

This analysis has highlighted the need for researchers, educators, and textbook authors to recognise and help overcome the challenges of developing STEM activities that illustrate the relevance and significance of mathematics in more open, real-world STEM contexts that include social issues. It is important for researchers and educational leaders to provide teachers with the necessary support and guidance to incorporate these elements since, as per our analysis, this task is difficult for teachers to undertake on their own. One approach for engaging teachers in the development of such activities is the use of lesson studies (Stepanek et al., 2006) in which teachers and researchers collaborate to create highly effective lessons that specifically integrate academic objectives and socio-scientific domains. Through various iterations of experimentation, it becomes possible to monitor whether genuine integration is occurring or if the work is progressing more in parallel.

Another implication is that school leaders must both assist teachers with delivering the opportunities and allocate time for teachers to develop such sequences while also allowing for post-reflection after teaching the activities. Many teachers have expressed that they are pressed for time and consequently do not have opportunities to develop STEM sequences or reflect on whether the sequences unfolded as intended. Here, another important consideration is that the schedule structure for traditional subjects in primary education often hinders the integration of subjects, which is a key challenge to address.

The third implication is that students can only benefit from such STEM activities if teachers have acquired the expertise to teach them. Teachers are not likely to already possess this expertise since the teacher education programme in Denmark does not focus on interdisciplinary teaching. Training in administering these activities can occur either during initial teacher education or in postgraduate training programmes. It is also possible for partnerships with universities, for instance, to play a pivotal role in reshaping teachers' beliefs and enhancing their capacities for teaching STEM activities.

Finally, there is a need for more research in this area. Further research efforts should delve into these educational environments while considering the epistemological variations across STEM disciplines and, specifically, the role of mathematics within these contexts. Future studies could start by identifying effective STEM problems within and across disciplines at various educational levels—for instance, by building upon the work of Reinholz et al. (2018)—while still focusing on the role of mathematics.

LabSTEM has recently received a new grant and is now referred to as LabSTEM+ . Moving forward, the project aims to build upon these insights and further investigate how educators can develop STEM sequences that integrate a focus on mathematics while simultaneously incorporating the first principle. It will address the possible concrete forms of such sequences as well as which knowledge and competencies educators must possess to craft such sequences within their own instructional contexts. Based on the analyses of LabSTEM and this chapter, LabSTEM + will focus on having laboratory instructors work much more closely with teachers in developing STEM activities to specifically help the teachers overcome the challenge of integrating the subjects and incorporating the first principle in a relevant and interconnected way.

References

Baldinger, E., Staats, S., Clarkson, L., Gullickson, E., Norman, F., & Akoto, B. (2020). A review of conceptions of secondary mathematics in integrated STEM education: Returning voice to the silent M. In J. Anderson & Y. Li (Eds.), *Integrated approaches to STEM education* (pp. 67–90). Springer. https://protect-au.mimecast.com/s/geILCP7LAXf09g7gouzRZKX?domain=doi.org

Becker, K., & Park, K. (2011). Effects of integrative approaches among science, technology, engineering, and mathematics (STEM) subjects on students' learning: A preliminary meta-analysis. *Journal of STEM Education: Innovations and Research, 12*(5–6), 23–36.

Bybee, R. W. (2018). *STEM education now more than ever*. National Science Teachers Association.

Christensen, M., Svabo, C., & Larsen, D. (2023). *STEM i erhversskolen: aktiviteter og ideer.* [STEM in vocational schools: Activities and ideas], Syddansk universitet. Laboratorium for STEM Udannelse og læring. Forskningscenter for Naturvidenskabelig Uddannelse og Formidling.

English, L. D. (2016). STEM education K–12: Perspectives on integration. *International Journal of STEM Education, 3*(3), 1–8.

Fitzallen, N. (2015). STEM education: What does mathematics have to offer? In M. Marshman (Ed.), *Mathematics education in the margins. Proceedings of the 38th annual conference of the Mathematics Education Research Group of Australasia* (pp. 237–244). MERGA.

Fitzmaurice, O., O'Meara, N., & Johnson, P. (2021). Highlighting the relevance of mathematics to secondary school students—Why and how. *European Journal of STEM Education, 6*(1), 07. https://doi.org/10.20897/ejsteme/10895

Honey, M., Pearson, G., & Schweingruber, H. (Eds.). (2014). *STEM integration in K–12 education: Status, prospects, and an agenda for research.* National Academies Press.

Hurley, M. M. (2001). Reviewing integrated science and mathematics: The search for evidence and definitions from new perspectives. *School Science and Mathematics, 101*(5), 259–268.

Kristensen, M. A., Larsen, D. M., Seidelin, L., & Svabo, C. (2024). The role of mathematics in STEM activities: Syntheses and framework from a literature review. *International Journal of Education in Mathematics, Science and Technology, 12*(2), 418–431. https://doi.org/10.46328/ijemst.3357

Larsen, D., Kristensen, M., Hjort, M., & Seidelin, L. (2022). STEM-didaktik–et internationalt, systematisk review om STEM undervisningens didaktik [STEM didactics—An international systematic review of STEM teaching didactics]. *MONA-Matematik-og Naturfagsdidaktik, 22*(1).

Larsen, D. M., Auning, C., Ellebæk, J. J., Bennedsen, K., & Svabo, C. (2023). *LabSTEM – STEM i grundskolen: aktiviteter og ideer* [LabSTEM - STEM in primary education: Activities and ideas]. Odense: Syddansk universitet. Laboratorium for STEM Uddannelse og læring. Forkningscenter for Naturvidenskabelig Uddannelse og Formidling.

Maass, K., Geiger, V., Ariza, M. R., & Goos, M. (2019). The role of mathematics in interdisciplinary STEM education. *ZDM Mathematics Education, 51*, 869–884.

Mariegaard, S., Jensen, B. L., Bennedsen, K., Foged, L., Larsen, D. M., & Svabo, C. (2023). *LabSTEM – STEM i dagtilbud: Aktiviteter og ideer [LabSTEM - STEM in early childhood education: Activities and ideas]*. Syddansk universitet. Laboratorium for STEM Uddannelse og læring.

Martín-Páez, T., Aguilera, D., Perales-Palacios, F. J., & Vílchez-González, J. M. (2019). What are we talking about when we talk about STEM education? *A Review of Literature. Science Education, 103*(4), 799–822.

Michelsen, C. (2016). Den matematiske og naturvidenskabelige dannelse [Mathematical and scientific formation]. In A. C. von Oettingen, C. Michelsen, J. J. Hansen, & J. Vinther (Eds.), *Almen dannelse: Dannelsesstandarder og fag* (pp. 37–80). Hans Reitzels Forlag.

Mujtaba, T., Sheldrake, R., Reiss, M. J., & Simon, S. (2018). Students' science attitudes, beliefs, and context: Associations with science and chemistry aspirations. *International Journal of Science Education, 40*(6), 644–667. https://doi.org/10.1080/09500693.2018.1433896

Niss, M., & Højgaard Jensen, T. (2011). *Competencies and mathematical learning: Ideas and inspiration for the development of mathematics teaching and learning in Denmark*. Roskilde University.

Oettingen, A. C., Michelsen, C., Hansen, J. J., & Vinther, J. (2016). *Almen dannelse: Dannelsesstandarder og fag* [General education: Educational standards and subjects]. Hans Reitzels Forlag.

Reinholz, D., Slominski, T., French, T., Pazicni, S., Rasmussen, C., & McCoy, B. (2018). Good problems within and across disciplines. *Journal of Research in STEM Education, 4*(1), 37–53. https://protect-au.mimecast.com/s/vz7cCQnMBZfXgV5V3IPuPpO?domain=doi.org

Schnack, K. (2003). Almendannelse som demokratisk dannelse [General education as democratic education]. *Uddannelse, 5*, 23–29.

Shaugnessy, M. (2013). By way of introduction: Mathematics in a STEM context. *Mathematics Teaching in the Middle School, 18*(6), 324.

Sjöberg, S. (2007). *Naturfag som almendannelse: En kritisk fagdidaktik* [Science education as general education: A critical pedagogy]. Klim.

Sjöström, J., Frerichs, N., Zuin, V. G., & Eilks, I. (2017). Use of the concept of *Bildung* in the international science education literature, its potential, and implications for teaching and learning. *Studies in Science Education, 53*(2), 165–192. https://doi.org/10.1080/03057267.2017.1384649

Skovsmose, O. (2023). *Critical mathematics education*. Springer International Publishing.

Stepanek, J., Appel, G., Leong, M., Mangan, M. T., & Mitchell, M. (2006). *Leading lesson study: A practical guide for teachers and facilitators*. Corwin Press.

Svendsen, M., Borch, K. B., Jepsen, K. N., Kristinsdottir, L., Larsen, D. M., & Svabo, C. (2023). *LabSTEM – STEM i gymnaiset: Aktiviteter og ideer [LabSTEM - STEM in secondary education: Activities and ideas]*. Syddansk universitet. Laboratorium for STEM Udannelse og læring.

Thibaut, L., Ceuppens, S., De Loof, H., De Meester, J., Goovaerts, L., Struyf, A., Boeve-de Pauw, J., Dehanene, W., Deprez, J. D., Cock, M., Hellinckx, L., Knipprath, H., Langie, G., Struyven, K., Van de Velde, D., Van Petergem, P., & Depaepe, F. (2018). Integrated STEM education: A systematic review of instructional practices in secondary education. *European Journal of STEM Education, 3*(1), 2.

Part III
Mathematics in the Development of 21st Century STEM Skills

Chapter 6
The Role of Mathematics in STEM for Developing Twenty-First Century Skills

Katie Makar

Abstract Twenty-first century skills are highly valuable in the workplace and for use in personal lives and citizenship. Internationally, school curriculum includes the development of critical thinking, creativity, collaboration and communication as an expectation across all content areas, including mathematics. Integrated tasks across science, technology, engineering and mathematics (STEM) are often seen as a vehicle to develop these skills, however STEM experiences in school often miss opportunities to engage mathematics in ways that enrich students' mathematical understanding. It can be difficult to find strong examples of mathematically-rich integrated STEM tasks. This chapter is divided into two parts. First, literature is reviewed that provides an overview of the specific contributions that mathematics makes in integrated STEM tasks to each of the four main twenty-first century skills. Then, an overview is given of the three chapters in this section and how they each illustrate specific examples of well-designed mathematically-rich integrated STEM tasks that have been used in classrooms to develop students' twenty-first century skills, with implications for practice.

Keyword Mathematically-rich integrated STEM tasks · Twenty-first century skills · problem solving · Critical thinking · Creativity · Collaboration · Communication

The Contribution of Mathematics to Developing Twenty-First Century Skills in STEM

In Sect. 1 of this book, we outlined cases of how mathematics is placed in integrated STEM tasks. The authors in this section demonstrated the challenges and opportunities for raising the profile of mathematics in integrated STEM tasks. These chapters illustrated the critical role of mathematics in STEM tasks, beyond their role as tools

K. Makar (✉)
The University of Queensland, St Lucia, QLD, Australia
e-mail: k.makar@uq.edu.au

J. Anderson and K. Makar (eds.), *The Contribution of Mathematics to School STEM Education*, https://doi.org/10.1007/978-981-97-2728-5_6

of computation (Goos et al., 2023). Integrated STEM tasks typically involve real-life problems, allowing them to be a vehicle for progressing reform ideas in mathematics education, particularly by enhancing problem solving.

Integrated STEM tasks often support a focus on common big ideas across the STEM areas, take a holistic approach to problem solving, and encourage transfer of learning across disciplines (Moore et al., 2020). The increased interest in STEM tasks in schools was partially a response to the need to embed twenty-first century (21C) skills into mathematics, science, technology and engineering. Early definitions of 21C skills focused on four skills: critical and creative thinking, collaboration, and communication (NRC, 2012; P21, 2009). Although more contemporary 21C skills have expanded beyond these four areas, with some descriptions including skills such as digital literacies, socio-emotional skills and citizenship, the four original skills remain a fairly constant presence, with consensus in a number of curricular documents. Each of these four areas have been interpreted in terms of the design of mathematically-rich STEM education experiences with important implications.

Critical Thinking, Creativity, Collaboration and Communication for Mathematics in STEM

Critical thinking is often classified as a cognitive process that emphasises domain-general skills such as complex problem solving, evidence-based reasoning and informed decision making (Lavi et al., 2021; NRC, 2012). Critical thinking is considered essential for information processing and analysis necessary for all aspects of life and work, such as making health-related decisions based on self-assessment, media reports, physician diagnosis and internet information gathering. In an information-age, the capacity to filter and critically evaluate large amounts of information to determine what is both relevant and credible is essential (van Laar et al., 2020). Critical thinking is the most apparent of the 21C skills across mathematically-rich STEM experiences as it highlights synthesis and selection of relevant information, explanation and evidence-based argumentation, transfer and application of disciplinary knowledge to an unfamiliar problem, and explicit expectations of complex problem solving and reasoning. Mathematics plays a key role in each of these elements, particularly in developing skills in explanation and evidence-based argumentation (Fielding-Wells, 2010).

Critical thinking is often highlighted as a more significantly perceived outcome by students of STEM experiences than other 21C skills (such as collaboration) or STEM-specific skills (Lavi et al., 2021). Because the focus of most STEM education learning experiences focus on complex problem solving, the implications of STEM education experiences for developing critical thinking is in providing students with the opportunity to practise a range of critical thinking skills. The contribution of mathematics to critical thinking is often in logical thinking and the way that evidence draws from mathematical concepts (e.g., data analysis, measurement). The authors in this section highlight multiple examples that emphasise critical thinking in mathematics within STEM experiences.

In contrast, creativity is a frequently over-looked 21C skill in STEM education. "Creativity is related to the production of new and useful ideas on products, services, or processes that are both novel and useful" (van Laar et al., 2020, p. 3). Complex problems in STEM often require solvers to generate, develop and design potential solution pathways (English, 2016). The capacity to do so is related to one's willingness to take intellectual risks because the solver must be willing to attempt a problem that they don't know how to solve (Allmond et al., 2016). In science and STEM areas, creativity involves students "… engaging in adaptive learning behaviors (sharing tentative ideas, asking questions, attempting to do and learn new things) … [and] highlights the fact that learning (like most any action) involves uncertainty and, therefore, some degree of risk" (Baghetto, 2009, p. 210).

Because creativity is often neglected as a skill relevant to mathematics, STEM experiences provide key opportunities for students to engage in the creative aspects of complex problem solving. The authors in this section demonstrate design elements that are more clearly linked with creativity, emphasising the mathematics in each case.

Most integrated STEM problems rely on teamwork to find solutions (Moore et al., 2020). Collaboration is one of the 21C skills that is an important driver for the rise in STEM education in schools, given that this skill is highly desired by employers. "Work is becoming more knowledge-based, interdisciplinary, and specialized. The complexity of tasks requires employees to collaborate, as individuals cannot possess all knowledge and skills" (van Laar et al., 2020, p. 3). Collaboration involves engaging with others to collectively produce a product. Working in collaboration with others develops leadership, goal setting, planning of tasks, reflection, interpersonal skills, and problem solving (Lavi et al., 2021; Stehle & Peters-Burton, 2019).

Embedding skills of collaboration in STEM education can also develop a more inclusive society by improving students' social skills, co-constructing and re-constructing ideas elicited from multiple perspectives that can greatly improve solutions, and build empathy in working collaboratively with others (Ah-Nam & Osman, 2017; Widya et al., 2019). However, STEM tasks may be designed with minimal levels of authentic collaboration, only putting students physically into groups without learning how to share responsibility, give and receive peer feedback, collectively make decisions or understand the nature and benefits of interdependence. In addition, when disciplinary experts collaborate to design integrated STEM tasks, there may be an imbalance in content that prioritises the learning goals of non-mathematics disciplines, re-enforcing the subservient role of mathematics in STEM (Goos et al., 2023). Mathematics can promote collaboration in integrated STEM tasks in multiple ways. For example, the emphasis on representations in mathematics provides a sharable artefact that can promote students' capacity to record and discuss observable patterns, relationships and key sources of evidence.

Finally, students' 21C skills of communication can be developed through mathematically rich integrated STEM education. Communication, "an ability to communicate effectively with a range of audiences", is often referred to as one of the 'soft skills' along with creativity and collaboration (Lavi et al., 2021, p. 10). Because inquiry and design challenges are common pedagogical approaches in STEM education, students have frequent opportunities to develop skills such as communication. For example, both inquiry and design models encourage collaboration, they provide opportunities to generate and articulate one's own ideas, listen and negotiate ideas with others, problem solve and defend their choices within the group, share interim progress and challenges and communicate findings appropriate to the audience including demonstrating how they have used evidence to reach their conclusion (Ah-Nam & Osman, 2017; Stehle & Peters-Burton, 2019). Mathematics disciplinary ways of knowing have been shown to support students' argumentation around complex STEM challenges; with the mathematical practices of representation, generalisation and justification identified as significant themes in the mathematically-rich integrated STEM education literature (Goos et al., 2023).

New Contributions of 21C Skills Embedded in Mathematically-Rich Integrated STEM Tasks

The chapters in this section each explain and illustrate ways that mathematics contributes to developing students' 21C skills in integrated STEM tasks. The classroom examples across the three chapters range from primary to secondary school and preservice teacher education.

In Chap. 7 "Mathifying STEM or STEMifying math? Challenges and possibilities for mathematics learning within integrated STEM contexts", Fitzpatrick et al. (2024) discuss the contributions of 21C skills in STEM and the dangers of neglecting disciplinary mathematics when designing integrated STEM tasks. Their study analysed integrated STEM tasks for middle school designed by preservice teachers. They found that even rich STEM lessons that integrated 21C skills in productive ways initially lacked appropriate mathematics. The authors proposed reasons why mathematics was often absent such as the novelty of engineering, the disconnect between the task and the curriculum and the difficulty in seeing opportunities for explicitly embedding mathematical content. The authors then used the design of a sustainable can holder before and after including a disciplinary mathematical focus as an example of how mathematics raised the quality of the task. By "mathifying" the task, they demonstrated how the mathematics enhanced the task beyond lower-order mathematical computations. Fitzpatrick et al. concluded that "attempts to *mathify* the STEM and *STEMify* the mathematics could uphold the integrity of disciplinary knowledge, while reaping the benefits of interdisciplinary approaches" (p. 127). The contrast of the quality of the task before and after mathematics was integrated contributes to an enriched understanding of the contributions that mathematics can have on integrated STEM tasks.

In Chap. 8, Crowder et al. (2024) investigate the benefits of developing 21C skills within a middle school problem-based curriculum. In their chapter, "Deepening mathematical understanding through a problem-based integrated STEM curriculum", the authors highlight the value of enriching a mathematics textbook series that "embeds mathematics in engineering-style design tasks" (p. 144). In the *Illustrative Mathematics* curriculum at the centre of their study, students repeatedly are asked to grapple with open-ended tasks across the range of content in the mathematics curriculum, ranging from simple warm up problems (estimating the height of a tree using an iconic building as a referent) to complex ones (designing a 4-person tent and creating a net to estimate materials needed).

In their study of four districts using the resource, interviews with teachers and administrators suggested that by valuing and supporting students through productive struggle (Warshauer, 2015), students appeared more engaged, demonstrated increased confidence in problem solving and reasoning, and developed 21C skills. The districts further perceived that when given time to develop their own professional practice with the materials, teachers' pedagogy became more student-centred. A significant element of the curriculum that encouraged change was a shift in focus from answers to more emphasis on reasoning and problem solving. For example, students were asked to "notice and wonder" throughout their work with the open-ended tasks. The authors provide several examples of the open-ended problems to illustrate and include excerpts from the interviews. Their work provides an example of opportunities that mathematically-rich integrated STEM tasks can bring to developing students' 21C skills in complex problem solving.

The last chapter in this section comes from Torralba and Rouse (2024), who introduce and illustrate the power of makerspace "as a vehicle for learning" (p. 155) to develop integrated STEM tasks with substantial and authentic development of mathematics. In "Doing mathematics in a school makerspace: Representing proportional reasoning through designing digital homes", the authors illustrate a rich integrated STEM design task in middle school that focuses on proportional reasoning (unit rates, percentage, scale), geometry (similarity) and the design cycle from engineering and technology. The middle school students in their study designed a dream home using digital tools to represent and explain their scaled design. The multi-modal approaches supported students to demonstrate their mathematical learning in ways that would not be possible in more traditional instruction and assessment formats. In the example provided, students share, interact, critique and offer feedback to their peers. Torralba and Rouse therefore highlight the opportunities of the makerspace movement to support teachers in designing units that allow students to demonstrate what they know through writing, talk and performance, thus developing 21C skills of communication, critical and creative thinking, and collaboration.

Across the three chapters in this section, the reader will gain valuable insights into ways that mathematically-rich integrated STEM tasks can enrich students' 21C skills. Across these chapters, the authors provide practical advice with specific examples of tasks that are classroom tested. All these tasks can develop students' mathematical content knowledge in ways that go beyond simple calculation. The authors also illustrate specific elements of 21C skills that are strengthened by engaging with these complex open-ended STEM tasks. The implications of their work are in demonstrating what mathematically-rich integrated STEM tasks look like in real classrooms and specific ways that 21C skills are enhanced. This can give teachers, curriculum designers and instructional leaders prototypes to design integrated STEM tasks that authentically draw on the power of mathematics.

References

Ah-Nam, L., & Osman, K. (2017). Developing 21st century skills through a constructivist-constructionist learning environment. *K-12 STEM Education, 3*(2), 205–216. https://www.learntechlib.org/d/209542

Allmond, S., Hillman, J., Huntly, K., Makar, K., & O'Brien, M. (2016). Assessing children's progress in taking intellectual risks in a mathematical inquiry classroom with a positive learning approach. In B. White et al. (Eds.), *Proceedings of the 39th annual conference of the Mathematics Education Research Group of Australasia* (pp. 94–101). MERGA.

Beghetto, R. A. (2009). Correlates of intellectual risk taking in elementary school science. *Journal of Research in Science Teaching, 46*(2), 210–223.

English, L. D. (2016). Advancing mathematics education research within a STEM environment. In K. Makar et al. (Eds.), *Research in Mathematics Education in Australasia 2012–2015* (pp. 353–371). Springer. https://doi.org/10.1007/978-981-10-1419-2_17

Fielding-Wells, J. (2010). Linking problems, conclusions and evidence: Primary students' early experiences of planning statistical investigations. In C. Reading (Ed.), *Proceedings of the 8th International Conference on Teaching Statistics* (ICOTS8). International Statistical Institute. http://icots.info/icots/8/cd/pdfs/invited/ICOTS8_2A3_FIELDINGWELLS.pdf

Goos, M., Carreira, S., & Namukasa, I. K. (2023). Mathematics and interdisciplinary STEM education: Recent developments and future directions. *ZDM–Mathematics Education, 55*(7), 1199–1217. https://doi.org/10.1007/s11858-023-01533-z

Lavi, R., Tal, M., & Dori, J. D. (2021). Perceptions of STEM alumni and students on developing 21st century skills through methods of teaching and learning. *Studies in Educational Evaluation, 70,* Article 101002. https://doi.org/10.1016/j.stueduc.2021.101002

Moore, T. J., Johnstone, A. C., & Glancy, A. W. (2020). STEM integration: A synthesis of conceptual frameworks and definitions. In C. C. Johnson, M. J. Mohr-Schroeder, T. J. Moore, & L. D. English (Eds.), *Handbook of research on STEM education* (pp. 3–16). Routledge.

National Research Council. (NRC) (2012). *Education for life and work: Developing transferable knowledge and skills in the 21st century.* National Academy Press. https://nap.nationalacademies.org/download/13398

Partnership for 21st Century Learning. (P21) (2009). *P21 framework for 21st century learning: 21st century student outcomes and support systems.* Partnership for 21st Century Learning.

Stehle, S. M., & Peters-Burton, E. E. (2019). Developing student 21st century skills in selected exemplary inclusive STEM high schools. *International Journal of STEM education, 6*, 39. https://doi.org/10.1186/s40594-019-0192-1

van Laar, E., van Deursen, A. J. A. M., van Dijk, J. A. G. M., & de Haan, J. (2020). Determinants of 21st-century skills and 21st-century digital skills for workers: A systematic literature review. *SAGE Open, 10*(1), 1–14. https://doi.org/10.1177/2158244019900176

Warshauer, H. K. (2015). Productive struggle in middle school mathematics classrooms. *Journal of Mathematics Teacher Education, 18*, 375–400. https://doi.org/10.1007/s10857-014-9286-3

Widya, Rifandi, R., & Rahmi, Y. L. (2019). STEM education to fulfil the 21st century demand: A literature review. *Journal of Physics: Conference Series, 1317*, 012208. https://doi.org/10.1088/1742-6596/1317/1/012208

Chapter 7
Mathifying STEM or STEMifying Math? Challenges and Possibilities for Mathematics Learning Within Integrated STEM Contexts

Michelle Fitzpatrick, Clodagh Cleary, and Aisling M. Leavy

Abstract Growing recognition of the importance of STEM education has led to an increase in the number of formal and informal STEM programmes. Efforts to evaluate the efficacy and reach of these programmes have not matched the growth in such initiatives and have become a cause for concern. As educational policy and curricular reform in Ireland increasingly advocate an integrated STEM approach, we sought to evaluate the efficacy of five integrated STEM activities designed by a team of mathematics educators and 30 preservice primary teachers as part of a semester-long mathematics specialism. Tasks were implemented across five days of intensive STEM workshops with a class of 11–12 year old students. We report on our analysis of video footage of the children's STEM engagement, alongside focus groups and interviews with the mathematics educators and preservice teachers. Our findings suggest that while STEM tasks were fertile ground for some disciplinary learning (in this case, science and engineering design process) and the development of certain Twenty-first-century skills, they failed to leverage curriculum-appropriate mathematics learning. We recommend that explicit mathematics disciplinary content be considered when planning STEM tasks to maximise the potential for developing children's grade-appropriate mathematical thinking and reasoning. We also present a series of recommendations and practical suggestions to support teachers in foregrounding mathematics when developing or modifying integrated STEM tasks and activities, offering a worked example of one of our case lessons (*The Harmless Can Holder*) to illustrate our findings throughout.

M. Fitzpatrick (✉) · C. Cleary · A. M. Leavy
Mary Immaculate College, University of Limerick, Limerick, Ireland
e-mail: Michelle.fitzpatrick@mic.ul.ie

J. Anderson and K. Makar (eds.), *The Contribution of Mathematics to School STEM Education*, https://doi.org/10.1007/978-981-97-2728-5_7

Introduction and Background to the Study

Similar to international trends, STEM education has become a focus of the policy discourse in Ireland. Economic imperatives emphasising the role of STEM in enhancing economic growth sparked initial interest in STEM (GOI, 2006), these were soon followed by efforts to enhance technological capacities in schools (GOI, 2009). A national STEM Education Review Group established in 2013 was tasked with reviewing STEM education in the Irish school system. STEM Education became a central theme in education policy (DES, 2017) and the formal introduction of the term 'STEM education' soon followed (GOI, 2017a, b). These policy documents advocate for an integrated approach to STEM education and recommend pedagogical approaches and classroom practices that support inquiry and nurture positive dispositions toward STEM. Possibly the most compelling 'call to action' is the publication of the *Primary Curriculum Framework* (NCCA, 2023) which introduces the curriculum area 'Science, Technology, Engineering and Mathematics Education' thereby requiring an integrated approach to STEM content provision. In parallel to this is the establishment of a STEM Education development group tasked with supporting the development of the curriculum specification for STEM Education. Given these fast-paced advances in STEM Education policy and curricular development, the authors were eager to explore the efficacy of integrated STEM tasks within the primary school setting.

Review of the Literature

Integrated STEM

There has been an impetus for more integrated approaches to STEM education, motivated by the opportunity to allow students to experience meaningful interaction between the STEM disciplines, often in response to real-world problems, and reflect the natural connections between STEM disciplines (Breiner et al., 2012; Honey et al., 2014). However, multiple different, and often contrasting conceptions, of integrated STEM exist. Some researchers define STEM as an approach that integrates the disciplines of Science, Technology, Engineering and Mathematics in a manner that authentically reflects the practice of STEM professionals currently working in STEM fields (Sanders, 2009). While others associate STEM education as a means of reforming from traditional didactic-based classrooms to a more constructivist approach encompassing problem solving and inquiry-based learning strategies (Breiner et al., 2012). Some proponents argue that integrated STEM requires the amalgamation of some or all four disciplines in one lesson (Moore et al., 2014) whereas others argue that integration between two or more STEM disciplines is sufficient (Kloser et al., 2018; Sanders, 2009). Efforts to evaluate the degree of integration of STEM have led to the development of frameworks that guide and support STEM integration. For example,

Vasquez et al. (2013) present integrated STEM as a continuum of increasing levels of interconnection among STEM disciplines. Other frameworks have been developed to identify characteristics for quality STEM approaches such as Moore et al.'s (2014) six major tenets for quality K-12 STEM education and Guzey et al.'s (2016) nine components for the evaluation of STEM integration curriculum. While the range of different interpretations of integrated STEM are problematic (English, 2016), the literature highlights some common key features of high- quality STEM education.

Characteristics of High-Quality STEM Education

A comprehensive review of the literature identifies seven key characteristics of high-quality STEM education common across many studies and frameworks: problem-solving design and approaches, engineering design and practices, the appropriate use of technology, the use of real-world contexts to situate learning, interdisciplinary knowledge and connections across STEM disciplines, the emphasis of pupil centred pedagogies and the development of twenty-first century skills and key competencies.

A common focus of integrated STEM education is to provide students with opportunities to apply mathematics and science understandings when approaching and *solving problems*. Problems used to contextualize STEM learning for students should incorporate *engineering design and practices*, frequently posed as an engineering design problem or challenge through the specific needs of a client (Moore et al., 2014; Sadler et al., 2000). Such tasks can encourage higher order critical thinking and problem-solving skills (Kelley & Knowles, 2016; Stohlmann et al., 2012), by engaging students in a problem-solving process involving open-ended questions that allow for multiple solutions. Students are thereby given the opportunity to use iterative test data to improve and refine their design solutions (Bryan et al., 2015; Moore et al., 2014) and justify for themselves scientific claims and design decisions with evidence and reasoning (Brown et al., 2011; Siverling et al., 2019). The appropriate *use and application of technology* is another key characteristic of integrated STEM education. Given the importance of modelling STEM careers as viable options (English, 2016; Herschbach, 2011; Kelley & Knowles, 2016) and engaging students in authentic engineering and scientific practices, it is particularly advantageous to integrate technology within STEM as that of the tools and practices utilised by STEM professionals (Ellis et al., 2020) wherein "technology enables the emergence of new scientific knowledge and assists in the design process" (Shahali et al., 2016, p. 6), technologies such as prototypes and products when they solve problems.

Engaging students in *real-world contexts* and problems is a key characteristic in the literature on integrated STEM (Blackley & Howell, 2019; Bybee, 2013; Kelley & Knowles, 2016; Moore et al., 2014; Shahali et al., 2016; Stohlmann et al., 2012). Tasks that have a familiar or real-life context are more meaningful and relevant for students and are shown to enhance student engagement, as they can draw on their everyday experiences to solve these types of tasks and are therefore more personally motivated to persevere with them (Blackley & Howell, 2019; Moll et al., 1992).

Learning experiences that engage students in authentic, real-world design challenges have been shown to foster the development and application of disciplinary knowledge and core skills across and between the combined STEM disciplines (Breiner et al., 2012; Butler et al., 2020; Kelley & Knowles, 2016; Moore et al., 2014; Sanders, 2009; Thibaut et al., 2018).

The importance of *disciplinary and interdisciplinary knowledge* being made explicit to students is also emphasized (English, 2016; Herschbach, 2011; Kelley & Knowles, 2016; Moll et al., 1992; Moore et al., 2014). There is evidence that learning is enhanced through integrative instruction (Sanders, 2009), as a cross-disciplinary approach allows students to transfer knowledge learned in one area and apply it to a different context which builds their understanding and expertise. Engineering design can, therefore, become the real-world, situated context and platform for STEM learning as it involves a process that can act as a powerful link between science, technology and mathematics (Kelley & Knowles, 2016; Moore et al., 2014). Acknowledgement of the critical role played by *pupil centred pedagogies* underpins the majority of STEM frameworks. Advocated pedagogies align with many universally recommended theories of teaching and learning and emphasise an amalgam of constructivist approaches that combine sociocultural perspectives. One recommended approach, and there are many more, that appears in the integrated STEM education literature is inquiry-based learning. Inquiry "motivates STEM learners to engage in problem stating and solving activities, which require planning, synthesis, and evaluation skills using relevant domain-specific content knowledge" (Duschl & Bismack, 2016, p.12).

Promoted pedagogies foster the development of *key competencies or twenty-first century skills* (Breiner et al., 2012; Dare et al., 2021; Sanders, 2009). Such skills are generally considered innovation and learning skills that emphasise the four C's, which according to Jackson et al. (2021) are essential skills to live a productive life in the twenty-first century and consist of critical thinking, creativity, collaboration and communication. When integrated STEM education is used with groups of learners, a robust environment for social interaction is provided, which is critical to the learning process (Sanders, 2009). Consequently, teamwork and communication are identified as significant components of these integrated STEM environments (Dare et al., 2021; Kelley & Knowles, 2016; Moore et al. 2014) where students are expected to work cooperatively and collaboratively within small groups to co-construct design solutions to real-world problems. In implementing this environment of social construction, students hone their communication, teamwork and critical thinking skills as they work with their peers towards a common goal and learn to communicate their thought processes in new ways. Collaboration focuses on sharing and contrasting ideas with other individuals engaged in a similar task with a similar aim (Blumenfeld et al., 2006). While such reflective classroom discourse environments are encouraged, Dare et al. (2021) advise, that given the ill-structured nature of real-world problems, students need to negotiate their understanding and decision-making. Doing so, they contend, requires a high level of pedagogical facilitation and support to ensure that all students' voices are equitably heard (Dare et al. 2021).

Challenges of Integrated Approaches to STEM Education: The Role of Mathematics

A major challenge of integrated approaches to STEM education is determining 'what' should be included. Of major consideration is which STEM disciplines should be integrated and in what ways. Attention to the traditional concerns of supporting scientific and mathematical literacy has broadened to consideration of ways to link these literacies to technology, engineering, design thinking, and computational thinking. Alongside this is an increasing recognition of the importance of developing 21st-century skills and attributes such as creative thinking, communicative, collaborative and problem-solving skills in addition to promoting the development of positive dispositions toward STEM. As these conceptualisations of STEM education broaden, there is the danger of moving towards a STEM "epistemic stew" (Lehrer, 2016). This leads to the significant challenge of maintaining a balance between different disciplines (English, 2016) alongside consideration of how to develop disciplinary knowledge and practices in tandem with incorporating integrated approaches to addressing real-world problems (Anderson et al., 2020).

A review of research by Anderson et al. (2020) provides valuable insights into integrated STEM learning particularly in relation to mathematics. Their review highlights the limited use of technology with mathematics emphasising tool use "rather than an integrated, essential component of an activity" (p. 38). With regard to science and mathematics integration, Anderson et al. (2020), alongside English (2016, 2017) refer to the dominance of science in many STEM programmes and highlight how the contribution of mathematics is frequently overlooked. Despite what the authors refer to as the limited attention to engineering in Australian primary schools, they report on the diversity of integrated STEM approaches being carried out by Australian mathematics educators, many of which position engineering design in the primary years and point to the importance of careful planning and the development of habits of mind to ensure focus on desired STEM concepts.

Others have also pointed to the underrepresentation of mathematics relative to other disciplines (Fitzallen, 2015; Martín-Páez et al., 2019) cautioning that, unless given significant attention, "the M is STEM is silent" (Shaughnessy, 2013, p. 324). Despite being generally accepted as the underpinning discipline in STEM (Maass et al., 2019), mathematics is often reduced to a service role in STEM tasks with little opportunity for fostering new mathematical insights (Tytler et al., 2019). Recently, Forde and colleagues (2023) examined video footage of over 2000 integrated STEM lessons. While their findings suggest that teachers' inclusion of mathematics in their lessons was related to increased cognitive engagement and depth of content integration, only 31% of the integrated STEM lessons observed included mathematical content. Indeed, they found that while there was, on many occasions, intent to include mathematical activity in the lesson, this intent was not always realised during implementation. Furthermore, they noted that the majority of tasks which included mathematics reported low levels of cognitive demand, failing to engage the students in understanding and exploring the nature of mathematical concepts, procedures and relationships.

There is, however, evidence of affective mathematical gains from integrated STEM. Lee et al. (2019) found that the 9th graders in their STEM Project-Based Learning intervention, demonstrated significantly greater affective mathematics engagement in terms of mathematical self-acknowledgement and value as compared to the non-STEM Project- Based Learning students. While Lee et al.'s (2019) study indicates the positive impact of highly situated and integrated instruction on affective mathematical engagement, more research is needed to investigate the impact of STEM on mathematical outcomes (Becker & Parks, 2011). Indeed, a review of integrated STEM curricula in the US (Honey et al., 2014, p. 3) revealed that, compared to science, while there was evidence of improvement in dispositions towards mathematics there was "less evidence of a positive impact on mathematics outcomes". Several reasons have been posited to account for such poor mathematical outcomes including the challenges for teachers in finding productive mathematics learning opportunities within STEM tasks (Tytler et al., 2019).

There is a diversity of interpretations of integrated STEM education and a corresponding lack of evidence-based research identifying the characteristics of high-quality programmes and pedagogies to support STEM learning. What is becoming more apparent is the lack of "clear advice for schools and school systems about STEM curriculum and pedagogy" (Anderson, 2020, p. 217) and a lack of conceptual clarity around what refer to as STEM ontologies and epistemologies (Chesky & Wolfmeyer, 2015). Even if identification of high-quality programmes is successful, there remains the challenge of supporting pre-service and in-service teachers to develop the knowledge and skills necessary to implement integrated STEM in classrooms (Anderson et al., 2020; Tytler et al., 2019). Furthermore, when considering the role of mathematics within integrated STEM, Anderson et al. (2020) refer to the need, not only to decide if mathematics should be foundational or incidental in integrated STEM education, they also highlight "the need to support teachers to know the difference between the two approaches and thus design learning activities accordingly" (p. 50). This study represents an effort to gain insights into opportunities and challenges for designing and implementing a series of integrated STEM tasks within the Irish educational system.

Methodology

Study Setting and Participants

The researchers were mathematics educators in the Department of STEM Education within a College of Initial Teacher Education in Ireland. The college certifies undergraduate (4-year degree) and postgraduate (2-year degree) students to teach children ages 4–12 in primary schools. STEM content is traditionally taught as discrete disciplinary areas within the school and college levels in Ireland; however, recent years have seen a shift to designing integrated STEM content across both settings.

Participants were 30 preservice teachers (PSTs) enrolled in a STEM education specialism module which met 3-h per week for a duration of 12 weeks. They were in the third year of their initial teacher education programme and had completed 5 mathematics education modules, 2 science modules and 2 ICT modules. In the previous semester, they had been introduced to STEM education via a 6-h online, asynchronous unit of work. PSTs were provided with input into the history and development of STEM education, models of integrated STEM, STEM task design and provided with opportunities to engage in collaborative STEM activities. Following these experiences, they contributed to the design and development of STEM tasks (see Table 7.1) and engaged in the implementation of these STEM activities in local primary school. A class of 5th grade children, and their teachers, in a local school also participated in the study. Participation in the research was voluntary and ethical approval was received from the institute's Research Ethics Committee.

Task Design and Development

Prior to the semester, five STEM tasks were selected and modified by the researchers based on a review of the literature and similar tasks. All tasks had the potential to incorporate the seven key characteristics identified in the review of literature. The tasks were: Harmless Can Holder, Weight-bearing Bridge, Earthquake Tower, Kicking-machine and Build a Rollercoaster (see Table 7.1).

During four of the in-college preparatory sessions PSTs were organised into five groups and each group was assigned one of the STEM tasks. The group completed

Table 7.1 Initial task outlines

STEM Challenge 1: ***Harmless Can Holder***
Design a holder for carrying a six-pack of cans that is harmless. The one that is currently used on packaging is made of plastic. Design an environmentally friendly holder to easily allow a person carry six full cans
STEM Challenge 2: ***Weight-bearing Bridge***
Children engineer a bridge in groups. They must design and build a bridge that spans a gap of 50 cm between two tables. Children will test their model bridges with weights
STEM Challenge 3: ***Earthquake Tower***
Design a model of a tower that could survive an 'earthquake' (the shake table)
STEM Challenge 4: ***Kicking-machine***
The children must design a kicking machine that can 'kick' a table tennis ball into a paper cup over a defined distance
STEM Challenge 5: ***Build a Rollercoaster***
Build a roller coaster from a limited amount of materials. Which group can design the roller coaster with the best turn/loop that safely delivers the marble into the paper cup

the task themselves and this experience provided further input into the lesson design with a view to its implementation within a primary classroom. Each group designed an introduction to the task. These introductions consisted of a driving context that provided a rationale for the task, alongside the selection of images, videos and other resources to support the task context. Materials were also identified, and task criteria and parameters were established. The PSTs completed a session on developing STEM talk moves to promote classroom discourse, and prepared a set of prompts and questions for their lesson based on expected student responses. PSTs were also supported in completing a STEM observation protocol while watching some STEM video footage, in an effort to develop their noticing skills. This work was overseen by the researchers.

Task Implementation

Tasks were implemented across five days with each task taking approximately 90 min. Each group of preservice teachers visited the school on an assigned day and implemented the task with a group of primary children. The children were in 5th grade (10–11 years old) and had no prior experience of integrated STEM education. Task implementation took place in the school recreational/sports hall as this provided adequate space for construction and use of materials. A large screen and audio-visual equipment were set up at the top of the hall. The children were organised into 6 groups with 4–5 children in each group. Group membership varied across the five tasks due to attendance and other management decisions. Each group was supported by one pre-service teacher. The pre-service teacher had two roles. They acted as a facilitator wherein they organised materials, answered questions and, where necessary, reminded the group of the task criteria and parameters (e.g. sketching a plan for the design). They did not contribute any creative ideas or contribute to the intellectual activity of the group. The second role was observer, noting the decision-making of children, recording the use of disciplinary knowledge, twenty-first century skills, alongside challenges and obstacles when engaging with tasks. All five activities were recorded by a professional video crew who used a roaming camera to gather footage of groups of children and their PST as they engaged in the each of the STEM activities.

Data Sources and Analysis

Using the seven *characteristics of high-quality STEM* education, as identified from the literature, as markers of success, the task design and implementation was evaluated through analysis of the data collected and presented in the table below. Data were collected at four different times: before, during, immediately following and one month after task implementation (identified as pre, 1, 2 and 3 respectively in

Table 7.2 Study data sources and phases

Data sources	When	Researchers	Pre-service teachers
Field notes of PST designing and testing tasks	Pre	✓	
Weekly reflective journal entries	Pre		✓
Field notes—observations of group work	1	✓	✓
Video recording of task implementation	1	✓	✓
Focus group discussion of preservice teachers and pre-service teacher group members	2	✓	✓
De-briefing conversation and reflection	2	✓	
Guided-reflective journal entry	3		✓
Final group presentations	3		✓

Table 7.2). The data consisted of fieldnotes, and observations made by the researchers of activities, presentations and discussions during the college preparatory sessions, digital video footage of the five STEM lessons and written documentation made by PSTs of their observations and reflections. The researchers incorporated reference to the seven characteristics of high quality STEM education when taking fieldnotes and during the focus group conversations with the pre-service teachers (Table 7.2).

Data were analysed using two different phases or passes through the data. Inductive analysis occurred first. It involved an open coding strategy which facilitated the emergence of codes and themes in an effort to generate meaning and establish patterns that would shed light on the integrated STEM lessons. One theme relating to mathematics emerged from the data. This theme, *Rich STEM lessons, with an emphasis on engineering and twenty-first century skills; but where's the mathematics?*, subsequently became the focus for the second pass through the data which was guided by a deductive analytic framework. The goal of the deductive analysis was to identify factors that contributed to, or inhibited, a focus on mathematics within the integrated STEM lessons. The reporting of the findings is structured according to these phases of data analysis, reporting firstly on the nature of the integrated STEM lessons and the role of mathematics, before discussing the factors that impacted the (limited) focus on mathematics.

Findings, Discussion and Recommendations

We present and discuss the findings in two parts. The first part reports briefly on our observations following the implementation of the STEM tasks, noting the strength of the task in leveraging some STEM learning (notably engineering and twenty-first century skills) and highlighting the absence of meaningful mathematics integration throughout the five tasks. The second section seeks to offer some retrospective insight into *why* mathematics appears to have been neglected in these lessons. We discuss the three factors that contributed to the neglect of mathematics and offer some recommendations on how we might address these shortcomings. We conclude with a discussion of a final theme that emerged from reflection on the lessons, with a view to recommending an approach to forefront mathematics within STEM units. In an effort to exemplify our findings and recommendations, the case of *The Harmless Can Holder*, the first of our five STEM tasks, is presented across both parts as a worked example. We use this example to illustrate the contrast between our initial task as it was implemented (see Fig. 7.1), and its potential, when modified, to serve as a mathematically-rich integrated STEM task (see Fig. 7.2), within a STEM inquiry cycle.

Setting the context: The PSTs facilitated a whole class discussion on river life, possible threats, and the issues of pollutants in our rivers. They explored the danger that plastics posed to fish, birds and other wildlife, in and on the riverbank.	
Presenting the problem: Students were given a letter from Limerick City and County Council outlining the problem of plastics in the local river and the risk to wildlife. The STEM challenge requires them to design and build an environmentally friendly holder that would easily allow a person to carry four full cans.	
Materials were distributed and students sketched a design for the *Harmless Holder.* Materials included: *cardboard, 4 cans, variety of tapes, string, scissors, lollipop sticks, hessian material, foil and paper.*	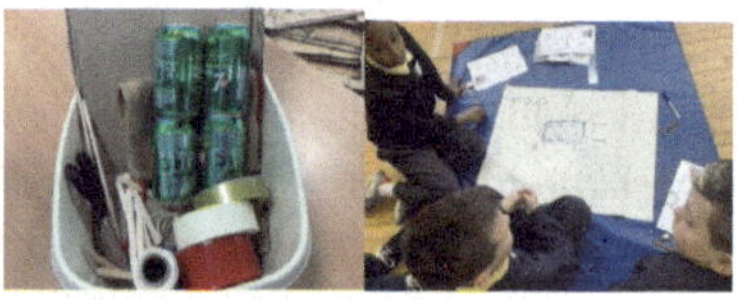
Building, testing, redesigning: Students built their holder, tested it, made amendments to improve their design and presented their prototype.	

Fig. 7.1 Case lesson: the harmless can holder lesson outline

Part One: Rich STEM Lessons, with an Emphasis on Engineering and Twenty-First Century Skills; but Where's the Mathematics?

> I just realised the power of STEM, you know? It can really bring through so much, all those 21st century skills. There was so much communicating, so much problem solving in just one lesson. [Joe, Harmless Holder]

The post-implementation data suggest that each of the five tasks designed and delivered, promoted key components of high-quality STEM education, as outlined in the literature review. The lessons were all centred around a real-world context that provided rich and authentic problem-solving opportunities. PSTs observed the effect this had on the children:

> It meant the kids were engaging in an activity that was relevant to their lives as it was the local environment. So, it was personally and socially relevant and that meant that they engaged in the task in a meaningful way. [Ciara, Harmless Holder]

The PSTs considered the tasks to be open-ended and inquiry based, with multiple solution paths, and were surprised by the uniqueness of the children's designs.

> Even though it was a prescribed task, there was no prescribed solution, or correct solution to the problem. So, the children had to work together in a meaningful way to discover their own creative solution to the problem. [Rebecca, Harmless Holder]
>
> With other lessons you'd have to go off and do your own work, or listen to the teacher the whole time, but with this it's like, them taking control. It's really student-led. [Melissa, Kicking Machine]
>
> ...and there was no right way of doing the task either. They could do it whatever way they wanted. There was no specific way that they had to do it, which was really good. And then everyone had very different designs at the end. [Claire, Harmless Holder]

The PSTs highlighted the centrality of the engineering design process in each lesson and recognised its role in the promotion of key competencies. Each group gave accounts of children demonstrating communication, collaboration, critical thinking and creativity. They were intrigued by the resilience of the children, and their attitudes toward failure.

> It was the engineering design process that fostered the 21st century skills. All through the process they were communicating (negotiating, discussing how they would go about this or that, or overcoming problems), collaborating (they were working together throughout) and thinking creatively to come up with a plan. Creativity was shown right through it [Joe, Harmless Holder]
>
> The kids were not afraid. If it fell apart the first time, they just fixed it and moved on... they just kept going... developing their problem-solving skills (Susan, Weight-bearing Bridge)
>
> Collaborative learning led to a lot of dialogue that allowed them [the children] to maximise their use of scientific language while developing higher order thinking skills [Ciara, Harmless Holder]
>
> None of them were afraid to do something wrong or failing. You know the way sometimes you're sitting in maths class and you can see they're afraid to get the wrong answer, whereas here, none of them were afraid to go wrong, because they knew they could redesign it. They

just kept persevering, which I thought was a great thing, because life, like you're going to have to alter and change things...you might not always get there the same way, so I liked the fact that the task leads them to redesign stuff [Claire, Harmless Holder]

When exploring the disciplinary nature of the task, PSTs and mathematics educators all agreed that engineering was the primary focus of each task. Although they played a *supporting* role in terms of content and context respectively, science and technology were evident across the five tasks. It was clear, however, that little to no mathematical learning took place over the five lessons. While some low-level mathematical *activity* was present amongst the students, such as measuring and computation, there was no evidence of planned, grade-appropriate mathematical teaching and learning across the five lessons. The mathematics involved was incidental for the most part and involved the children utilizing known skills and knowledge, far below their grade level expectations. When asked about the mathematical content of the lessons, PSTs reported on the presence of mathematics such as:

the weight element...they [the children] were saying it's too heavy (Oonagh, Harmless Holder)

counting in 10's... measuring 50 cm... using the weights at the end (Susan, Weight-bearing Bridge)

measuring, things like that... totting up the creativity points at the end (Valerie, Rollercoaster).

It was evident that the PSTs had difficulty in recognising the mathematical deficiencies of the tasks, initially viewing these unambitious and incidental activities as authentic mathematics integration. Furthermore, they struggled to identify meaningful opportunities to explore rich mathematical content to promote mathematical thinking and required extensive scaffolding from the mathematics educators in recognising opportunities for appropriately challenging mathematics integration. This was surprising given that this PST cohort had completed all core mathematics education modules and was undertaking this STEM elective as part of a mathematics specialism.

Part Two: Why Was There so Little Evidence of Mathematical Learning?

An Over-Focus on Engineering from the Beginning

The pace of STEM policy and curricular change in Ireland gives impetus to exploring ways of preparing our future teachers for interdisciplinary approaches. As discussed previously, STEM content has traditionally been taught as discrete disciplines across the sectors. Given that integrated STEM is a nascent approach, preservice teachers have limited experiences as STEM learners themselves. As expected, while they communicated confidence in teaching mathematics and science, the PSTs in this study reported unfamiliarity with engineering practices and were concerned about how such practices could be incorporated into the primary classroom.

Given the novelty of engineering, focus was placed on the engineering design process. STEM tasks modelled in class sought to allow the PSTs to experience engineering practices through the eyes of the learner, while at the same time modelling innovative STEM pedagogies they could adopt when teaching. While prioritizing the engineering design process, however, we underestimated the need to maintain the spotlight on mathematics. In the developing enthusiasm for designing and making, the engineering design process slowly became the overall focus of the tasks and mathematics became almost 'a given'. This finding lends support to the caution provided by Shaughnessy (2013) of the risk of mathematics becoming the silent 'M' in STEM and emphasizes the challenges identified by English (2016) in maintaining a balance between the different STEM disciplines. We also made assumptions about the ease with which PSTs would apply prior learning in mathematics education modules to the design and delivery of their integrated STEM task.

Recommendation: Integrated STEM tasks must contain reference to relevant explicit mathematics disciplinary content.

No Curriculum-Based Learning Outcomes Were Identified

Owing to this over-focus on the engineering design process, there were no explicit objectives for mathematical learning in the task design. While the potential for mathematics was identified and discussed when planning for the lesson, specific objectives from the curriculum documents were not identified. This proved insufficient in ensuring that mathematics learning was translated into the lesson. PSTs were initially satisfied with the level of mathematics involved in each task and, as already outlined, offered examples of menial mathematical activities as evidence of integration and referred to children's enjoyment engaging in these mathematics tasks. These observations support the finding, arising from a review of US integrated STEM curricula, of improvement in dispositions towards mathematics and the absence of evidence for gains in mathematics learning outcomes (Honey et al., 2014). When probed, however, on the mathematics component and the suitability of such experiences for the grade level, PST began to reflect on the appropriateness of the mathematics involved:

> It was weak enough for their class level like you know, just informal measuring and stuff. No… there was nothing concrete. Say if we went to the curriculum now, I'd say we'd struggle to pull out their learning objectives. So as a task, I'd say it wasn't the best from a maths point of view [Joe, Harmless Holder]

PSTs identified their unfamiliarity with the class as a reason for the lack of mathematical challenge as indicated by Ciara in the following extract:

> It would have been beneficial if we had known pupils' maths ability prior to the lesson, because then we could have integrated more meaningful maths at a developmentally appropriate level for them (Ciara, Harmless Holder).

The standalone nature of the teaching experience and the unfamiliarity with the class were certainly challenges in designing a meaningful STEM lesson. That being said, the mathematics evident in the tasks did not reflect expectations for the fifth class in the national curriculum. It appears that PSTs had difficulty in identifying appropriately challenging objectives from the mathematics curriculum. Consequently, PSTs were unprepared to deal with potentially rich mathematics when they arose, as recognized by Evan in the following focus group extract:

> There was definitely some opportunities for integration that we've missed, you know, they were coming out with like, isosceles triangles, one of them said that, and I just, I certainly didn't anticipate a lot of the things they were coming out with. [Evan, Earthquake Tower]

This challenge is not unique to PSTs and was also identified by Tytler et al., (2019, p. 63) who reported the challenge for practicing teachers in "extracting meaningful mathematics from cross-disciplinary topics in ways that can extend students' mathematical thinking".

Recommendation: Integrated STEM tasks must incorporate grade-appropriate learning objectives for the development of mathematics skills and understanding.

The Task Criteria and Materials Available Did not Stimulate Mathematical Thinking

Following each task implementation, each group of PSTs reported that they would adjust their criteria and materials used to improve the quality of the STEM task more generally, to assist the running of the lesson from a classroom management perspective, and to enhance and support the mathematical nature of the task. Two groups (*Earthquake Tower* and *Rollercoaster*) felt they had too many, overly restrictive criteria that stymied the creativity of the children and their designs. The remaining three groups reported that the criteria they provided could have been clearer, more detailed, and included specific requirements that would have compelled their students to consider the mathematics involved in the task. Following the *Kicking Machine* task, it was suggested that amending the criteria to focus on the distance it could travel, rather than hitting a target (i.e. getting the ball in the cup), could have promoted mathematical thinking by encouraging the students to investigate angles, heights, and lengths. Meanwhile, in the *Earthquake Tower* task, the group felt that their challenge description (*to build the tallest earthquake resistant tower*) did not reflect the direction of the lesson, whereby the children had to concentrate on creating *a* tower that would stand at all, thereby shifting the focus from *height* to *stability*, and changing the mathematic lens from *measures* to *geometry*.

> Kathryn: We did bring maths in with the models and explaining shape, but we could have focused on it more...
>
> Aine: I thought we'd be focusing on height and so the emphasis would have been on measure, but shape was more the focus

Some groups also considered the materials and resources they distributed and how that could encourage or curtail mathematical thinking and reasoning. The *Weight-bearing Bridge* group, for example, stated that if reteaching the lesson, they would give the children access to weights throughout the activity to encourage more authentic measuring opportunities. The *Kicking Machine* and *Rollercoaster* groups could have encouraged the exploration of angles by providing angle checkers, anglegs, geostrips and protractors (an opportunity that was capitalized on by one pupil, who ran back to his class for his maths set during the rollercoaster activity).

Recommendation: 'Mathify' the parameters and criteria of existing integrated STEM tasks to maximise the mathematics potential and ensure the provision of multiple opportunities to engage in grade-appropriate mathematical thinking and reasoning.

Mathifying *a STEM Task: The Case of* the Harmless Can Holder

During the focus group discussions and in follow-up module activities, PSTs and mathematics educators reflected on the role of mathematics in the lesson and considered possible adaptions they could make to reposition the mathematics within the STEM task. We present one case, *The Harmless Can Holder* task, as an example of the PSTs' efforts to redesign the STEM task with an enhanced mathematical focus (see Fig. 7.2).

Spotlighting and Positioning Mathematics in STEM Inquiry Cycles

Outside of possible adaptions to the STEM task itself, PSTs discussed other means of incorporating meaningful mathematics integration. Given that this was an isolated STEM teaching block, there was little opportunity to build upon discrete mathematics learning. PSTs identified future potential for building specific STEM tasks into their long-term planning. In the case of the *Harmless Can Holder*, for example, we explored the value of this STEM task as a rich context to build upon discrete mathematics lessons on the properties of the circle, in which student would apply their developing geometric understandings in their designs and constructions. Arguments were also made concerning the task's capacity to act as a springboard to introduce children to explorations on the properties of the circle.

Arising from a review of Australian STEM learning initiatives, Tytler et al. (2019) recognised 'the application of mathematics to projects that were 'authentic' and meaningful to students and involved them either in developing mathematics that was new, or applying known mathematics in new ways' as the 'fundamental principle of STEM-focused mathematics' (p. 77). Based on our findings, we have offered some recommendations, or considerations, to forefront mathematics teaching and learning when designing or modifying a STEM task. When attempting to reposition the 'M' in STEM, we should consider *where* we place specific STEM tasks in the mathematics learning sequence; that is, we need to make conscious decisions about *when* and

Image 1

Image 2

Image 3

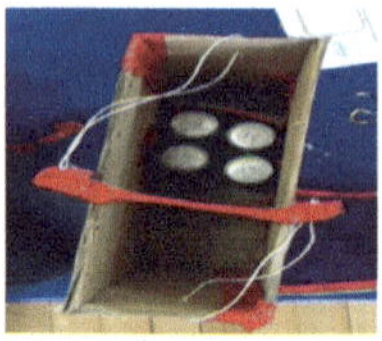

Image 4

Observations from lesson	**Suggested adaptions to promote mathematics within this task**
Observation 1: Little to no mathematical thinking was involved in this task. Mathematical content was incidental and required skills and knowledge explored in the early years (such as non-standard units [see *image 2*])	**Identify appropriate curriculum objectives** • select and use appropriate instruments of measurement • estimate and measure length/weight using appropriate metric units • renaming units of length, *mm* to *cm* • identify the properties of the circle • construct a circle of a given radius or diameter
Observation 2: The task was open-ended and allowed for creativity. However, the vague criteria did not stimulate mathematical thinking	**Purposefully building mathematics into the criteria and constraints** • e.g. *"It must be designed to hold four cans only"*, *"cans must remain upright"* and *"using the least cardboard possible"*. • This would necessitate more precise measuring, as many of the containers designed gave no regard to the dimensions of the container and could hold much more than four cans [see *images 3 and 4*]
Observation 3: The materials provided did not stimulate grade-appropriate mathematics or mathematical inquiry. Giving the cans out during the planning and building phases removed much of the potential mathematical challenge (tracing the outline of the can [see *image 1*] and using the cans themselves to measure [see *image 2*]). Also, the lack of mathematics resources discouraged the children from considering the role of mathematics within the task.	**Careful consideration of the materials provided to encourage/compel mathematical inquiry** • Only give the dimensions of the can (height, diameter, weight, etc.) and not the can itself, to encourage the students to apply curriculum-based mathematical knowledge and skills. • Distribute one can during the planning phase, allowing the children to determine the dimensions themselves for their designs. • The cans themselves could be accessed, early on, perhaps, as a differentiated support during the planning and making phase if needed, and later for groups to test their designs. • Include a range of mathematics resources and instruments to prompt and allow for meaningful mathematical activity (e.g. compass, rulers, weights, scales)

Fig. 7.2 Case lesson: The Harmless Can Holder lesson observations and suggested adaptions

Table 7.3 Mathematics *for* STEM and Mathematics *from* STEM, planning prompts for teachers

When thinking about planning mathematics *for* STEM, teachers might ask; • What are the key mathematical concepts and skills needed to complete this authentic task? • What mathematics should I explore in my discrete mathematics so children can apply this knowledge during the task? • Where would this task most meaningfully fit into my long-term mathematics plan/ scheme of work? Teachers might also consider opportunities for mathematical inquiry that may arise *from* a specific STEM task; • What new mathematical knowledge or skills could be developed during the task? • Are there potential mathematical discoveries or 'aha' moments within the task? • What mathematical unknowns might arise that could be used as a springboard into a future (discrete) mathematics lesson? • Is there an obvious follow-up mathematics lesson, for which this STEM task could serve as a rich context?

where we position both mathematics and STEM in our teaching. Perhaps we might consider **mathematics *for* STEM** and **mathematics *from* STEM** (see Table 7.3). By mathematics *for* STEM, we refer to the specific disciplinary knowledge and skills, developed in preceding discrete mathematics lessons, to be utilised and developed in new ways during the STEM task. Mathematics learning *from* STEM, on the other hand, refers to new mathematical knowledge and skills developed during the STEM task. Furthermore, it may also refer to new mathematical concepts that the STEM task presents. In this case, the STEM task acts as both a rich context and a springboard for future mathematics inquiries.

Conclusions

The STEM tasks explored in this study were effective in terms of the key characteristics identified in the review of literature. Each of the five tasks presented an engaging engineering design task, rooted in a real-world context. Students were challenged to use and develop key twenty-first century competencies in solving an authentic problem and were deeply engaged in their work. Our findings, however, support the work of Lehrer (2016, 2017), who contends that, while engaging for students, integrated STEM projects can fall short of engaging students in deeper disciplinary practices, failing to promote a curriculum agenda that would characterise a coherent knowledge progression (Tytler, 2020; Tytler et al., 2019). While some lower order, procedural mathematical activity was observed, the lessons failed to leverage rich skills and proficiencies promoted by mathematics education research and advocated in reform curricula (NCTM, 2014; NRC, 2001). As Baldinger et al. (2020) caution, although the experiential, materials-based engineering design challenge approach can be highly creative, "it may predispose students to use mathematics in a manner that could be restricted to measurement and computational skills", and may in fact "inadvertently equate mathematics with measurement and routine, procedural work" (pg. 69). Therefore, we need to ensure that rich curriculum-appropriate mathematics are incorporated in our STEM tasks from the outset. In our attempts to introduce and promote engineering practices and twenty-first century competencies, we let the mathematics become a service tool, and failed to position it as a site of learning itself (Baldinger et al., 2020). In our eagerness to develop aspects of twenty-first century skills, we lost the emphasis on disciplinary-specific knowledge that is incorporated in influential frameworks of twenty-first century competencies (Beswick & Fraser, 2019). Our retrospective analysis however, identified ways in which we might mitigate this shortcoming and develop rich mathematical learning opportunities that support the promotion of key competencies.

We need, therefore, to carefully consider how best to balance the demands of disciplinary knowledge development with the opportunities for interdisciplinary mathematics. Developing rich mathematical learning opportunities in STEM tasks is a recognised challenge for teachers. Our recommendations offer practical guidance for maintaining the spotlight on mathematics in STEM. Reflecting on opportunities for mathematics before, during and after STEM tasks, in the planning phase, could support teachers in recognising opportunities for meaningful inquiry cycles, whereby discrete disciplinary learning could inform meaningful interdisciplinary work, and vice versa. Such attempts to *mathify* the STEM and *STEMify* the mathematics could uphold the integrity of disciplinary knowledge, while reaping the benefits of interdisciplinary approaches.

References

Anderson, J. (2020). The STEM education phenomenon and its impact on school curriculum. *Curriculum Perspectives, 40*(2), 217–225.

Anderson, J., English, L., Fitzallen, N., & Symons, D. (2020). The contribution of mathematics education researchers to the current STEM education agenda. In: J. Way, C. Attard, J. Anderson, J. Bobis, H. McMaster, K. Cartwright (Eds.), *Research in mathematics education in Australasia 2016–2019* (pp. 27–57). Springer, Singapore. https://doi.org/10.1007/978-981-15-4269-5_3

Baldinger, E. E., Staats, S., Covington Clarkson, L. M., Gullickson, E. C., Norman, F., & Akoto, B. (2020). A review of conceptions of secondary mathematics in integrated STEM education: Returning voice to the silent M. In J. Anderson & Y. Li (Eds.), *Integrated Approaches to STEM Education: An International Perspective* (pp. 67–90). Springer International Publishing. https://doi.org/10.1007/978-3-030-52229-2_5

Becker, K., & Park, K. (2011). Effects of integrative approaches among science, technology, engineering, and mathematics (STEM) subjects on students' learning: A preliminary meta-analysis. *Journal of STEM Education, 12*(5/6), 23.

Beswick, K., & Fraser, S. (2019). Developing mathematics teachers' 21st century competence for teaching in STEM contexts. *ZDM Mathematics Education, 51*, 955–965.

Blackley, S., & Howell, J. (2019). The next chapter in the STEM education narrative: Using robotics to support programming and coding. *Australian Journal of Teacher Education (online), 44*(4), 51–64.

Blumenfeld, P. C., Kempler, T. M., & Krajcik, J. S. (2006) *Motivation and cognitive engagement in learning environments* (pp. 475–488). na.

Breiner, J. M., Harkness, S. S., Johnson, C. C., & Koehler, C. M. (2012). What is STEM? A discussion about conceptions of STEM in education and partnerships. *School Science and Mathematics, 112*(1), 3–11.

Brown, R., Brown, J., Reardon, K., & Merrill, C. (2011). Understanding STEM: Current perceptions. *Technology and Engineering Teacher, 70*(6), 5.

Bryan, L. A., Moore, T. J., Johnson, C. C., & Roehrig, G. H. (2015). Integrated STEM education. In C. C. Johnson, E. E. Peters-Burton, & T. J. Moore (Eds.), *STEM road map: A framework for integrated STEM education* (pp. 23–38). Routledge.

Butler, D., McLoughlin, E., O'Leary, M., Brown, M. & Costello, E. (2020). Towards the ATS STEM conceptual framework. http://doras.dcu.ie/25151/1/STEM%20Conceptual%20Framework.pdf

Bybee, R. W. (2013). *The case for STEM education: Challenges and opportunities*. NSTA Press.

Carpenter, T. P., Romberg, T. A. & Carpenter, S. (2004). *Powerful practices in mathematics & science*. National Center for Improving Student Learning and Achievement in Mathematics and Science.

Chesky, N. Z., & Wolfmeyer, M. R. (2015). *Philosophy of STEM education: A critical investigation*. Palgrave Macmillan.

Dare, E. A., Hiwatig, B., Karatithamkul, K., Ellis, J. A., Roehrig, G., Ring-Whalen, E. A., Rouleau, M., Faruqi, F., Rice, C. & Titu, P. (2021). Improving STEM education: The design and development of a K-12 STEM observation protocol (STEM-OP). *ASEE Annual Conference proceedings*.

Department of Education and Skills. (2017). *Action Plan for Education*, Department of Education and Skills: Dublin, available at: https://www.education.ie/en/Publications/Corporate-Reports/Strategy-Statement/Action-Planfor-Education-2017.pdf

Duschl, R. A. & Bismack, A. S. (2016). *Reconceptualizing STEM education: The Central Role of Practices*.Routledge.

Ellis, J., Wieselmann, J., Sivaraj, R., Roehrig, G., Dare, E., & Ring-Whalen, E. (2020). Toward a productive definition of technology in science and STEM education. *Contemporary Issues in Technology and Teacher Education, 20*(3), 472–496.

English, L. D. (2016). STEM education K-12: Perspectives on integration. *International Journal of STEM Education, 3*(1), 1–8.

English, L. D. (2017). Advancing elementary and middle school STEM education. *International Journal of Science and Mathematics Education, 15*(Suppl 1), S5–S24.

Fitzallen, N. (2015). STEM Education: What Does Mathematics Have to Offer? *Mathematics Education Research Group of Australasia*, 237–244.

Forde, E. N., Robinson, L., Ellis, J. A., & Dare, E. A. (2023). Investigating the presence of mathematics and the levels of cognitively demanding mathematical tasks in integrated STEM units. *Disciplinary and Interdisciplinary Science Education Research, 5*(1), 3–18. https://doi.org/10.1186/s43031-022-00070-1

National Council for Curriculum and Assessment. (2023). *Primary Curriculum Framework*, available: https://www.gov.ie/en/publication/0db24-primary-curriculum-framework/

Government of Ireland. (2006). *Strategy for science, technology and innovation 2006–2013*. Dublin: Stationary Office.

Government of Ireland. (2009). *SMART SCHOOLS = SMART ECONOMY Report of the ICT in Schools Joint Advisory Group to the Minister for Education and Science*. Dublin: Stationary Office.

Government of Ireland. (2017a). *STEM Education Policy Statement 2017–2026*. Dublin: Stationary Office.

Government of Ireland. (2017b). *STEM Education Implementation Plan 2017–2019*. Dublin: Stationary Office.

Guzey, S. S., Moore, T. J., Harwell, M., & Moreno, M. (2016). STEM integration in middle school life science: Student learning and attitudes. *Journal of Science Education and Technology, 25*(4), 550–560.

Herschbach, D. R. (2011). The STEM initiative: Constraints and challenges. *Journal of STEM Teacher Education, 48*(1), 96–122.

Honey, M., Pearson, G., & Schweingruber, H. (2014). *STEM integration in K-12 education: Status, prospects and an agenda for research.* National Academies Press.

Jackson, C., Mohr-Schroeder, M. J., Bush, S. B., Maiorca, C., Roberts, T., Yost, C., & Fowler, A. (2021). Equity-oriented conceptual framework for K-12 STEM literacy. *International Journal of STEM Education, 8*(1), 1–16.

Kelley, T. R., & Knowles, J. G. (2016). A conceptual framework for integrated STEM education. *International Journal of STEM Education, 3*(1), 1–11.

Kloser, M., Wilsey, M., Twohy, K. E., Immonen, A. D., & Navotas, A. C. (2018). 'We do STEM': unsettled conceptions of STEM education in middle school S.T.E.M Classrooms. *School Science and Mathematics, 118*, 335–347.

Lee, Y., Capraro, R. M., & Bicer, A. (2019). Affective mathematics engagement: A comparison of STEM PBL versus non-STEM PBL instruction. *Canadian Journal of Science, Mathematics and Technology Education, 19*(3), 270–289. https://doi.org/10.1007/s42330-019-00050-0

Lehrer, R. (2016, October). *Perspectives on elementary STEM education.* Keynote address at the forum: Putting STEM education under the microscope. Melbourne: Deakin University. https://blogs.deakin.edu.au/steme/wp-content/uploads/sites/39/2017/04/Lehrer-Deakin_STEM_2_reduced.pdf

Lehrer, R. (2017, September). *Considering epistemological junctions when designing for interdisciplinary learning.* Paper presented at the conference of the European Association for Research in Learning and Instruction (EARLI), Tampere, Finland.

Maass, K., Geiger, V., Ariza, M. R., & Goos, M. (2019). The role of mathematics in interdisciplinary STEM education. *ZDM Mathematics Education, 51*(6), 869–884. https://doi.org/10.1007/s11858-019-01100-5

Martín-Páez, T., Aguilera, D., Perales-Palacios, F. J., & Vílchez-González, J. M. (2019). What are we talking about when we talk about STEM education? A Review of Literature. *Science Education, 103*(4), 799–822.

Moll, L. C., Amanti, C., Neff, D., & Gonzalez, N. (1992). Funds of knowledge for teaching: Using a qualitative approach to connect homes and classrooms. *Theory into Practice, 31*(2), 132–141.

Moore, T. J., Stohlmann, M. S., Wang, H. H., Tank, K. M., Glancy, A. W. & Roehrig, G. H. (2014). Implementation and integration of engineering in K-12 STEM education In Purzer, S., Strobel, J. & Cardella, M.E. (Eds.), *Engineering in pre-college settings: Synthesizing research, policy, and practices* (pp. 35–60). Purdue University Press.

National Council of Teacher of Mathematics. (2014). *Principles to actions: Ensuring mathematical success for all.* The National Council of Teachers of Mathematics, Inc. https://pubs.nctm.org/view/book/9780873539043/9780873539043.xml

National Research Council. (2001). *Adding it up: helping children learn mathematics.* National Academies Press. https://doi.org/10.17226/9822

Sadler, P. M., Coyle, H. P., & Schwartz, M. (2000). Engineering competitions in the middle school classroom: Key elements in developing effective design challenges. *The Journal of the Learning Sciences, 9*(3), 299–327.

Sanders, M. E. (2009). Stem, stem education, stemmania. *The Technology Teacher, 68*, 20–26.

Shahali, E. H. M., Halim, L., Rasul, M. S., Osman, K., & Zulkifeli, M. A. (2016). STEM learning through engineering design: Impact on middle secondary students' interest towards STEM. *EURASIA Journal of Mathematics, Science and Technology Education, 13*(5), 1189–1211.

Shaughnessy, M. (2013). Mathematics in a STEM context. *Mathematics Teaching in the Middle School, 18*(6), 324.

Siverling, E. A., Suazo-Flores, E., Mathis, C. A., & Moore, T. J. (2019). Students' use of STEM content in design justifications during engineering design-based STEM integration. *School Science and Mathematics, 119*(8), 457–474.

Stohlmann, M., Moore, T. J. & Roehrig, G. H. (2012). Considerations for teaching integrated STEM education. *Journal of Pre-College Engineering Education Research (J-PEER), 2*, 28–34

Thibaut, L., Ceuppens, S., De Loof, H., De Meester, J., Goovaerts, L., Struyf, A., Boeve-de Pauw, J., Dehaene, W., Deprez, J., & De Cock, M. (2018). Integrated STEM education: A systematic review of instructional practices in secondary education. *European Journal of STEM Education, 3*, 1–12.

Tytler, R., Williams, G., Hobbs, L., & Anderson, J. (2019). Challenges and opportunities for a STEM interdisciplinary agenda. In B. Doig, J. Williams, D. Swanson, R. Borromeo Ferri, & P. Drake. (Eds.), *Interdisciplinary mathematics education: State of the art and beyond* (pp. 51–84). Springer Open.

Tytler, R. (2020). STEM education for the twenty-first century. In J. Anderson & Y. Li (Eds.), *Integrated approaches to STEM education: an international perspective* (pp. 21–43). Springer International Publishing. https://doi.org/10.1007/978-3-030-52229-2_3

Vasquez, J. A., Sneider, C., & Comer, M. (2013). *STEM Lesson Essentials, Grades 3–8: Integrating Science, Technology, Engineering, and Mathematics*. Heinemann.

Chapter 8
Deepening Mathematical Understanding Through a Problem-Based Integrated STEM Curriculum

Anita Crowder, Rayna Flye, and Kristin Umland

Abstract While much STEM education research has focused on mathematics as a tool for STEM, this chapter will identify ways that STEM was used to deepen the knowledge, use, and enjoyment of mathematics via the use of the freely available, widely-adopted middle school curriculum from Illustrative Mathematics (IM). IM purposely designed its research-based K-12 mathematics curriculum to invite all students into mathematics by leveraging contexts through a culturally relevant, problem-based approach. Researchers analyzed interview data in a multiple-case study from four diverse school districts across the United States. The findings suggest that aspects of the curriculum's instructional model, such as collaborative real-world and mathematical problem-solving and explicit mathematical discourse structures, impacted shifts in educator classroom practice, perceived student engagement and achievement in mathematics, and long-term valuing of problem-based instruction. The outcomes have practical implications for supporting teachers to understand how mathematical understanding can be enriched through STEM problems.

Keywords Problem-based learning · Mathematics curriculum · Real-world problem solving · Middle school

Introduction

The decades-long "STEM shortage," the insufficient number of workers with the skills required to fill available science, technology, engineering, and mathematics (STEM) jobs, has continued unabated (Camilli & Hira, 2019; Smith & White, 2019). As technology has advanced and become an increasingly integral part of society, the impact of the STEM shortage has threatened to persist and intensify. Preparing today's youth for tomorrow's occupations and increasing interest and engagement in STEM has become necessary to expand the workforce. In the United States,

A. Crowder (✉) · R. Flye · K. Umland
Illustrative Mathematics, Tucson, AZ, USA
e-mail: crowdera@vcu.edu

J. Anderson and K. Makar (eds.), *The Contribution of Mathematics to School STEM Education*, https://doi.org/10.1007/978-981-97-2728-5_8

researchers have partially explained the STEM shortage by the rapidly changing advances that often make skills learned in one year obsolete by the next (Deming & Noray, 2018). Therefore, future occupations will require STEM knowledge and twenty-first century critical thinking, collaboration, creativity, and communication skills to accommodate a quickly changing technological landscape (Rotherham & Willingham, 2010). Mathematics can be considered the language of STEM because of its ability to describe and explain underlying relationships between phenomena in the world. Further, foundational mathematics skills, including critical thinking, analysis, and problem-solving, lend themselves to all other content areas. Decomposing complex problems into manageable parts, sorting and analyzing data, and comparing and contrasting strategies to find solutions are all vital for daily problem-solving.

Many factors can deter students from studying STEM, such as perceived academic difficulty and an unwelcoming culture (van den Hurk et al., 2019). In the United States, the middle school period, when students are between the ages of eleven and fourteen, has been identified as a critical point when students develop perceptions of their academic ability. Researchers have recognized middle school as when students decide to pursue higher-level mathematics courses in later grades and, later, STEM fields in postsecondary education (Blotnicky et al., 2018; Falco, 2020; Kang et al., 2019).

To help encourage all youth to engage in STEM, many education researchers have found that engaging students in hands-on, personally relevant, integrated STEM activities in informal settings increased affect and interest in STEM courses and careers (Baran et al., 2019; Crowder & Whittle, 2022; Siregar et al., 2019). However, for many of these experiences, mathematics was used as a tool for other STEM disciplines without surfacing critical mathematical ideas and concepts. In their comprehensive framework for integrated STEM education, Roehrig et al. (2021) wrote that, to be effective vehicles for learning, integrated STEM activities required learners to see the explicit connections to mathematics and science.

This chapter will share the results of a multiple-case study from four districts using Illustrative Mathematics (IM), a problem-based integrated STEM mathematics curriculum. The authors analyzed interview transcripts and identified recurring themes to uncover commonly reported impacts. The chapter will first describe the background of STEM education in the United States and the approaches researchers and educators have taken to address the persistent underrepresentation of students from diverse communities in STEM, including problem-based instruction. Next, the authors will detail how IM developed the standards-based resource based on recognized best teaching practices, culturally responsive pedagogy, and student discourse as vehicles for effective, inclusive mathematics instruction and how this approach supports the twenty-first century skills students need in a rapidly changing society. The chapter will then share a few examples from the IM curriculum designed for students to uncover mathematical concepts by utilizing STEM as the context for collaborative exploration and discovery. Finally, the authors will report on the methodology of data collection and findings on the impact of the IM curriculum on educators and students.

Background

Mathematics has always been a foundation for STEM, but mathematical ideas and concepts are only sometimes explicitly addressed in STEM activities (Stohlmann, 2018). In their meta-analysis of mathematics achievement in integrated STEM education studies, Siregar et al. (2019) found that while ten out of 17 of the examined studies showed a statistically significant positive impact on mathematics achievement, there was still much research needed to understand the most effective method to leverage STEM as a vehicle to teach mathematics. Crowder and Whittle (2022) described a four-day informal STEM experience for middle school girls where each day focused on a different subject area (science, technology, engineering, or mathematics). Students at the camp engaged in various STEM activities, including robotics, but the mathematical concepts necessary to fully understand the tasks were not made explicit. For example, in the robotics activity, students were asked to program a robot to follow a path and use angles to control direction. However, there was no discussion of circles, measurement, or rotation to link the activity to mathematical ideas. Although participants expressed that the four-day experience increased their interest in pursuing STEM careers, they also reported that the last day of the camp, dedicated to stand-alone mathematics concepts, was the least popular. Crowder and Whittle (2022) surmised this response was because mathematics was presented out of context, lessening the content's perceived utility.

Similarly, Baran et al. (2019) discussed an informal integrated STEM program designed to meet science and mathematics standards. While the program results showed significant improvement in participants' attitudes toward science, there was no significant positive change in attitudes toward mathematics in the context of STEM. These findings led the authors to determine that further examination of the mathematics content of the program was needed.

Most of the STEM experiences described above followed a problem-based instructional (PBI) model. In a literature review of problem-based learning in science and mathematics, Merritt et al. (2017) reported that the problem-based approach consisted of several common pillars: presenting students with a problem before formal instruction; providing opportunities for small group collaboration; and teachers acting in the role of facilitator rather than instructor. Research has shown that the PBI model positively impacted students' critical thinking skills (Nurazmi & Bancong, 2021). However, this approach could directly conflict with the mathematics instruction experienced by many mathematics teachers (King & Smith, 2020). The traditional mathematics classroom in the United States has relied on direct instruction and rote memorization of procedures. This approach often does not facilitate a deep understanding of underlying concepts, meaning that students may be able to perform the mechanics of problem-solving but have little grasp on mathematical ideas and the relationships between them (Ferrini-Mundy, 2000). The traditional approach to teaching mathematics frequently teaches concepts disconnected from real-life applications, leading to students having a low utility value for the content (Blumenfeld et al., 1991; Hacker, 2016).

Additionally, studies show that rote learning of procedures can produce limited problem-solving ability because students are rarely encouraged to develop and critique multiple strategies to come to a solution (Fuson et al., 2005). Finally, instead of valuing and elevating the process of student thinking, success in most traditional mathematics classrooms is measured by how quickly and accurately students solve problems, with the most emphasis on obtaining the correct answer. This approach means there is a very narrow definition of "success" in this approach to teaching and learning, which may feel exclusionary to many students from diverse backgrounds and cultures (Assouline & Lupkowski-Shoplik, 2021). Further, Usher (2009) found that the way instruction was structured played a factor in the development of middle school students' mathematics self-efficacy, the belief in one's own capability, and attitudes toward mathematics.

Although a PBI model in the formal classroom has proven effective for mathematics achievement and self-efficacy (Clarke et al., 2004), educators have faced challenges to implementation. Using a PBI model has often required a change in teaching practices from traditional methods teachers may have experienced, deep professional learning, and curricular supports as teachers shift their classroom instruction (Schettino, 2016).

Illustrative Mathematics

In 2011, IM designed an openly available set of tasks to illustrate the new Common Core State Standards in Mathematics (CCSSM) to meet the needs of educators. These tasks evolved into free K-12 mathematics curricula grounded in a well-defined, learnable PBI model. The approach of the IM curriculum has been to place mathematics into a real-world context while elevating mathematics itself. Currently, the curricula are in use at over 1800 school districts across the United States, and more than 40,000 educators and 2 million students access them. To be effective, a PBI model for mathematics instruction requires that educators believe that all students have the potential and ability to engage with and understand mathematical concepts. IM's middle school mathematics curriculum has been designated a high-quality instructional resource by EdReports (2019), a national non-profit organization that evaluates educational materials to help educators choose appropriate resources.

The power of a PBI model rests in shifting inquiry and knowledge from teacher to student. Following the recommendations of Li and Schoenfeld (2019), IM curriculum developers approached mathematics as a sense-making activity rather than memorizing content and learning procedural steps (Illustrative Mathematics, 2021a). As such, lessons were structured to invite students to create their own meaning and understanding of concepts when presented with mathematical and real-world applications. IM's curriculum developers explicitly incorporated the twenty-first century skills of critical thinking, collaboration, creativity, and communication to address student mathematics identity and achievement by designing the curriculum to leverage best practices for teaching mathematics identified by the National Council of Mathematics Teachers: learning goals; student thinking; multiple representations; student discourse; purposeful questions; understanding before procedures; building persistence; and evidence-based assessment (Leinwand, 2014).

The design of the IM curriculum is consistent with aspects of the engineering design process following the definition of Shaughnessy (2013), who described STEM as a multidisciplinary approach to solving problems iteratively and collaboratively. Each IM lesson encourages the engineering design process and engineering habits of mind outlined by Wheeler et al. (2019). IM curriculum developers designed lessons to support student creativity, divergent thinking, systems thinking, optimism, collaboration, communication, and ethics.

Developers leveraged collaboration and mathematics discourse as pathways to deeper conceptual understanding by presenting a coherent, repeated structure to a lesson. The structure of each lesson includes a warm-up activity; one or more instructional activities; a synthesis; and a cool-down. The warm-up activity serves as an invitation to the mathematical work of the day, engaging students and providing access to the mathematics of the lesson. The instructional activities provide students an opportunity to grapple with problems that help advance their understanding of the mathematics identified in the learning goals. The synthesis builds the link between the activities and the mathematical concepts and helps students consolidate what they have learned. The cool-down is an exit ticket that allows students to apply what they have learned and teachers with a quick formative assessment to guide planning for future instruction.

STEM in the Curriculum

This section will detail two examples of how IM's middle school curriculum offered students opportunities to reason about mathematical concepts and then apply them to an engineering problem. The first example of embedded mathematics was Lesson 1 in Grade 6 Unit 3, in a section on Units of Measurement. The student learning goals identified for this lesson are to "evaluate (orally) the usefulness of calculating a rate per 1 when solving problems involving unfamiliar rates" and "explain (orally, in writing, and through other representations) how to solve a problem involving rates in a less familiar context, e.g., minutes per window" (Illustrative Mathematics, 2021b, p. 27). The lesson warm-up has students use the Think-Pair-Share instructional routine in a height estimation activity based on Fig. 8.1.

The curriculum does not expect students to be exact in their estimations in the design of the activity. The purpose is to support discourse by encouraging students to clarify their thoughts and compare their estimates and the strategies used to determine them.

The next activity, "Window Washing," places the students in the context of a real-world situation. The lesson shows students a photo of the Burj Khalifa, the world's tallest building. The lesson guides teachers to utilize the Notice and Wonder instructional routine with the whole class, in which students share what they notice

Fig. 8.1 Estimating relative height. From "Lesson 1: The Burj Khalifa," by Illustrative Mathematics, Grade 6 Unit 3 Teacher Guide, p. 29, CC BY 4.0. (Illustrative Mathematics, 2021b)

and wonder about the photo and the building itself. Afterward, the lesson asks how long it might take to wash all the building's windows, given that it takes 15 min for a crew to wash 18 windows. The problem provides no other information, pushing students to think about what details they might need to answer the question.

The structure of this lesson encourages students to wrestle with the problem independently first, affording opportunities to find different strategies to answer the question at hand. During the activity synthesis, the lesson again calls for facilitating discourse to help students arrive at different unit rates (e.g., time/window, hours/workday). The teacher's guide states (Illustrative Mathematics, 2021b, p. 30):

> Students…try to determine how long it would take to wash all of its 24,348 windows given a specific rate of work by a window-washing crew. The fact that the total number of windows is not a multiple of the corresponding value in the given ratio (15 windows in 18 minutes) motivates students to identify the rate for 1 and then scale that number to answer the question.

Next, students tackle the rate of someone climbing the building, given the following information: "In 2011, a professional climber scaled the outside of the Burj Khalifa, making it all the way to 828 m (the highest point on which a person can stand) in 6 h" (ibid). Students then apply their knowledge of calculating unit rates to different contexts. For example, given the proportions of weight on Earth to weight on Mars and the moon, students calculate how much a person would weigh in those places given their weight on Earth.

The cool-down for this lesson asks, "The fastest elevators in the Burj Khalifa can travel 330 feet in just 10 s. How far does the elevator travel in 11 s? Explain your reasoning" (ibid). The most important part of the student response for a teacher is not the numerical answer but how the student explains the reasoning behind the answer.

Additionally, the lesson includes seven practice problems using the concepts of unit rates in varying contexts, including how long it takes to service tables in a restaurant, how much a babysitter is paid by the hour, and the number of copies of an image that could be made given a total number of shapes. All these activities elevated the mathematical concepts as they were being used to solve different real-world problems, purposefully supporting students to think critically and broadly about the ideas. Each unit in the curriculum culminates in a project called "Let's Put it to Work" so that students can apply concepts they have learned throughout the unit to a real-world problem. IM has included this type of engineering design task throughout the curriculum.

The second example is also in the Grade 6 curriculum, Lesson 19 in Unit 1. This unit focuses on teaching the concepts of area and surface area using aspects of mathematical modeling (Kaiser, 2020). This task is an end-of-unit lesson project in which sixth graders design a tent to meet specific criteria (Illustrative Mathematics, 2021c, p. 377): "Your task is to design a tent to accommodate up to four people, and estimate the amount of fabric needed to make your tent. Your design and estimate must be based on the information given and have mathematical justification." The learning goals were to "Apply understanding of surface area to estimate the amount of fabric in a tent and explain (orally and in writing) the estimation strategy. Compare and contrast (orally) different tent designs [and] Interpret information (presented in writing and through other representations) about tents and sleeping bags" (p. 377). For this task, students are given the measurements for a standard sleeping bag and drawings of various tent designs. Students are encouraged to use the given information to create a tent in a style of their choice. This project builds on earlier lessons in this unit, where students create three-dimensional nets to calculate surface area. They are first required to use three-dimensional nets to represent their tent designs. While the lesson has students work in small groups to clarify specifications, each student is responsible for designing a tent individually. This purposeful approach of scaffolded problem-based instruction leverages student discourse, curiosity, and creativity to help students create their own understanding of mathematical concepts and relationships.

Methodology

To understand how the IM curriculum and its supports have impacted student and educator learning, as well as identify existing challenges to implementation, this collective analysis combined data from four case studies conducted with districts and schools that were implementing the IM curriculum. Because IM was interested in the "how" of in-school implementation, the case study approach was deemed appropriate for data collection (Yin, 1994). Case studies have proven to offer the opportunity to conduct in-depth exploration and provide rich data that can offer insights into a complex phenomenon like curriculum implementation (Brown, 2008; Rowley, 2002). Case studies have also helped mediate the ethical issues of short-term research that can arise with vulnerable populations, particularly young people from underserved communities, by developing meaningful and lasting partnerships with districts, schools, educators, and students (Hays, 2003; Swanborn, 2010). Furthermore, case studies have allowed for understanding phenomena in context, something difficult to accomplish using quantitative methods alone. This deeply contextual insight has helped researchers to build a holistic picture of implementation.

Proposition

The proposition for this analysis was IM's theory of action, where the hypothesis was that the combination of high-quality curriculum, professional learning, and school supports would positively impact teaching practices and student mathematics affect, self-efficacy, and achievement. The case study approach was helpful to support or to refute IM's theory of action by providing real-world examples and evidence. In this context, authors defined mathematics affect as the expression of positive emotion, attitude, and belief by students or teachers (Jacobson & Kilpatrick, 2015; Leder & Forgasz, 2002; Zan et al., 2006). Mathematics education researchers have long studied the role of mathematics self-efficacy in mathematics achievement (e.g., Hackett & Betz, 1989). Similar to mathematics affect, researchers have found that self-efficacy in mathematics and problem-solving impacts the direction young people choose to follow in their scholarly and career pursuits (Lent et al., 1991). For this analysis, the authors defined self-efficacy in teacher and student reflections of positive mastery experiences that helped support their belief in their ability to achieve in future mathematics activities (Usher, 2009). Achievement in this study was both actual and perceived. Standardized or classroom assessment reflected actual achievement, while teacher observation or experience represented perceived achievement.

Cases

Districts for the four case studies were recruited and selected based on several criteria, including implementation of the IM curriculum, completion of IM professional learning, openness to data collection, and availability of site visits. The four school districts included in this analysis were varied regarding region of the United States, community context, and student demographics. As described below, the four districts represented urban, suburban, and rural community contexts, as well as the south, southwest, western, and midwestern regions of the United States.

District A At the time of the case study, District A was a suburban district in the western United States with 13 schools serving pre-kindergarten to eighth grade. The student population was diverse: 7% Asian/Pacific Islander; < 1% American Indian/ Alaskan Native; 6% Black/African American; 44% Hispanic; 37% White; and 5%

multiracial. 47% of the students were designated as economically disadvantaged, defined as the eligibility for the federal Free and Reduced Price Meal program, where family income is at or below 185 percent of the federal poverty guidelines (Legislative Analyst's Office, 2009). The student body contained 11% multilingual learners and 8% of students in special education programs.

District B District B was a rural county in the midwest of the United States. At the time of the case study, this district was composed of three schools with the following student demographics: 1% Asian; < 1% American Indian/Alaskan Native; < 1% Black; 6% Hispanic; 91% White; and 1% multiracial. Nearly 49% of the students were designated as economically disadvantaged, 1% were multilingual learners, and 13% were enrolled in special education programs.

District C Located in the southwest United States, District C was a large, urban district with 25 schools. The following are the demographic statistics of District C during the case study: < 1% Asian/Pacific Islander; 5% American Indian/Alaskan Native; 3% Black/African American; 84% Hispanic; and 8% White. Approximately 80% of the students were eligible for free/reduced-price meals, 17% were multilingual learners, and 13% were enrolled in special education programs.

District D District D was a pre-kindergarten through eighth-grade charter school system in an urban southern United States region. During the school year of the case study, the demographic makeup of the student body was 1% Asian/Pacific Islander, < 1% American Indian/Alaskan Native, 48% Black/African American, 12% Hispanic, 29% White, and 12% multiracial. Over 73% of students qualified for free/reduced-price meals, 6% were multilingual learners, and 14% were students with disabilities.

Data Collection and Analysis

Data collection methods for the full case studies included informal interviews, focus groups, classroom observations, and interactions with administrators, educators, and students. The findings reported in this chapter focused on the analysis of transcripts from educators to identify common themes across districts. Data were iteratively coded for reported impacts on teacher attitudes, classroom practices, and perceptions of student engagement and achievement, then clustered into categories to capture recurring themes (Ezzy, 2013).

Results

Despite the contextual differences between the school districts, the authors identified several recurring themes in the data. For educator interactions in the case studies, the data revealed several important themes: shift in practice, perceived student achievement, and appreciation of the long-term value of problem-based mathematics instruction. Collectively, the data highlighted the importance of ongoing teacher professional development and support in effectively implementing the new curriculum and promoting a deeper understanding of mathematics among students.

Shift in Practice

The data suggested a significant shift in educators' reporting of their approach to mathematics teaching, moving away from traditional methods characterized by rote memorization and individual work. Instead, there is a focus on problem-based and collaborative learning, emphasizing conceptual understanding, critical thinking, and real-world applications.

> District A Administrator: Our teachers are changing their mindsets, too. Before, when teachers were planning lessons, they didn't necessarily step back and look at the big picture of how the lessons they taught in September would connect to lessons they taught in December or March. Thanks to the coherence of the 6–8 Math curriculum, they now have a better understanding of these connections and about how mathematics is about building mastery over time, rather than simply mastering one skill or concept and then moving to the next one.
>
> District B Teacher: I tried a few units last year to see if I liked the curriculum and I did. I liked the coherence from unit to unit and the built-in instructional routines for student engagement. I also liked how it helped students understand abstract concepts with concrete, hands-on activities.
>
> District C Administrator: The curriculum is also changing teachers' perceptions of what math teaching and learning should look like.
>
> District D Administrator: In terms of transitioning to the IM curriculum, teachers need time and support as they explore the curriculum in the classroom, and figure out what it should feel like.

The quotations above also illustrated educators' need for time with the curriculum to explore and experience in a classroom and to facilitate a change in attitude toward mathematics instruction. The quotes also alluded to the necessity of learning a new way of envisioning how a mathematics classroom is supposed to "look" and "feel." These statements suggested that the problem-based approach may be a new experience for many teachers as both learners and instructors. Across districts, educators expressed that they required more than one academic year to situate themselves in the curriculum. Educators shared that the curriculum's purposeful ties to mathematics standards and overall coherence led to confidence and comfort in delivering the content.

Perceived Student Achievement

According to educators in the case studies, early data and observations suggested that the IM 6–8 Mathematics curriculum positively affected student performance. The cases reported that students who experienced the problem-based and collaborative learning approach showed improved preparation and performance in mathematics compared to those exposed to traditional methods.

> District A Teacher: My students are grasping concepts more quickly. Because there's coherence from unit to unit, I haven't seen the dips in their scores that I used to see. I've seen increased scores on the same learning targets as compared to our previous math curriculum, and I've seen an increase in student performance overall.
>
> District B Teacher: For students, there's a nice balance between problem-based tasks and procedures. It's like a lightbulb has come on for them. Now they're not just giving me the answer; they're thinking about where that answer came from. Instead of just memorizing a formula, such as for the circumference of a circle, they're seeing the steps and asking, 'How we did [sic] we get this formula?
>
> District C Administrator: When we look at the teachers who were the early adopters of the 6–8 curriculum—the ones who jumped in at the very beginning and said 'I'm going to do this and trust in the process'—their students' data shows better than that [of] their colleagues.
>
> District D Teacher: The conceptual understanding students get from IM is much stronger.

Data collected from the participating districts also reported positive outcomes for achievement. Educators in the district case studies referred to increased assessment results and a perception that students seemed more engaged and enthusiastic about learning mathematics than they had with previous curricula. For example, a principal of one of the implementing elementary schools attributed the following to IM's focus on student engagement in mathematics content:

> We conducted a study last year over our behavior, and when I came here four years ago, we had significant behaviors, almost 700 behavioral referrals in a year. Last year, we were down to below 300. But we were not getting them during math when we first implemented IM, because kids were actually engaging in the work, and they were expected to engage.

Other behavioral indicators the districts reported included students expressing anticipation for mathematics class and excitement about interacting with their peers when discussing mathematics and concepts placed in the context of real-world problems.

Long-Term Value of IM's PBI Approach

Educators noted that, over time, the benefits of using the IM 6–8 Mathematics curriculum became evident, with year-to-year improvement in student preparedness and outcomes. Although transitioning to a new approach presented challenges, teachers believed that the investment justified the rewards of deeper understanding and increased student engagement.

District A Teacher: One of the biggest benefits of 6–8 Math is that our students now see math differently. They see that it is problem-based, that mistakes can be valuable in their learning, and that it's about the process, not just the result. This is not only important for their success in math but in life.

District B Teacher: We do standards-based grading and I've seen a noticeable improvement compared to previous years.

District C Administrator: At the beginning of 2017, our seventh grade teachers said that their incoming seventh graders—who had used the 6–8 curriculum in sixth grade—were more prepared. And at the beginning of 2018, teachers again saw an improvement over the previous year. So we're starting to see that return on investment that comes with using the IM curriculum over time.

District D Teacher: I feel so comfortable doing it in year three, versus year one, where I think it was more of a challenge for me. But once you get the hang of it, it makes sense, and it clicks, and it just comes second nature to you. Now, I could look at a lesson and say, 'OK, this is what I need to do, this is what the outcome is.' It all makes sense. So, it feels good.

The data analyzed from the districts reported that IM's curriculum gave students exposure to content multiple times throughout the learning progression, introducing a concept, and then returning to that concept as a scaffold to learn a new one. Therefore, they found that students not only became accustomed to embedded instructional routines in the classroom but also had the opportunity to recognize and build upon concepts learned in previous lessons, units, and grade levels. As the case study participants expressed, students who experienced the IM curriculum in the sixth grade seemed more ready to tackle seventh-grade mathematics content than those who experienced a different sixth-grade curriculum. Educators also shared that the IM curriculum appeared to offer learning that extended beyond traditional mathematics content. These skills included generalized problem-solving skills, comparing and critiquing different strategies, and understanding that making mistakes is part of the learning process.

Practical Implications

The practical implications of these findings could be far-reaching. Because mathematics has played a critical role in students' decision-making about pursuing further mathematics education in secondary and postsecondary school, and mathematics achievement has been linked to affect and engagement, the results painted a compelling picture of the IM curriculum's potential long-term effects on these middle school teachers and their students. If, as observed by case study participants, students felt more engaged in mathematics and learned concepts more deeply than traditional instruction, this may have impacted both their mathematics affect and self-efficacy.

Shifts in teacher attitudes toward problem-based mathematics instruction have traditionally been challenging to achieve (Herbel-Eisenmann et al., 2006; Jao, 2017; Merritt et al., 2017). However, cumulative results from these four case studies have shown that implementing a high-quality curriculum, which embeds mathematics in engineering-style design tasks and appropriate professional learning support for educators, helped facilitate that shift. Allowing teachers the time and space to deeply engage with the problem-based content and offering curricular and instructional support increased teachers' willingness and confidence to let students grapple with mathematical ideas and find answers for themselves. Although researchers collected no data on behavioral changes in the case studies analyzed for this report, data from a case study IM conducted with a district using IM's elementary school (K–5) curriculum did produce an interesting finding.

One of the limitations of this report was that all participating districts in the case studies were implementing the middle school IM curriculum, which meant that they were satisfied with the curriculum and its impact on students and educators. It is also necessary to engage in longitudinal research to track IM students' choices in their academic journeys as they enter secondary school and college. The findings show that the problem-based approach of the IM curriculum could offer an important gateway for future STEM professionals. Further large-scale studies are underway to provide evidence for the possible relationships between the effective implementation of the IM curriculum and student achievement, particularly those students from underrepresented populations in STEM.

References

Assouline, S. G., & Lupkowski-Shoplik, A. (2021). *Developing math talent: A comprehensive guide to math education for gifted students in elementary and middle school*. Routledge.

Baran, E., Canbazoglu Bilici, S., Mesutoglu, C., & Ocak, C. (2019). The impact of an out-of-school STEM education program on students' attitudes toward STEM and STEM careers. *School Science and Mathematics, 119*(4), 223–235.

Blotnicky, K. A., Franz-Odendaal, T., French, F., & Joy, P. (2018). A study of the correlation between STEM career knowledge, mathematics self-efficacy, career interests, and career activities on the likelihood of pursuing a STEM career among middle school students. *International Journal of STEM Education, 5*, Article 22. https://doi.org/10.1186/s40594-018-0118-3

Blumenfeld, P. C., Soloway, E., Marx, R. W., Krajcik, J. S., Guzdial, M., & Palincsar, A. (1991). Motivating project-based learning: Sustaining the doing, supporting the learning. *Educational Psychologist, 26*(3–4), 369–398.

Brown, P. A. (2008). A review of the literature on case study research. *Canadian Journal for New Scholars in Education/Revue canadienne des jeunes chercheures et chercheurs en education, 1*(1), 1–13. https://journalhosting.ucalgary.ca/index.php/cjnse/article/view/30395

Camilli, G., & Hira, R. (2019). Introduction to special issue—STEM workforce: STEM education and the post-scientific society. *Journal of Science Education and Technology, 28*, 1–8. https://doi.org/10.1007/s10956-018-9759-8

Clarke, D., Breed, M., & Fraser, S. (2004). The consequences of a problem-based mathematics curriculum. *The Mathematics Educator, 14*(2), 7–16.

Crowder, A., & Whittle, K. (2022). Increasing middle school girls' STEM self-efficacy with soldering, robots, and Mobius strips. *International Journal of Gender, Science and Technology, 14*(2), 232–246.

Deming, D. J., & Noray, K. L. (2018). *STEM careers and the changing skill requirements of work* (No. w25065). National Bureau of Economic Research.

EdReports. (2019). *Kendall Hunt's illustrative mathematics 6–8 math*. www.edreports.org/reports/overview/kendall-hunts-illustrative-mathematics-6-8-math-2019
Ezzy, D. (2013). *Qualitative analysis*. Routledge.
Falco, L. D. (2020). An intervention to support mathematics self-efficacy in middle school. In L. M. Harrison, E. Hurd, & K. Brinegar (Eds.), *Integrative and interdisciplinary curriculum in the middle school* (pp. 37–66). Routledge.
Ferrini-Mundy, J. (2000). Principles and standards for school mathematics: A guide for mathematicians. *Notices of the American Mathematical Society, 47*(8).
Fuson, K. C., Kalchman, M., & Bransford, J. D. (2005). Mathematical understanding: An introduction. In M. S. Donovan & J. D. Bransford (Eds.), *How students learn: Mathematics in the classroom* (pp. 217–256). National Academy Press.
Hackett, G., & Betz, N. E. (1989). An exploration of the mathematics self-efficacy/mathematics performance correspondence. *Journal for Research in Mathematics Education, 20*(3), 261–273.
Hacker, A. (2016). *The math myth: And other STEM delusions*. The New Press.
Hays, P. A. (2003). Case study research. In K. B. de Marrais & S. D. Lapan (Eds.), *Foundations for research* (pp. 233–250). Routledge.
Herbel-Eisenmann, B. A., Lubienski, S. T., & Id-Deen, L. (2006). Reconsidering the study of mathematics instructional practices: The importance of curricular context in understanding local and global teacher change. *Journal of Mathematics Teacher Education, 9*, 313–345.
Illustrative Mathematics. (2021a). *Illustrative mathematics design principles—teachers*. Kendall Hunt. https://im.kendallhunt.com/k5/teachers/teacher-guide/design-principles.html
Illustrative Mathematics. (2021b). *Illustrative mathematics grade 6 unit 3 teacher guide*. Kendall Hunt.
Illustrative Mathematics. (2021c). *Illustrative mathematics grade 6 unit 1 teacher guide*. Kendall Hunt.
Jacobson, E., & Kilpatrick, J. (2015). Understanding teacher affect, knowledge, and instruction over time: An agenda for research on productive disposition for teaching mathematics. *Journal of Mathematics Teacher Education, 18*, 401–406.
Jao, L. (2017). Shifting pre-service teachers' beliefs about mathematics teaching: The contextual situation of a mathematics methods course. *International Journal of Science and Mathematics Education, 15*, 895–914.
Kaiser, G. (2020). Mathematical modelling and applications in education. In S. Lerman (Ed.), *Encyclopedia of mathematics education* (pp. 553–561). Springer.
Kang, H., Calabrese Barton, A., Tan, E., Simpkins, S. D., Rhee, H. Y., & Turner, C. (2019). How do middle school girls of color develop STEM identities? Middle school girls' participation in science activities and identification with STEM careers. *Science Education, 103*(2), 418–439.
King, B., & Smith, C. (2020). Using project-based learning to develop teachers for leadership. *The Clearing House: A Journal of Educational Strategies, Issues and Ideas, 93*(3), 158–164.
Legislative Analyst's Office. (2009). *Improving academic success for economically disadvantaged students*. Author. https://lao.ca.gov/2009/edu/academic_success/academic_success_0109.aspx
Leder, G. C., & Forgasz, H. J. (2002). Measuring mathematical beliefs and their impact on the learning of mathematics. In G. C. Leder, E. Pehkonen, & G. Törner (Eds.), *Beliefs: A hidden variable in mathematics education* (pp. 95–113). Kluwer Academic Publishers.
Leinwand, S. (2014). *Principles to actions: Ensuring mathematical success for all*. National Council of Teachers of Mathematics.
Lent, R. W., Lopez, F. G., & Bieschke, K. J. (1991). Mathematics self-efficacy: Sources and relation to science-based career choice. *Journal of Counseling Psychology, 38*(4), 424.
Li, Y., & Schoenfeld, A. H. (2019). Problematizing teaching and learning mathematics as "given" in STEM education. *International Journal of STEM Education, 6*. Article 44.
Merritt, J., Lee, M. Y., Rillero, P., & Kinach, B. M. (2017). Problem-based learning in K–8 mathematics and science education: A literature review. *Interdisciplinary Journal of Problem-Based Learning, 11*(2).

Nurazmi, N., & Bancong, H. (2021). Integrated stem-problem based learning model: its effect on students' critical thinking. *Kasuari: Physics Education Journal (KPEJ), 4*(2), 70–77.

Roehrig, G. H., Dare, E. A., Ring-Whalen, E., & Wieselmann, J. R. (2021). Understanding coherence and integration in integrated STEM curriculum. *International Journal of STEM Education, 8*, 1–21.

Rotherham, A. J., & Willingham, D. T. (2010). "21st-century" skills: Not new, but a worthy challenge. *American Educator, 34*(1), 17–20.

Rowley, J. (2002). Using case studies in research. *Management Research News, 25*(1), 16–27.

Shaughnessy, J. M. (2013). Mathematics in a STEM context. *Mathematics Teaching in the Middle School, 18*(6), 324.

Schettino, C. (2016). A framework for problem-based learning: Teaching mathematics with a relational problem-based pedagogy. *Interdisciplinary Journal of Problem-Based Learning, 10*(2). Article 12.

Siregar, N. C., Rosli, R., Maat, S. M., & Capraro, M. M. (2019). The effect of science, technology, engineering and mathematics (STEM) program on students' achievement in mathematics: A meta-analysis. *International Electronic Journal of Mathematics Education, 15*(1), 1–12.

Smith, E., & White, P. (2019). Where do all the STEM graduates go? Higher education, the labour market, and career trajectories in the UK. *Journal of Science Education and Technology, 28*, 26–40.

Stohlmann, M. (2018). A vision for future work to focus on the "M" in integrated STEM. *School Science and Mathematics, 118*(7), 310–319.

Swanborn. (2010). *Case study research: What, why, and how?* (1st ed.). SAGE Publications.

Usher, E. L. (2009). Sources of middle school students' self-efficacy in mathematics: A qualitative investigation. *American Educational Research Journal, 46*(1), 275–314.

van den Hurk, A., Meelissen, M., & van Langen, A. (2019). Interventions in education to prevent STEM pipeline leakage. *International Journal of Science Education, 41*(2), 150–164.

Wheeler, L. B., Navy, S. L., Maeng, J. L., & Whitworth, B. A. (2019). Development and validation of the classroom observation protocol for engineering design (COPED). *Journal of Research in Science Teaching, 56*(9), 1285–1305.

Yin, R. K. (1994). Discovering the future of the case study. Method in evaluation research. *Evaluation Practice, 15*(3), 283–290.

Zan, R., Brown, L., Evans, J., & Hannula, M. S. (2006). Affect in mathematics education: An introduction. *Educational Studies in Mathematics, 63*(2), 113–121. https://www.jstor.org/stable/25472116

Chapter 9
Doing Mathematics in a School Makerspace: Representing Proportional Reasoning Through Designing Digital Homes

Juan Torralba and Rob Rouse

Abstract In this pilot study, we extend existing research in the field of mathematics-focused maker-based STEM education. Specifically, we report on the ways in which seventh-grade students represented their knowledge of proportional reasoning as they created digital 3-dimensional models of their dream homes in a school makerspace STEM integrated activity. Using a directed content analysis, we examined written, verbal, and visual data collected from the students across the project. Results indicated that students represented their understanding of proportional reasoning through open-ended written responses more accurately than through verbal descriptions or digital artifacts. In addition, the geometric and numeric dimensions of proportional reasoning, and their respective components were represented more often by students than any other dimension of proportional reasoning across the data. This suggests a maker-based instructional approach to teaching proportional reasoning in the middle grades may help students gain a solid foundation in those mathematical components and twenty-first century skills such as communication and collaboration. Implications and recommendations for practice and research are discussed.

Keywords Makerspace · Proportional reasoning

Introduction

Engaging students in STEM (science, technology, engineering, and mathematics) activities during the early grades can positively impact their motivation and increase their interest in taking advanced STEM courses in secondary school (Blotnicky et al.,

J. Torralba (✉)
Department of Teaching and Learning, University of Miami, Florida, USA
e-mail: Jxt677@miami.edu

R. Rouse
School of Education, University of Queensland, Brisbane, Australia
e-mail: R.rouse@uq.edu.au

J. Anderson and K. Makar (eds.), *The Contribution of Mathematics to School STEM Education*, https://doi.org/10.1007/978-981-97-2728-5_9

2018; Dabney et al., 2012; Harackiewicz et al., 2012). In addition, students who demonstrate interest in the STEM fields by the eighth grade are two to three times more likely to pursue, persevere, and complete a degree in a STEM subject (Sadler et al., 2012). These findings demonstrate the importance of fostering an interest in STEM in students in the early and middle grades to support more students advancing to participate in STEM careers in the future.

Although the acronym "STEM" is generally used to address all four subjects in silos (i.e., science, technology, engineering, and mathematics), learning STEM through the integration of its content, overarching concepts, and practices, has been found to promote student stronger interest, motivation, and participation in STEM activities (Blikstein, 2013; Guzey et al., 2017; Honey et al, 2014; Martin, 2015; Ramey et al., 2020). Mathematics performance in STEM activities, in particular, has been found to impact students' ability to enter and follow a STEM career path (Blotnicky et al., 2018; Marshall et al., 2016; Martinez Ortiz, 2015). Furthermore, students' understanding of middle grades mathematics content (specifically ratios, proportion, and proportional reasoning), has been identified as a critical factor in their ability to succeed in secondary school mathematics, as such content serves as foundational for overarching mathematical concepts found in geometry, algebra, and calculus (Lobato et al., 2010). Therefore, with the goal of more effectively preparing students for secondary school mathematics and beyond, it is important to continue to investigate the ways in which students in the middle grades can develop a robust understanding of proportional reasoning.

In this chapter, we contribute to the literature on middle grades STEM learning and, more specifically, students' learning of proportional reasoning in mathematics. We report findings from a pilot study in which we investigated seventh-grade students' representations of proportional reasoning following a mathematics-focused STEM integrated activity in which they designed and built digital versions of their dream home. Uniquely, our study took place in the school's makerspace, and all instruction engaged students in maker-based learning experiences, which have been found to improve student understanding of STEM content (Chu et al., 2017; Ke et al., 2019). The study was guided by the question: How, and to what extent, do students represent their knowledge of proportional reasoning through designing digital dream homes in a school makerspace?

In what follows, we first provide background information on learning experiences in mathematics, making as a vehicle for learning, and STEM and mathematics learning through making. Next, we present information on the participants and procedures in our pilot study, followed by our methods for collecting and analyzing data. After presenting study results, we discuss the implications of our findings for learning mathematics through making and STEM integration. Through this work, our hope is to provide further evidence and support for using school makerspaces as venues for teaching standards-based middle school mathematics content.

Background

Providing meaningful and positive student learning experiences in mathematics classrooms has become a strong focus for educators, not only because of the potential positive impact on students' mathematics performance, but also because of the potential of such experiences to spark students' interest in mathematics, strengthen their confidence in mathematics performance, and foster the development of a mathematics identity (Anderson, 2006; Boaler & Selling, 2017; Cobb, 2007). In the past, researchers have investigated mathematics learning experiences in middle grades to understand their impact on student academic performance (Martínez Ortiz, 2015), mathematical competence (Boesen et al., 2014), and mathematics skills and learning attitudes (Furukawa, 2018; Weinberg et al., 2011). However, much remains unexplored related to student mathematics learning experiences that are guided by a standards-based curriculum, embrace innovative experiential learning opportunities, and assess students using a variety of modes (including via writing, talk, and student-created artifacts as evidence of learning).

More recently, studies on experiential learning, as an instructional approach for mathematics learning have made several recommendations. First, to develop mathematics learning experiences that promote equity and engagement through non-traditional learning approaches that activate students' engagement through open-ended and visual mathematics tasks (Boaler et al., 2017). Second, to engage students in meaningful mathematical experiences in middle and secondary school that may have a long-term impact on students' identities and professional decisions in their adult lives (Boaler & Selling, 2017). Third, to stimulate student excitement and develop their curiosity through discovery in mathematics (Tong et al., 2020). Fourth, to provide opportunities for students to reflect, relearn, rethink, reapply, and reexperience as they identify and gain new knowledge (Chan, 2023). Overall, the literature on student success in mathematics suggests that an experiential approach has the potential to meaningfully engage students in mathematics learning and increase their interest in higher-level mathematics.

Making as a Vehicle for Learning

One novel approach that has gained interest within the field of education is the idea of learning through making. This approach has its roots in a concept called the maker movement (Bevan, 2017; Breaux, 2017; Hsu et al., 2017; Martin, 2015; Papavlasopoulou et al., 2017). The maker movement emerged from the "do-it-yourself" (DIY) movement in the early 2000s and focuses on the designing, building, and sharing of physical and digital artifacts with others (Dougherty, 2012; Halverson & Sheridan, 2014). The DIY movement generally focused on individuals creating their own digital or physical products/artifacts, instead of purchasing them, turning individuals into producers rather than consumers. The maker movement, on the other hand, goes beyond working individually and instead fosters establishing physical and virtual spaces for individuals to create and share their physical or digital artifacts, as well as to learn from each other, share equipment, and build a community (Breaux, 2017).

There are several compelling reasons for integrating maker education (i.e., the name for instruction based on principles of the maker movement) into schools. First, because maker education is somewhat ill defined, teachers who adopt maker education can more easily connect to a broad set of instructional goals. For example, teachers can connect to academic goals (e.g., STEM standards), twenty-first century skills (e.g., collaboration, creativity), or specialized goals (e.g., culturally relevant instruction). Second, maker education shares attributes with several established instructional models such as experiential learning and project-based learning (PBL). Like experiential learning, maker education focuses on students learning-by-doing and reflecting on their experiences (Kolb, 2014). Similarly, maker education and PBL are student-centered, position the teacher as a facilitator, emphasize process over product, and lead to students creating personalized products (Chounta et al., 2017; Marshall & Harron, 2018; Hsu et al., 2017).

While there is currently no universally-agreed-upon model for how to enact maker education in schools, one place to integrate maker education is in the school's makerspace. Makerspaces are sites where instruction is aligned with the principles of the maker movement. That is, they promote sharing and collaboration, and they often have low- and high-tech tools useful for students prototyping and creating. Some makerspaces are large and hold complex machinery while others are mobile and stocked with comparatively simple tools and materials (Craddock, 2015; Krishnan, 2015; Wong & Partridge, 2016). Regardless of how they look, as school-based makerspaces become more popular and gain traction in schools, it is imperative to investigate how students learn best in these spaces.

STEM and Mathematics Learning Through Making

Previous studies have found that participation in maker-based STEM integrated activities in schools can increase student interest in the STEM fields (Tofel-Grehl et al., 2017), engage those who are typically underrepresented in STEM (Holbert, 2016), and promote voices and power for those who are typically excluded from the maker movement outside of schools (Calabrese Barton & Tan, 2018; Calabrese Barton et al., 2017). Research has also shown that students who participate in maker-based activities develop twenty-first century skills such as collaboration, critical thinking, and creativity (Hughes, 2017; Hughes et al., 2019); shape stronger STEM identities (Davis et al., 2017; Torralba & Rouse, 2019); and, to some extent, increase student understanding of STEM content (Litts et al., 2017; Torralba, 2019). However, the extant literature on learning through making mostly focuses on, science, technology, and engineering content-focused STEM integration, with comparatively little focus on mathematics learning (Hsu et al., 2017; Papavlasopoulou et al., 2017; Ventä-Olkkonen et al., 2019). Considering the identified benefits in the literature on STEM learning through integration and making and the role student understanding of proportional reasoning in middle school play in their success in upper-level mathematics, science, physics, geography, and economics (Ben-Chaim et al., 2012), we sought to investigate how middle school students demonstrated their understanding of the two main dimensions of proportional reasoning through making; namely, numerical proportionality (e.g., percentages, rates, and ratios) and geometric proportionality (e.g., similarity and attributes of similarity) (Lobato et al., 2010).

Method

Participants

Thirty-four seventh-grade students from a public urban K-8 school in the southeast United States participated in this study. Unfortunately, due to many absences across the term, of these students, only eight students completed all parts of the project. These eight students (five female and three male) comprised our sample. The sample was representative of the classroom.

Instructional Design

Students worked individually on the dream home project, the focal instructional activity integrating mathematics and technology in this study, for seven weeks, meeting twice-a-week for 2 h per session. We designed the project to facilitate students demonstrating what they knew about proportional reasoning by engaging with dual 2D/3D modeling technology and completing design-based activities. In addition, we designed the project to facilitate students growing twenty-first century skills as many of the activities required them to think critically, collaborate, and communicate with their classmates. All parts of the project occurred during the normal school day. During the implementation of the project, the classroom teacher led the class through the project activities, while the first author served as a support facilitator by clarifying expectations for each task, troubleshooting technology issues, and providing feedback and assistance to students when they had questions. To complete the dream home project, the teacher instructed the students to begin by sketching what they envisioned as their future dream homes using pencil and paper. After receiving feedback from their classmates on the paper sketch, the teacher directed students to create digital models of their sketches, using an online digital modeling program called sweethome3d.com. This program allowed students to replicate their paper sketch on a digital 2D sketchpad that then created a digital 3D model of the sketch. The digital models used the same units of measurement as the paper sketch and included the same details (e.g., windows, bedroom items, living room items). The digital 3D model also allowed students to incorporate 3D models of furniture, fixtures, and outdoor items (e.g., cars, pools) into their sketches. After completing their digital models, the teacher instructed students to use wix.com, a free online website builder, to create digital portfolios showcasing the work they completed during the design of their digital dream home models. Students' portfolios displayed screenshots and interactive 2D and 3D models of their homes, along with written descriptions of how they came up with their designs and ultimately built their dream homes. The portfolio served as a tool to help students reflect on their digital design work. After completing their portfolios, students presented their completed projects to the class and then individually wrote responses to prompts in which they described how they used mathematics concepts in the construction of their digital models.

Data Collection

To answer our research question, (How, and to what extent, do students represent their knowledge of proportional reasoning through designing digital dream homes in a school makerspace?), we drew on four data sources aligned with different phases of the project: (a) digital artifacts, (b) portfolios, (c) verbal presentations, and (d) written explanations. Digital artifacts consisted of the digital models students created in sweethome3d.com. Portfolios included students' individual websites displaying their completed dream homes as well as their written descriptions related to the design and construction tasks in which they engaged throughout the project. For these written descriptions, the teacher asked students to make comprehensive notes related to the decisions they made at different stages of the project, the problems they experienced, and how they solved those problems. Verbal presentations included students' video-recorded presentations that they gave to their classmates at the end of the project. Presentations ranged in length from 5 to 6 min. Written explanations included students' written responses to three brief open-ended questions. These questions asked students to describe the mathematics concepts they used to complete the project and build their dream homes and were the same for all students. These questions included: "Describe the process you used to create your dream home," "What proportional reasoning knowledge did you use to create your dream home?" and "How did you ensure the measurements in your work were accurate?"

Analysis

We used a directed content analysis (Hsieh & Shannon, 2005) to determine how students represented their knowledge of proportional reasoning after completing the different tasks in the dream home project. This type of analysis relies on identifying key concepts ahead of time (i.e., concepts related to proportional reasoning in this case), creating a coding scheme based on operationalized definitions, and then coding the available data. Accordingly, our analysis took place in three parts. First, we identified dimensions of proportional reasoning to guide our creation of a coding scheme. Second, we created a coding scheme. Third, we coded the data and looked for themes.

Identifying Dimensions of Proportional Reasoning

Given that we sought to investigate how and to what extent students represented their knowledge of proportional reasoning after participating in the dream home project, we aimed to create a coding scheme targeting proportional reasoning. We did this by searching the appropriate seventh-grade standards document (in this case, we searched the Texas Essential Knowledge and Skills (TEKS) for middle school mathematics) and identifying the extant concepts related to proportional reasoning. We took this step because this activity took place within a school in which teachers were required to connect instruction to the state standards. Thus, we sought to make concrete connections to the appropriate state standards for mathematics. Our search led us to identify two overarching dimensions of proportional reasoning, *geometric* and *numeric*, within the state standards. We used these two overarching dimensions as broad categories, adding distinctions to each category. Within the *geometric* category, the components of proportional reasoning included *attributes of similarity* and *similarity*. Within the *numeric* category, the components of proportional reasoning included *percents, rates, and ratios* and *unit rate*. It is important to note that standards documents developed and enacted in different locations may contain additional categories or subcategories related to concepts of proportional reasoning.

Creating a Coding Scheme

Once we identified the concepts related to proportional reasoning present in the seventh-grade standards documents, we operationalized each concept based on definitions within the standards documents. In this way, we were able to develop a coding scheme using well-established mathematical terminology for proportional reasoning. The result was a coding scheme composed of two main categories each with two sub-categories (Table 9.1). Once we established our coding scheme, we entered it into NVivo, a coding software package.

Table 9.1 Targeted coding scheme for proportional reasoning

Category	Operational definition	Example
Geometric		
Attributes of similarity	The student generalizes the critical attributes of similarity, including ratios within and between similar shapes	First, I drew out a floor plan that I later used to design the house through SweetHome3D.com. Then, after creating the "perfect" floor plan, I realized that my garage was big enough to fit about 5 cars! So I had to change the design to fit 2 cars and making each room properly to fit the furniture
Similarity	The student solves mathematical and real-world problems involving similar shape and scale drawings	I used a specific process to draw up the blueprint for my dream house, using an aerial view and graphing paper and 'built' my house using sweeth ome3d.com
Numeric		
Percents, rates, and ratios	The student solves problems involving percents, rates, and ratios, including multi-step problems involving percent increases and percent decreases	To determine the cost of my house, I used the total area of the house and went onto Zillow.com to find the price of similar houses in different areas
Unit rate	The student calculates unit rates from rates in mathematical and real-world problems	The way I determined the cost of my house was by getting the square feet and multiplying it by the amount of money it was for one square foot

Coding the Data and Looking for Themes

After entering our coding scheme into NVivo, we uploaded our data from each source (i.e., digital artifacts, portfolios, verbal presentations, written explanations) into the software and coded them accordingly. When coding, we took several steps to avoid overrepresenting our codes within each data source. For example, because digital artifacts were represented by 2D sketches and by 3D models, we determined the unit of analysis to be one layout of a home and coded only one of the digital artifacts. We took similar steps to avoid overrepresenting codes between data sources. For example, although students' portfolios included both screenshots of their digital representations of their dream home and written reflections of the tasks they completed throughout the project, we coded only the written reflections and not the screenshots as we had already coded them as digital artifacts. To ensure coding accuracy, an outside researcher who was not involved in the study double coded a subset of the data ($\approx 33\%$). Upon completion, the first author met with the outside researcher to examine and discuss the double-coded data. During this examination, the first author and the outside researcher discussed coding disagreements and clarified questions related to the unit of analysis. Disagreements were resolved through discussion and commenced until reaching 100% agreement on all codes.

Results

We report our results in two parts. In part one, we present broad themes that emerged in aggregate across the data. Examining the data in aggregate allowed us to answer the first part of our research question (How, and to what extent, do students represent their knowledge of proportional reasoning through designing digital dream homes in a school makerspace?) and to determine if students were able to represent their knowledge of proportional reasoning better in one part of the dream home project than in another. In this section, we also identify the modality by which students represented their knowledge of proportional reasoning most frequently.

In the second part, we report how students represented their knowledge of proportional reasoning across the individual categories of our coding scheme. Doing this allowed us to understand how students represented specific aspects of proportional reasoning in their digital artifacts, portfolios, verbal presentations, and written explanations, answering the "to what extent" portion of our research question. All student names used in this section are pseudonyms.

Part I: Aggregate-Level Representations of Proportional Reasoning

Students represented their knowledge of proportional reasoning across all four data sources—although not equally. In addition, students represented their knowledge of proportional reasoning through three distinct modalities: orally through their verbal presentations, visually through their digital artifacts, and in writing through their portfolios and written explanations. Table 9.2 summarizes the aggregate-level representations of proportional reasoning for each data source and modality.

Part II: Instance-Level Representations of Proportional Reasoning

The two overarching dimensions of proportional reasoning guiding our coding scheme were *geometric* and *numeric*. The *geometric* dimension included two sub-categories, *attributes of similarity* and *similarity*. The *numeric* dimension also included two sub-categories, *percents, rates, and ratios*, and *unit rate*. Across our data set, students represented their knowledge of the geometric dimension of proportional

Table 9.2 Aggregate-level representations of proportional reasoning

Data source	Number of instances	Number of students
Representations by source		
Digital artifact	10	6
Portfolio	14	5
Verbal presentation	16	6
Written explanation	21	8
Representations by modality		
Oral	16	6
Visual	10	6
Written	35	8

reasoning 39 times (64%) compared to 22 times (36%) for the numeric dimension. Table 9.3 summarizes the instance-level representations of proportional reasoning. In the following sections, we examine these distinctions more closely.

Representations of Proportional Reasoning by Geometric Dimension: Within the geometric dimension, students represented their knowledge of proportional reasoning 39 times across the data. Students mostly represented their knowledge of *attributes of similarity* and *similarity* in writing, with students representing their knowledge 14 times in their portfolios or written explanations, accounting for 48% of the instances. All students represented knowledge of proportional reasoning through the geometric dimension on one task or another.

Attributes of Similarity. Students demonstrated knowledge of attributes of similarity by generalizing the critical attributes of similarity, including ratios within and between similar shapes (see Table 9.1 for an example of when a student demonstrated knowledge of attributes of similarity). With 29 total instances occurring across the data set, attributes of similarity was the most common type of proportional reasoning represented by students. Students mostly represented attributes of similarity in writing (48%) through their written explanations and portfolios. For example, in their written explanation, one student noted:

Table 9.3 Instance-level representations of proportional reasoning

Mode	Geometric proportionality		Numeric proportionality	
	Attributes of similarity	Similarity	Percents, rates, ratios	Unit rate
Digital artifact	9	1	0	0
Portfolio	4	1	9	0
Verbal presentation	6	2	7	1
Written explanation	10	6	4	1

> I made sure to include a big pool, huge outdoor entertainment space, and lots of fun toys for my kids I hope to have. I made sure that my measurements were proportional and accurate by comparing the square footage of my home to other homes similar to mine.

Relatedly, another student represented *attributes of similarity* in her portfolio when she reported:

> My backyard has an outdoor eating area and a playground while my front yard simply has a driveway. Building my first floor proved to be somewhat difficult at times because I didn't want to have a very big house, yet all the rooms I had made my area extremely large. I made my rooms proportional by comparing the length of my walls to real life things, such as a six-foot-tall person.

Students also represented *attributes of similarity* orally (21%) through their verbal presentations of their projects. For example, in his verbal presentation, one student stated:

> First, I sketched out the house on a piece of paper. And then I went on to the website to build the wall, which was one of the hardest parts because you're making the base, like the foundation of the house. And it takes longer because sometimes you can change your design and rearrange things on the paper.

Finally, students represented *attributes of similarity* visually (31%) through their digital artifacts. Figures 9.1 and 9.2 depict two students' digital artifacts.

In both Figs. 9.1 and 9.2, students represented *attributes of similarity* by including rooms in which the furniture, avatars, and outdoor items were proportional to the room size.

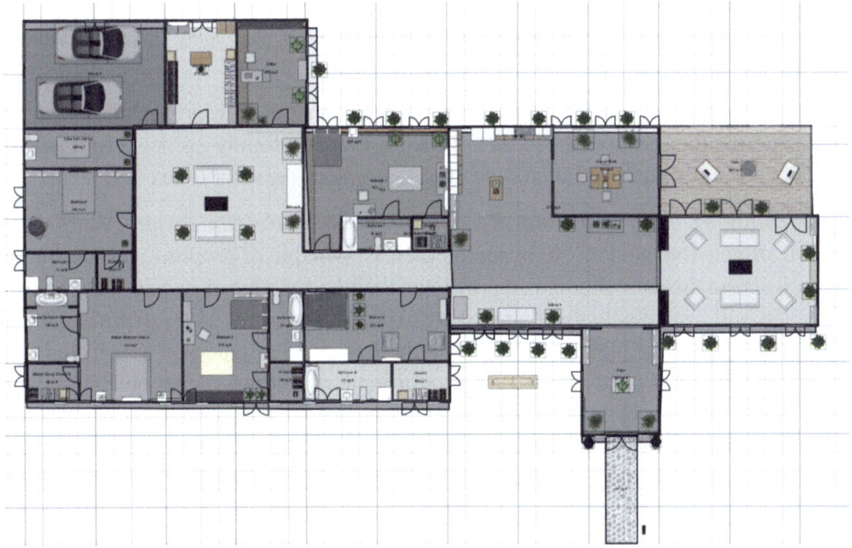

Fig. 9.1 Kanu's digital artifact

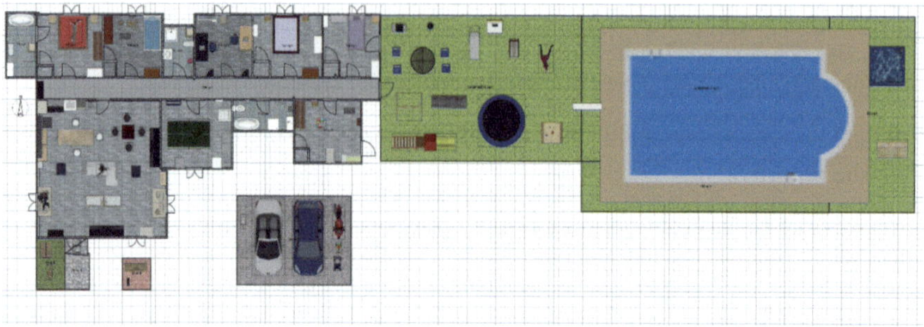

Fig. 9.2 Melanie's digital artifact

Similarity. Students demonstrated knowledge of similarity by solving mathematical and real-world problems involving similar shape and scale drawings (see Table 9.1 for an example of when a student demonstrated knowledge of similarity). With 10 total instances occurring across the data set, similarity was the third-most common type of proportional reasoning represented by students. Students' written explanations revealed resemblances in the processes that they used to create their dream homes. For example, one student wrote:

> To add floors or stories, you click the "add floor" button and you make walls the same way as the first floor, and apply furniture same [as] the first floor. Repeat this same mechanism on and on and on until you've reached [the correct] amount of floors.

Another student shared their process, stating:

> First, I sketched the house that I wanted to build... Then I found the area for each room to get the total area of the first floor. I later on repeated these steps while doing the second floor of my house.

These reports highlight the idea that in many cases students represented *similarity* by considering the layout of the of their dream home, developing a process for making the rooms proportional, and then repeating that process several times.

Representations of Proportional Reasoning by Numeric Dimension: Within the numeric dimension, students represented their knowledge of proportional reasoning 22 times across the data. Students represented their knowledge of *percents, rates, and ratios* and *unit rate* mostly through writing, with 14 instances (64%) stemming from portfolios or written explanations. Notably, there were no instances in which students represented the numeric dimension of proportional reasoning visually through their digital artifacts. This is because the final versions of their digital artifacts included only static measurements of the parts of the dream home. All students represented knowledge of proportional reasoning through the numeric dimension on one task or another.

Percents, Rates, and Ratios. Students demonstrated their knowledge of the percents, rates, and ratios subcategory of the numeric dimension of proportionality by solving problems involving percents, rates, and ratios, including multi-step problems involving percent increases and percent decreases (see Table 9.1 for an example of when a student demonstrated knowledge of percents, rates, and ratios). With 20 total instances occurring across the data set, percents, rates, and ratios was the second most common type of proportional reasoning represented by students.

Students mostly represented percents, rates, and ratios through writing. For example, to explain how he determined the cost of his home and to ensure it was accurate, Edgar wrote:

> The way I determined the cost of my house was by getting the square feet and multiplying it buy the amount of money it was for one square foot and comparing it on Zillow.com with other houses with similar square feet in the area.

Students also represented percents, rates, and ratios orally through their verbal presentations. For example, to describe the process she used to create her home, Ayanna explained:

> [I] looked at the coordinates and saw that every square was 6 feet, and therefore, I made accurate measurements to have a proportional and nice-looking modern-day house. My house was a 3-bedroom two-bathroom one-story house, with a living room, kitchen, utility room, dining room, and a backyard with a pool included. the total square footage of the house was 2938 square feet.

In both of these instances, students referenced comparing in-project measurements and standards to out-of-project measurements and standards to bridge the gap between what worked in the project and how it compared to the outside world.

Unit Rate. Students demonstrated knowledge of attributes of unit rate by calculating unit rates from rates in mathematical and real-world problems (see Table 9.1 for an example of when a student demonstrated knowledge of unit rate). With only two total instances occurring across the data set, unit rate was the least common type of proportional reasoning represented by students by far. One instance occurred as a written explanation and the other as a verbal presentation.

Discussion and Recommendations

In this chapter, we reported the results of a pilot study in which we investigated how and to what extent seventh-grade students demonstrated their understanding of proportional reasoning when digitally designing dream homes in the school's makerspace. This work was motivated, in part, by past research indicating that providing STEM integrated learning experiences to students in the early and middle grades can increase their interest in taking more STEM courses in high school (Blotnicky et al., 2018; Harackiewicz et al., 2012), motivative them to pursue a STEM career in college (Dabney et al., 2012; Sadler et al., 2012), and help them persevere and excel in STEM courses in college (Marshall et al., 2016). More importantly, positive learning experiences and success in middle school mathematics activities have been identified as critical in fostering student interest and motivation to pursue a STEM career later in college (Boaler & Selling, 2017). We grounded our study on this potential and developed a project to investigate students' representations of mathematics knowledge through experiential learning activities that integrated the application of mathematics concepts and digital modeling technology. Across the project, students created digital models of their dream homes, built digital portfolios in which they documented and described parts of the project, made presentations to their peers in which they displayed and described their finished dream homes, and wrote explanations of how they used mathematics to create their dream homes.

We found that students represented their knowledge of proportional reasoning visually, orally, and through writing during the dream home project. However, students represented their knowledge of proportional reasoning most frequently through writing tasks. These tasks included creating the digital portfolio and recording written explanations. In addition, we also found that students represented specific facets of proportional reasoning throughout the project—although unevenly. For example, students represented the *geometric* dimension of proportional reasoning more frequently than the *numeric* dimension of proportional reasoning. Furthermore, within the *geometric* dimension, students represented *attributes of similarity* far more than they represented *similarity*. Likewise, within the *numeric* dimension of proportionality, students represented *percents, rates, and ratios* much more frequently than they represented *unit rate.*

Implications

Taken as a whole, this study contributes several notable findings to the emerging literature on learning mathematics through making in school makerspaces. First, our findings support the notion that activities enacted in school makerspaces can connect to STEM content in a standards-based curriculum (Chu et al., 2017; Ke et al., 2019). By demonstrating that it is possible for students to represent knowledge of proportional reasoning throughout the various tasks that occurred during the dream home project, we also made progress towards addressing a significant tension related to incorporating makerspaces into schools—the idea that it may be difficult to align the goals of makerspaces instruction with the content in a standards-aligned curriculum (Godhe et al., 2019; Halverson & Sheridan, 2014; Rouse & Gillespie Rouse, 2022). While this tension is certainly still a concern, we contend that it is possible to pursue standards-based instruction in a school makerspace given careful planning and the development of tasks that allow students to demonstrate what they know in a variety of ways. Practically, this finding lends support to teachers who currently use makerspaces to enact innovative STEM instruction. Makerspaces are places where teachers can connect to standards while at the same giving students new and dynamic learning experiences. Second, our findings support the practice of designing tasks that allow students to create multimodal representations to demonstrate evidence of learning during activities enacted in a school makerspace at the elementary (Chu et al., 2017), secondary (Litts et al., 2017; Tofel-Grehl et al., 2017), and post-secondary school levels (Saorín et al., 2017). This finding is particularly important for mathematics teaching and learning, as it supports previous studies suggesting that students who generally struggle in traditional learning environments often find it easier to express their knowledge through multimodal avenues (Hughes, 2009). The real-world implication of this finding is that teachers doing work in makerspaces should continue to design instruction so that students have a variety of ways to demonstrate what they know, including through writing, talk, and performance. Importantly, because demonstrating knowledge through multiple modes may require students to share, interact, offer critiques, and provide feedback to one another, this type of instruction is likely to support teachers incorporating instruction related to twenty-first century skills such as communication, collaboration, and teamwork.

Limitations

There are several limitations to this pilot study. First, the number of participants was small, and attrition was high. Out of the 34 students in the class, only eight students completed all four tasks within the dream home project. In future iterations, it will be important to increase the number of participants and implement supports that ensure all students complete the different parts of the project. Second, the uneven way in which students represented their knowledge of proportional reasoning raises questions about if students indeed had opportunities to represent their knowledge of the different dimensions of proportional reasoning that made up our coding scheme. In the case of coding students' digital artifacts for evidence of the *numeric* dimension of proportional reasoning, we know our method to be problematic—students couldn't represent what they knew because the digital artifacts were static. In future studies, we intend to include interviews with students in which they answer questions about their digital artifacts coupled with computer-aided design (CAD) user interaction data—shedding more light on the specific facets of proportional reasoning they used to create their artifacts. We also believe it would be helpful to develop performance tasks that would allow student to record on paper how they used proportional reasoning to make decisions or solve problems like those they experienced in the dream home project. Relatedly, we intend to ask a similar question related to the results from analyzing the written tasks. Is it possible that these tasks were targeted in such a way that allowed students to disproportionally represent their knowledge of proportional reasoning compared to the other tasks? To solve this potential issue in future studies, we plan to better align our coding scheme with the activities and prompts in the various project tasks, making sure to give students opportunities to represent proportional reasoning at each stage of the project.

Although not a limitation in the strictest sense, it is also important to acknowledge that the scope of our data collection and analysis did not allow us to investigate social components of learning through making during the dream home project. Without collecting and analyzing data from social interactions amongst students, we cannot make claims about students' collaborative learning and if or how the unique environment and instructional style within the makerspace may have influenced students' ability to represent their understanding of proportional reasoning. Pursuing investigations such as these in the future would be worthwhile.

Next Steps

As the number of schools adopting makerspaces as places for innovative STEM-focused instruction continues to increase, there will be an additional need for research related to learning through making in school makerspaces. Broadly, the results of this additional research will help to better inform teachers and administrators how to get the most out of their school's makerspace instruction. Specifically, and related to the results of our pilot study, there is a need for additional research related to designing and implementing school makerspace instruction that connects to and integrates the content in standards-based curriculum, resulting in the alignment of activities happening in school makerspaces with STEM content found in the curriculum.

In the future, it will also be important to continue to conduct research related to how students can best show what they know by participating in school makerspace activities. Practically, this research will lead to teachers understanding how best to assess their students when participating in a makerspace activity. Our findings showed that students were able to represent knowledge of proportional reasoning visually, orally, and in writing. This raises the question of whether providing a wider variety of outlets (e.g., video games, interactive demonstrations, self-made videos), might impact the extent to which students demonstrate what they know. Regarding writing, we found that students mostly demonstrated their understanding of proportional reasoning in written tasks. This finding aligns with previous research suggesting writing-to-learn approaches can facilitate student learning of mathematics content (Graham et al., 2020; Lopas et al., 2021). This raises the question of whether researchers can use writing-to-learn approaches to measure students' learning during their participation in school makerspace activities. No matter the next investigation, school makerspaces offer researchers ample opportunities to connect to the next wave of in-school STEM learning opportunities.

References

Anderson, R. (2006). *Mathematics, meaning, and identity: A study of the practice of mathematics education in a rural high school* (Unpublished doctoral dissertation). Portland State University, Oregon.

Ben-Chaim, D., Keret, Y., & Ilany, B. (2012). *Ratio and proportion: Research and teaching in mathematics teachers' education (pre- and in-service mathematics teacher of elementary and middle school classes).* Sense Publishers.

Bevan, B. (2017). The promise and the promises of making in science education. *Studies in Science Education, 53*(1), 75–103.

Blikstein, P. (2013). Digital fabrication and 'making' in education: The democratization of invention. In J. Walter-Herrmann & C. Büching (Eds.), *FabLabs: Of machines, makers and inventors* (pp. 203–221). Transcript Publishers.
Blotnicky, K. A., Franz-Odendaal, T., French, F., & Joy, P. (2018). A study of the correlation between STEM career knowledge, mathematics self-efficacy, career interests, and career activities on the likelihood of pursuing a STEM career among middle school students. *International Journal of STEM Education, 5*(1), 1–15. https://doi.org/10.1186/s40594-018-0118-3
Boaler, J., & Selling, S. K. (2017). Psychological imprisonment or intellectual freedom? A longitudinal study of contrasting school mathematics approaches and their impact on adults' lives. *Journal for Research in Mathematics Education, 48*(1), 78–105. https://doi.org/10.5951/jresematheduc.48.1.0078
Boaler, J., Munson, J., & Williams, C. (2017). *Mindset mathematics: Visualizing and investigating big ideas, grade 4.* https://ebookcentral.proquest.com
Boesen, J., Helenius, O., Bergqvist, E., Bergqvist, T., Lithner, J., Palm, T., & Palmberg, B. (2014). Developing mathematical competence: From the intended to the enacted curriculum. *Journal of Mathematical Behavior, 33*(1), 72–87.
Breaux, C. (2017). Why making? *Computers and Composition, 44*, 27–35. https://doi.org/10.1016/j.compcom.2017.03.005
Calabrese Barton, A., & Tan, E. (2018). A longitudinal study of equity-oriented stem-rich making among youth from historically marginalized communities. *American Educational Research Journal, 55*(4), 761–800.
Calabrese Barton, A. C., Tan, E., & Greenberg, D. (2017). The makerspace movement: Sites of possibilities for equitable opportunities to engage underrepresented youth in STEM. *Teachers College Record, 119*(6), 1–44.
Chan, C. K. Y. (2023). *Assessment for experiential learning.* Taylor & Francis.
Chounta, I., Manske, S., & Hoppe, H. (2017). From making to learning: Introducing dev camps as an educational paradigm for re-inventing problem-based learning. *International Journal of Educational Technology in Higher Education, 14*(1), 1–15.
Chu, S. L., Deuermeyer, E., & Quek, F. (2017). Supporting scientific modeling through curriculum-based making in elementary school science classes. *International Journal of Child-Computer Interaction, 16*, 1–8.
Cobb, G. (2007). One possible frame for thinking about experiential learning. *International Statistical Review, 75*(3), 336–347.
Craddock, I. (2015). Makers on the move: A mobile makerspace at a comprehensive public high school. *Library Hi Tech, 33*(4), 497–504.
Dabney, K., Tai, R., Almarode, J., Miller-Friedmann, J., Sonnert, G., Sadler, P., & Hazari, Z. (2012). Out-of-school time science activities and their association with career interest in STEM. *International Journal of Science Education, Part B, 2*(1), 63–79. https://doi.org/10.1080/21548455.2011.629455
Davis, D., Mason, L., Poling, A., & Lundervold, D. A. (2017). A behavioral phenomenological inquiry of maker identity. *Behavior Analysis: Research and Practice, 17*(2), 174–196.
Dougherty, D. (2012). The maker movement. *Innovations: Technology, Governance, Globalization, 7*(3), 11–14.
Furukawa, L. (2018). Trajectory of learning experience from the performance of Canada's youth in mathematics. *International Journal of Innovation in Science and Mathematics Education, 26*(6), 62–75.
Godhe, A. L., Lilja, P., & Selwyn, N. (2019). Making sense of making: Critical issues in the integration of maker education into schools. *Technology, Pedagogy and Education, 28*(3), 317–328.
Graham, S., Kiuhara, S. A., & MacKay, M. (2020). The effects of writing on learning in science, social studies, and mathematics: A meta-analysis. *Review of Educational Research, 90*(2), 179–226.

Guzey, S., Harwell, M., Moreno, M., Peralta, Y., & Moore, T. J. (2017). The impact of design-based STEM integration curricula on student achievement in engineering, science, and mathematics. *Journal of Science Education and Technology, 26*(2), 207–222.

Halverson, E. R., & Sheridan, K. M. (2014). The maker movement in education. *Harvard Educational Review, 84*(4), 495–504.

Harackiewicz, J., Rozek, C., Hulleman, C., & Hyde, J. (2012). Helping parents to motivate adolescents in mathematics and science: An experimental test of a utility-value intervention. *Psychological Science, 23*, 899–906. https://doi.org/10.1177/0956797611435530

Holbert, N. (2016). Leveraging cultural values and "ways of knowing" to increase diversity in maker activities. *International Journal of Child-Computer Interaction, 9–10*, 33–39.

Honey, M., Pearson, G., & Schweingruber, H. (2014). *STEM integration in K-12 education: Status, prospects, and an agenda for research.* National Academies Press.

Hsieh, H.-F., & Shannon, S. E. (2005). Three approaches to qualitative content analysis. *Qualitative Health Research, 15*(9), 1277–1288. https://doi.org/10.1177/1049732305276687

Hsu, Y., Baldwin, S., & Ching, Y. (2017). Learning through making and maker education. *TechTrends, 61*(6), 589–594.

Hughes, J. (2009). New media, new literacies and the adolescent learner. *E-Learning and Digital Media, Special Issue: Digital Literacies, 6*(3), 259–271.

Hughes, J. M. (2017). Digital making with "at-risk" youth. *International Journal of Information and Learning Technology, 34*(2), 102–113.

Hughes, J. M., Morrison, L. J., Kajamaa, A., & Kumpulainen, K. (2019). Makerspaces promoting students' design thinking and collective knowledge creation: Examples from Canada and Finland. In A. Brooks, E. Brooks, & C. Sylla (Eds.), *Interactivity, game creation, design, learning, and innovation* (pp. 343–352). Springer International Publishing.

Ke, F., Clark, K. M., & Uysal, Ş. (2019). Architecture game-based mathematical learning by making. *International Journal of Science and Mathematics Education, 17*, 167–184.

Kolb, D. A. (2014). *Experiential learning: Experience as the source of learning and development.* Pearson FT Press.

Krishnan, G. (2015). *Designing a mobile makerspace for children's hospital patients: Enhancing patients' agency and identity in learning.* https://etd.library.vanderbilt.edu//available/etd-08182015-034719/

Litts, B., Kafai, K., Lui, Y., Walker, B., & Widman, D. (2017). Stitching codeable circuits: High school students' learning about circuitry and coding with electronic textiles. *Journal of Science Education and Technology, 26*(5), 494–507.

Lobato, J., Ellis, A., & Charles, R. (2010). *Developing essential understanding of ratios, proportions, and proportional reasoning for teaching mathematics in grades 6–8 (Essential understanding series).* National Council of Teachers of Mathematics.

Lopas, C., Zygouris-Coe, V. I., Grysko, R. A., & Gao, S. (2021). Writing to learn in science: Accommodations to support English-language learners' writing skills and science content learning in grade 5. *The Reading Teacher, 74*(5), 617–630. https://doi.org/10.1002/trtr.1979

Marshall, M. M., Carrano, A. L., & Dannels, W. (2016). Adapting experiential learning to develop problem solving skills in deaf and hard of hearing engineering students. *Journal of Deaf Studies and Deaf Education, 21*(4), 403–415. https://doi.org/10.1093/deafed/enw050

Marshall, J. A., & Harron, J. R. (2018). Making learners: A framework for evaluating making in STEM education. *Interdisciplinary Journal of Problem-Based Learning, 12*(2). Article 3. https://doi.org/10.7771/1541-5015.1749

Martin, L. (2015). The promise of the maker movement for education. *Journal of Pre-College Engineering Education Research (J-PEER), 5*(1), 30–39.

Martínez Ortiz, A. (2015). Examining students' proportional reasoning strategy levels as evidence of the impact of an integrated Lego robotics and mathematics learning experience. *Journal of Technology Education, 26*(2), 46–69.

Papavlasopoulou, S., Giannakos, M., & Jaccheri, L. (2017). Empirical studies on the maker movement, a promising approach to learning: A literature review. *Entertainment Computing, 18*, 57–78.

Ramey, K. E., Stevens, R., & Uttal, D. H. (2020). In-FUSE-ing STEAM learning with spatial reasoning: Distributed spatial sensemaking in school-based making activities. *Journal of Educational Psychology, 112*(3), 466–493.

Rouse, R., & Gillespie Rouse, A. (2022). Taking the maker movement to school: A systematic review of preK-12 school-based makerspace research. *Educational Research Review, 35*. https://doi.org/10.1016/j.edurev.2021.100413

Sadler, P., Sonnert, G., Hazari, Z., & Tai, R. (2012). Stability and volatility of STEM career interest in high school: A gender study. *Science Education, 96*(3), 411–427.

Saorín, J. L., Melian-Díaz, D., Bonnet, A., Carbonell Carrera, C., Meier, C., & De La Torre-Cantero, J. (2017). Makerspace teaching-learning environment to enhance creative competence in engineering students. *Thinking Skills and Creativity, 23*, 188–198.

Tofel-Grehl, C., Fields, D., Searle, K., Maahs-Fladung, C., Feldon, D., Gu, G., & Sun, C. (2017). Electrifying engagement in middle school science class: Improving student interest through e-textiles. *Journal of Science Education and Technology, 26*(4), 406–417.

Tong, D. T., Loc, N. P., Uyen, B. P., & Cuong, P. H. (2020). Applying experiential learning to teaching the equation of a circle: A case study. *European Journal of Educational Research, 9*(1), 239–255.

Torralba, J. (2019). A mixed-methods approach to investigating proportional reasoning understanding in maker-based integrative STEAM projects. *Proceedings of FabLearn 2019: 8th annual conference maker education* (pp. 81–88). The Association for Computing Machinery.

Torralba, J., & Rouse, R. (2019). Developing an engineering identity through immersive design challenges in academic makerspaces: A qualitative case study. In *Proceedings of the 126th American society of engineering education annual conference exposition*. https://peer.asee.org/32636

Ventä-Olkkonen, L., Hartikainen, H., Norouzi, B., Iivari, N., & Kinnula, M. (2019). A literature review of the practice of educating children about technology making. In D. Lamas, F. Loizides, L. Nacke, H. Petrie, M. Winckler, & P. Zaphiris (Eds.), *Human-Computer Interaction—INTERACT 2019* (pp. 418–481). Springer International Publishing.

Weinberg, A. E., Basile, C. G., & Albright, L. (2011). The effect of an experiential learning program on middle school students' motivation toward mathematics and science. *RMLE Online: Research in Middle Level Education, 35*(3), 1–12.

Wong, A., & Partridge, H. (2016). Making as learning: Makerspaces in universities. *Australian Academic & Research Libraries, 47*(3), 143–159.

Part IV
Mathematical Modelling and the Development of STEM Competencies

Chapter 10
The Role of Mathematical Modelling for Developing STEM Competencies

Katie Makar

Abstract STEM tasks have typically neglected mathematics beyond a servant role of computation. Mathematical modelling is a content-rich mathematical process that involves solving complex problems through the creation, testing and refinement of representations (models) that capture the mathematical structures and key characteristics within phenomena of interest. In this chapter, the role of mathematical modelling is investigated to embed mathematically-rich contributions to the solution of complex STEM-based problems. The three chapters in this section are summarised to illustrate practical examples of mathematical modelling within integrated STEM tasks. Many of these tasks also emphasised design-thinking, a strategy increasingly used in STEM education. Implications for practice are highlighted.

Keywords Mathematical modelling · STEM education · Design-thinking in STEM · Mathematics in integrated STEM tasks

Contributions of Mathematical Modelling to Integrated STEM Education

> Mathematical models and modelling thinking [is] central to developing design solutions before prototyping stages. … It follows, that discussing the notion of an authentic STEM education relies heavily on considering the pivotal functions of models and modelling (Hallström & Schönborn, 2019, p. 3).

Maass et al. (2019) set out three interdisciplinary approaches in STEM education: Twenty-first century skills, mathematical modelling and responsible citizenship. Other sections of this book (Anderson & Makar, 2024) focus on the first and third element, while this chapter outlines ways that mathematical modelling can be a productive tool for enriching both the mathematics in STEM tasks and to develop

K. Makar (✉)
The University of Queensland, Brisbane, Australia
e-mail: k.makar@uq.edu.au

J. Anderson and K. Makar (eds.), *The Contribution of Mathematics to School STEM Education*, https://doi.org/10.1007/978-981-97-2728-5_10

STEM competencies. Hestenes (2010) considers a model as a way of representing structure within a system. Models are more than representations, however, as they carry both their history and their future use as cognitive and cultural activity (Doerr et al., 2017). Mathematical modelling specifically seeks to represent relationships between real-world phenomena and the abstract world of mathematics (Niss et al., 2007). Modelling is a critical practice in STEM because each of its separate disciplines are intertwined through the practice of modelling, yet mathematical modelling in particular is typically interdisciplinary in nature (Hallström & Schönborn, 2019; Maass et al., 2019). In addition, mathematics provides unique disciplinary tools to modelling that not only advance STEM but also enrich mathematical understanding.

Mathematical modelling therefore focuses holistically on solving a problem rather than using modelling only as motivation to practice skills. Drawing on Papert's modelling of a LOGO turtle with technology, for example, Sengupta et al. (2018) contended that modelling with the LOGO turtle was not about the coding or the linear equations that modelled the turtle's behaviour. Instead, the LOGO activities sought to design learning opportunities in which students could have the "experience of *thinking like the turtle* and *thinking with the turtle* … that is anti-thetical to *learning as usual* in K-12 science (and beyond)" (p. 54, emphasis in original). The point here is that often in STEM contexts, the focus can be on the goal of a particular disciplinary law and neglect important *qualitative reasoning* that integrates contextual problem solving within and across STEM.

A key contribution of mathematical modelling is therefore the notion of *mathematising*, which seeks to capture qualitative reasoning with quantitative measures (cf. Wheeler, 2001). Freudenthal (1981) identified mathematising as one of the most important skills in problem solving, arguing "that in the learning process mathematising a situation deserves priority over solving word problems" (p. 145). Mathematising is a modelling practice across science, technology, engineering and mathematics in which one seeks to identify the measures and mathematical structures that quantify important qualitative characteristics of a situation (e.g., change in the volume of a dam when torrential rains hit). Mathematics provides a unique window therefore into developing the data, measures and evidence within phenomenon that may not be obvious in the problem (Sherwood, 2023). In STEM contexts, modelling encompasses the assumptions, measures, estimates, tools and mathematical structures that can develop learners' capacity to transfer learning to unfamiliar problems (phenomena with isomorphic structures that may not be apparent in the problem description). For example, in observing simulations of how viruses spread in a population, if students recognise the multiplicative nature of (exponential) growth over time in the phenomenon as an underlying mathematical structure, then it can provide an important starting point to creating a mathematical and scientific model of the phenomenon.

English (2016a) recognising that mathematics is often neglected in STEM tasks, argued that mathematical modelling was critically valuable to lift the profile of mathematics in STEM education. In particular, she advocated three suggestions that could promote mathematical modelling in STEM education: (1) developing mathematical content through 'idea-generating' problems in STEM; (2) engaging students in context-rich problems that showcase mathematics; and (3) explicitly teaching general skills and processes in mathematics and STEM that highlight the mutually beneficial contributions that they each make. Mathematical modelling provides an important perspective that can greatly enhance STEM education. In particular, mathematical modelling places the focus of problem solving on addressing a real-life context with the mathematics authentically woven into the context like a "tapestry" rather than using the context as decorative "wrapping" or a "border" around a conventional textbook-like problem (Stillman, 1998). Weaving the mathematics authentically through a problem with modelling provides opportunities for students to draw on their personal knowledge of the context to enrich their use and understanding of the content in a STEM problem.

STEM Competencies and Design-Thinking in Integrated STEM Tasks

STEM competencies are considered as those practices and skills needed to address STEM tasks, but more broadly as skills that "young people will need to meet the demands of their futures—personal, civic and workplace" (Maass et al., 2019, p. 870). Twenty-first century skills—critical thinking, creativity, collaboration and communication—are regularly listed among skills considered essential in STEM education (English, 2016b). Digital literacies, complex problem-solving, systems thinking, social and intercultural competence, self-regulation and project management are often listed as well (Jang, 2016). It is beyond this chapter to delve into these elements individually. Other chapters in this volume focus more specifically on twenty-first century skills (Makar, 2024) and responsible citizenship (Anderson, 2024). However, one can think of the design-thinking element of mathematical modelling as partially addressing these competencies.

Design science, when applied to mathematics education, involves investigation of complex dynamic systems, creation of products of innovation that are socially influenced, purposeful outcomes that are sharable, reusable, enhanced by interdisciplinarity and can be improved through iteration of revisions (Lesh & Sriraman, 2005).

> Design thinking comprises a set of activities that broadly include problem definition, brainstorming, and planning followed by iterative cycles of creation, testing, and evaluation. Though more typical of life beyond school, many educators embrace design thinking as an antidote to pedagogies that emphasize single correct answers and narrow definitions of academic achievement. … Design-based activities appear in science, [technology], math, and engineering as well as more interdisciplinary, project-based learning. (Chin et al., 2020, p. 339)

Design-thinking orients students to seek feedback, which encourages revision and improvement, and ultimately better learning outcomes (Cutumisu et al., 2020). Next, the three chapters that follow this one are summarised in how they focus on the application of mathematical modelling to integrated STEM tasks.

New Contributions of Mathematical Modelling to Integrated STEM Tasks

The authors in Sect. 10.3 showcase ways that mathematical modelling enhance STEM education tasks, with each providing innovative examples of integrated STEM tasks with rich mathematics content and mathematical modelling at the forefront.

In Chap. 11, English (2024) illustrates students' involvement in two integrated STEM design tasks that ask them to create models. Her chapter, "Design-based mathematical modelling within STEM contexts" emphasises how design thinking can encourage students' innovative ideas when engaging with integrated STEM tasks. Within the context of the author's study, the participants used design thinking in complex STEM tasks that allowed them to identify the problem, brainstorm solution pathways, develop prototypes, and refine outcomes through iterations of improvement. In addition, she argued that design thinking encouraged qualities such as empathy that focus on outcomes that consider the needs of others.

Through two cases involving classes of primary children (aged 10–11), the author showcases applications of student-generated design-based models within integrated STEM tasks that were mathematically rich. The first case focused on the design of a mini-golf course model that involved application of perimeter and area, two measures in mathematics that students often confuse. The lessons included learning about the history and construction of golf courses, and students' personal and social experiences with mini-golf. In the second case, students designed a model of a sustainable town that included environmental (e.g., renewable energy) and infrastructure considerations, and mathematical evidence in their justifications. Peer feedback and critique encouraged iterative improvements of the designs. These two cases provide practical insight into classroom-tested integrated STEM tasks involving design of models and mathematical evidence.

Fermi problems have recently been popularised as an approach to support students with addressing complex, open-ended modelling problems with mathematical evidence in a low-stakes environment (see for example Ärlebäck, 2009; Segura & Ferrando, 2023). In a Fermi problem, students develop mathematical models to generate a reasoned estimation of a quantity in a real situation. In Chap. 12, Ärlebäck and Albarracín (2024) demonstrate the opportunities of applying a mathematics education perspective to STEM education. The authors introduce a Fermi Problem Activity Template to help teachers support their students in structuring Fermi problems. They then illustrate the template in the context of a STEM-oriented task involving rich application of estimation and mathematical modelling. In addition to mathematics, the "Tin foil boat challenge" problem draws on Archimedes' principle in science as well as design processes in engineering. The findings from their chapter emphasise the opportunities of mathematical modelling for creating mathematically rich STEM tasks. This is important given the phenomenon of mathematics as the "silent M" in STEM education.

In Chap. 13, Kawakami and Saeki (2024) investigate and extend a data-driven modelling framework to capitalise on STEM's interdisciplinary opportunities. Emphasising the contributions of mathematics (measurement) and statistics, the authors illustrate the extended framework with two tasks for primary children involving the design of physical and mathematical models of paper helicopters and the dispersal system of tree seeds. Design is again at play in their chapter where children iteratively design, test and evaluate prototypes using mathematical and statistical evidence; they use their findings to make predictions about the comparative flights of the paper helicopters with different wing spans and alternative designs of tree seed dispersals. The chapter looks in depth at one student's work to explain how the framework illuminates her reasoning across different STEM disciplines. The findings of their study can support teachers to design high-quality interdisciplinary data-driven modelling tasks that provide authentic mathematical and statistical content.

References

Anderson, J. (2024). How mathematics in STEM can contribute to responsible citizenship education in schools. In J. Anderson & K. Makar (Eds.), *The contribution of mathematics to school STEM education: Current understandings*. Springer.

Anderson, J., & Makar, K. (Eds.) (2024). *The contribution of mathematics to school STEM education: Current understandings*. Springer.

Ärlebäck, J. B. (2009). On the use of realistic Fermi problems for introducing mathematical modelling in school. *The Mathematics Enthusiast, 6*(3), 331–364. https://doi.org/10.54870/1551-3440.1157

Ärlebäck, J. B., & Albarracín, L. (2024). Lessons learned from research on Fermi problems and mathematical modelling: Theoretical and practical implications for school STEM education. In J. Anderson & K. Makar (Eds.), *The contribution of mathematics to school STEM education: Current understandings*. Springer.

Chin, D. B., Blair, K. P., Wolf, R. C., Conlin, L. D., Cutumisu, M., Pfaffman, J., & Schwartz, D. L. (2020). Educating and measuring choice: A test of the transfer of design thinking in problem solving and learning. *Journal of the Learning Sciences, 28*(3), 337–380. https://doi.org/10.1080/10508406.2019.1570933

Cutumisu, M., Schwartz, D. L., & Lou, N. M. (2020). The relation between academic achievement and the spontaneous use of design-thinking strategies. *Computers & Education, 149*, 103806. https://doi.org/10.1016/j.compedu.2020.103806

Doerr, H. M., Delmas, R., & Makar, K. (2017). A modeling approach to the development of students' informal inferential reasoning. *Statistics Education Research Journal, 16*(2), 86–115.

English, L. D. (2016a). Advancing mathematics education research within a STEM environment. In K. Makar et al. (Eds.), *Research in mathematics education in Australasia 2012–2015* (pp. 353–371). Springer. https://doi.org/10.1007/978-981-10-1419-2_17

English, L. D. (2016b). STEM education K-12: Perspectives on integration. *International Journal of STEM Education, 3*(3), 1–8.

English, L. D. (2024). Design-based mathematical modelling within STEM contexts. In J. Anderson & K. Makar (Eds.), *The contribution of mathematics to school STEM education: Current understandings*. Springer.

Freudenthal, H. (1981). Major problems in mathematics education. *Educational Studies in Mathematics, 12*, 133–150.

Hallström, J., & Schönborn, K. J. (2019). Models and modelling for authentic STEM education: Reinforcing the argument. *International Journal of STEM Education, 6*. Article 22. https://doi.org/10.1186/s40594-019-0178-z

Hestenes, D. (2010). Modeling theory for math and science education. In R. Lesh, P. L. Galbraith, C. R. Haines, & A. Hurford (Eds.), *Modeling students' mathematical modeling competencies: ICTMA 13* (pp. 13–41). Springer.

Jang, H. (2016). Identifying 21st century STEM competencies using workplace data. *Journal of Science Education and Technology, 25*(2), 284–301. https://doi.org/10.1007/s10956-015-9593-1

Kawakami, T., & Saeki, A. (2024). Extending data-driven modelling from school mathematics to school STEM education. In J. Anderson & K. Makar (Eds.), *The contribution of mathematics to school STEM education: Current understandings*. Springer.

Lesh, R., & Sriraman, B. (2005). Mathematics education as a design science. *ZDM Mathematics Education, 37*(6), 490–505. https://doi.org/10.1007/BF02655858

Makar, K. (2024). The role of mathematics in STEM for developing 21st century skills. In J. Anderson & K. Makar (Eds.), *The contribution of mathematics to school STEM education: Current understandings*. Springer.

Maass, K., Geiger, V., Ariza, M. R., & Goos, M. (2019). The role of mathematics in interdisciplinary STEM education. *ZDM, 51*, 869–884. https://doi.org/10.1007/s11858-019-01100-5

Niss, M., Blum, W., & Galbraith, P. (2007). Introduction. In W. Blum, P. L. Galbraith, H.-W. Henn, & M. Niss (Eds.), *Modelling and applications in mathematics education* (pp. 3–32). Springer.

Segura, C., & Ferrando, I. (2023). Pre-service teachers' flexibility and performance in solving Fermi problems. *Educational Studies in Mathematics, 113*, 207–227. https://doi.org/10.1007/s10649-023-10220-5

Sengupta, P., Dickes, A., & Farris, A. (2018). Toward a phenomenology of computational thinking in STEM education. In M. S. Khine (Ed.), *Computational thinking in the STEM disciplines: Foundations and research highlights* (pp. 49–72). Springer. https://doi.org/10.1007/978-3-319-93566-9_4

Sherwood, C. (2023). *Contextualised storytelling: Broadening the notion of what is data and data-ing*. Paper presented at the Thirteenth International Forum for Research in Statistical Reasoning, Thinking, and Literacy (SRTL-13), 25 June–1 July 2023, Maleny QLD.

Stillman, G. (1998). The emperor's new clothes? Teaching and assessment of mathematical applications at the senior level. In P. Galbraith, W. Blum, G. Booker, & D. Huntley (Eds.), *Mathematical modelling: Teaching and assessment in a technology-rich world* (pp. 243–253). Horwood.

Wheeler, D. (2001). Mathmatisation as a pedagogical tool. *For the Learning of Mathematics, 21*(2), 50–53. http://www.jstor.org/stable/40248362

Chapter 11
Design-Based Mathematical Modelling Within STEM Contexts

Lyn English

Abstract With the current global upheavals and continued disruptions, the need for adaptive and innovative problem solvers has never been greater. This chapter advocates that the foundations for developing future-oriented problem solvers begin in the primary school. One approach to such development is through implementing design-based mathematical modelling within STEM contexts. Consideration is given to features of design-based mathematical modelling, which contrast with the requirements of traditional problem types. Examples of students' responses to two design challenges are presented, namely, designing a mini-golf course and a new sustainable town. Both cases enabled multiple entry and exit points, various approaches to product creation, and opportunities for the innovative application of existing knowledge, thus providing agency for advancing STEM learning for all students.

Keywords Design-based mathematical modelling · Problem solving · Design thinking

Introduction

With the prevailing emphasis on STEM integration, the power of mathematical problem solving—transferable across the four disciplines, and beyond—has been eclipsed. Over two decades we have witnessed a decline in research on mathematical problem solving, and on students' mathematical thinking and learning (Lester & Cai, 2016). This is of major concern, especially when work and social life increasingly demand resources beyond standard "textbook" problem solving (Chin et al., 2019). Such changing demands could not have been more starkly exposed than in the COVID-19 crisis, when modelling and statistical analyses led the search for strategies to "flatten the curve" with minimal social or economic detriment. Unfortunately, the modelling scenarios especially in the early stages of the pandemic yielded unreliable predictions and consequent inappropriate recommendations. Projections from the

L. English (✉)
Queensland University of Technology, Brisbane, Australia
e-mail: l.english@qut.edu.au

J. Anderson and K. Makar (eds.), *The Contribution of Mathematics to School STEM Education*, https://doi.org/10.1007/978-981-97-2728-5_11

modelling thus became a major source of controversy (Rhodes & Lancaster, 2020), indicating little societal appreciation of the critical nature of mathematical models (Barbosa, 2006) and how assumptions in the modelling process can sway decisions.

This heightened visibility of mathematics in society sparked major questions about the repercussions for mathematics and STEM education and research (Bakker & Wagner, 2020). In a similar vein, concerns regarding the presence of mathematics in STEM projects continue to be raised, with Martinovic and Milner-Bolotin (2022) commenting that "Very often STEM projects in K-12 classrooms pay less attention to mathematics and engineering, while focusing on science and technology, thus STEM becomes Stem or even S" (p. 156). Krajcik and Delen (2017) expressed similar concerns, noting that our society requires resourceful and STEM capable individuals, who possess a basic knowledge of scientific (and mathematical) and engineering ideas and practices, as well as creative problem-solving skills and effective communication. Indeed, we cannot ignore the importance of STEM disciplinary knowledge in students' problem-solving toolkit. Becoming a creative, capable, and adaptive problem solver cannot occur with a limited or depleted knowledge base. As educators, we must ensure all students have both the knowledge and the thinking skills to navigate the increasing challenges in society. Importantly, we need to nurture individuals who can develop new and adaptable knowledge, rather than apply limited procedures or strategies. This is not an easy task, especially when compounded with the need for diverse learning experiences—both cultural and social—to develop students' identity as STEM learners and reflective community members (Bybee, 2013; Couso et al., 2022).

One approach to advancing mathematics within STEM lies in design-based mathematical modelling, including situations where individual and community issues arise (Couso et al., 2022; English, 2023; Maass et al., 2019, 2022; Makar et al., 2023). As the term implies and as used in this chapter, design-based mathematical modelling incorporates a focus on design, which is posited as a sub-component of planning within a modelling cycle. Both modelling and designing have flourished in recent times across many fields of society, with each having significant implications for STEM education (Bhandari, 2023; English, 2021, 2023). As highlighted by the National Council of Supervisors of Mathematics and Teachers of Mathematics (NCSM-NCTM, 2018), engineering design can support students' development of problem-solving skills, reinforce and extend their thinking processes, and provide tools to tackle the types of problems appearing in both their personal and professional lives. Hjalmarson et al. (2020) addressed modelling as a design practice in engineering, illustrating the importance of the discipline's realistic contexts including constraints. As modelling is "the connection between representations and the structure of the real world, engineering is an opportunity for the theoretical to meet the practical in terms of modeling" (pp. 227–228). Other research applying engineering design ideas to modelling includes Carriera et al.'s (2020) study of Year 9 students' and university students' development of a hand biometric recognition system. There has been limited attention, however, on design-based mathematical modelling and how it can enhance young children's problem-solving skills where the solution extends beyond a single answer to a numerical problem. This chapter addresses design-based mathematical modelling where primary school students develop their own models in response to a challenging real-world situation.

Design-Based Mathematical Modelling in STEM Contexts

> The importance of mathematical modelling in the school curriculum is clear. It both demonstrates the widespread applicability of mathematics and enhances mathematical understanding through inquiry. It serves as a powerful corrective to those who view mathematics as a set of discrete facts and procedures to be taught and learned. (Burkhardt, 2018, p. 74)

As indicated by Burkhardt, mathematical modelling displays the richness and versatility of the discipline. Numerous definitions of models and modelling have appeared in the literature (e.g., Abassian et al., 2020; English, 2013; Lesh & Harel, 2003; Meyer & Lima, 2023; Stillman, 2019). As applied to the studies reported in this chapter, Hestenes' (2010) definition is adopted, namely, "A model refers to a representation of structure in a given system", with a system considered a "set of related objects, which may be real or imaginary, physical or mental, simple or composite" (p. 17). For STEM contexts, this chapter views models as systems comprising important mathematical and STEM-based concepts and/or relationships that represent the structural characteristics of a relevant, real-world scenario (English, & Mousoulides, 2011; Lesh & Doerr, 2003; Lesh & Harel, 2003). Unfortunately, nothing even preparatory to modelling of this nature is apparent in typical primary school curricula—it is largely absent from the current Australian Mathematics Curriculum (ACARA, 2022) until the secondary years. This is despite modelling being a powerful enabler of early readiness for advanced mathematical thinking (Burkhardt, 2018; English, 2021; Lesh et al., 2013).

In contrast to traditional classroom problem solving, design-based mathematical modelling activities engage students in (a) posing questions about stimulating, open-ended situations, (b) exploring and identifying assumptions to be considered in the design process, (c) seeing alternative perspectives and meanings, (d) taking into account given constraints in working with simple variables in the design process, and (e) adapting modelled solutions to new situations (cf., Suh & Seshaiyer, 2017). Effective modelling requires moving beyond the conventional ways of thinking applied in typical school problems (Lesh et al., 2013) to include contextual and critical analysis.

In today's world, critiquing the origin of a model, what it yields, and what is learned, is of increasing social importance (Barbosa, 2006; English, 2023; Maass et al., 2022; Sánchez Aguilar & Castaneda, 2021). As a twenty-first century skill, critical thinking is increasingly recognised as essential in mathematics and STEM education but is sadly lacking in many education systems and in society. Given the increased societal use of mathematical modelling, an appreciation of the critical nature of models and how assumptions in the modelling process can sway decisions is essential (cf. Barbosa, 2006; Skovsmose et al., 2023). As Abassian et al. (2020) pointed out, because mathematical modelling plays a significant role in making societal decisions, those who can both create and critique models become powerful members of society.

Students' skills in asking critical questions as they work with data in constructing and improving a model, reflecting on what their models convey, contemplating consequences, and justifying and communicating their conclusions require nurturing throughout school (Ärlebäck & Doerr, 2018; English, 2016; Gibbs & Young Park, 2022; Maass et al., 2019). With the escalation of statistical data from the mass media, it is imperative that the foundations of mathematical modelling begin early, including exposing young students to real-world contexts that challenge their thinking rather than reinforce stereotyped procedures.

Modelling and Design

The links between modelling and design, including design thinking, have received limited research in STEM education, especially in the primary grades. This lack of attention is problematic for future citizens (Ericson, 2022), given that modelling and product design can have both positive and negative impacts on communities, including economic growth and the natural environment. For example, current and future implications of the global rush towards renewable energy sources include problems with power reliability and hardware waste disposal (e.g., Ateyo, 2022).

Expressed simply, design may be defined as dealing with "the ways in which human beings modify their environments to better satisfy their needs and wants" (Kangas & Seitamaa-Hakkarainen, 2018, p. 597). Design thinking, an important component of complex problem solving, involves the processes of problem identification, brainstorming ideas for solution, creating prototypes, and testing and refining outcomes (Chin et al., 2019; English et al., 2020). Design is currently practised in qualitatively different ways across a variety of disciplines (van Kuijk et al., 2019) and is considered to be multifaceted. The educational implications of these developments warrant a closer look.

Envisioning new strategies and/or new directions to transform existing (or possibly potential) situations into ones that are more future-oriented and sustainable, is a key feature of today's design thinking (Gruber et al., 2015). In line with mathematical modelling, design demands thinking "outside-the-box" in generating ideas for a desired outcome—one that not only meets minimum requirements but also incorporates innovative features that enhance and distinguish the outcome from others (cf., Carlgren et al., 2016; Micheli et al., 2019). At the same time, both modelling and design involve identifying and challenging existing assumptions especially when looking towards changing one's direction. Several examples can again be drawn from society's reaction to Covid—innovative companies were forced to challenge long-standing assumptions in their manufacturing repertoires and thus change directions in their product creations to meet consumer demands. One manufacturer turned from gin creation to incorporate hand sanitisers; another company from packaging to face masks; and yet another, from car manufacturing to incorporate air purifying respirators.

Designing and modelling share several other features including their open-ended and challenging nature, their user-centredness and empathy, and their fostering of creativity and innovation (idea generation). Importantly, the comprehensive nature of design-based mathematical modelling calls for critical analysis during product development and refinement. In contrast to the demands of traditional problem types, these experiences draw upon a different array of skills, ones that more closely align with today's requirements for effective problem solvers.

In recent years, problem solvers' centredness and empathy, together with their awareness of citizen and consumer needs, have gained prominence in both the modelling and design literatures (e.g., English, 2023; Maass et al, 2023; Micheli et al., 2018; Skovsmose et al., 2023). Some design researchers consider empathy as the "core value" of centredness (Carlgren et al., 2016, p. 51), where the problem solver takes the perspective of others in a problem situation. Similarly, modelling problems require students to take the perspective of a client in a problem scenario (e.g., English & Watters, 2005), or as a problem solver faced with a challenging situation in which many courses of action are possible (e.g., an explorer determining a suitable site for building a new town, English, 2022).

In contrast to traditional expectations, design-based mathematical modelling assumes that learners can move beyond their existing competence to generate new knowledge in solving unanticipated problems. Even when students achieve average results on standardised tests, they can display advanced mathematical thinking not traditionally expected (Maas et al., 2019)—especially when common practices emphasise number skills at the expense of problem solving and reasoning with numbers (Kazemi, 2021). Experiences with design-based modelling warrant consideration in mathematics curriculum development—as factors that can both facilitate the creation of a wider range of classroom problems and can nurture students' competencies in tackling them. Despite promoting a wider range of problem-solving skills, such experiences tend to be neglected in STEM curricula. This chapter provides two examples of how this situation could be addressed, namely, designing a model of a mini-golf course and designing a model of a new sustainable town. In both cases, the models created are 2-D representations of how students would structure their golf course and sustainable town respectively.

Applying Design-Based Modelling in the Classroom

Designing and Creating a Model of a Mini-Golf Course

Measurement is a major component of mathematics education, yet there remains limited research on the learning and teaching of measurement despite the many difficulties primary school students experience (Sarama et al., 2021). Such difficulties include a lack of comprehension of the fundamental concepts of spatial measurement, such as confusing procedures and formulas for measuring length, area, and volume (Tan-Sisman & Aksu, 2016). Contributing to this confusion is a failure to appreciate conservation of area and difficulty in seeing how length units produce area units. A poor understanding of the structure of arrays and grids, coupled with a limited comprehension of the two-dimensional structure of area (Larsen, 2006, p. 41), compounds the confusion between perimeter and area. One of the main causes of these difficulties lies in the teaching of these measurements as sets of procedures, rather than in a meaningful way through contextualised problems such as the present example (Moyer, 2001; Tan-Sisman & Aksu, 2016).

The STEM-based activity, "Putt-Putt", was implemented in two fifth-grade classes ($n = 47$; aged 10–11 years) in an Australian state primary school. Students designed and constructed 2-D models of a mini-golf course, working with cut-outs of squares, rectangles, and composite shapes they had created on square centimetre grid paper. Exhibiting features of interdisciplinary real-world problems, the activity was developed and refined in collaboration with the teachers, who had requested that we incorporate the students' recent study of squares and rectangles. The activity was implemented across five days, for a total of five hours, with additional time spent completing introductory experiences beforehand. These experiences included: (a) a history of mini-golf courses, their locations, the nature of their construction, and their different designs; (b) discussions on the students' experiences with mini-golf courses; and (c) viewing of mini-golf YouTube snippets, with students describing important information that they had learned.

Applying and consolidating their geometric knowledge, together with spatial and visualisation skills, students first determined the areas and perimeters of their shapes, using whatever approaches they chose. They then explored how their measurements impacted on the positioning of the shapes on their golf course (created on large card paper). Students were not given specific directions on approaches to completing the activity. After experimenting with their various squares, rectangles, and composite shapes, students were to first individually select eight shapes—two of which were to be a square and a rectangle, with the remaining six shapes comprising compositions of squares and rectangles. A final set of eight shapes from those created by each group member was chosen for the group golf course model, together with a contextual theme. Each shape was to comprise a tee and a golf hole, together with obstacles that aligned with their chosen mini-golf course theme. At the end of the activity, students rated their golf course models, and indicated what improvements they would make. Group presentations were subsequently undertaken.

Each student completed their own activity workbook in which they sketched their initial designs, recorded their measurement calculations, and responded to specific questions about the activity. The students were not given any specific teacher directions, rather, the workbooks served to scaffold the students' learning by asking a range of questions about their creations. Consideration is given here to a selection of these questions, namely, the design processes student groups applied including how they positioned and coordinated the shapes to create their models, their group's main ideas in completing the activity, their approaches to determining the positioning and nature of obstacles on the shapes, and whether their mini-golf course met their criteria. Group ratings of their model, the changes they would make to improve it, and how they were using their disciplinary knowledge were recorded. From their recorded responses, together with transcripts of their group discussions and final group reports, insights into students' design-based modelling were gleaned. To provide an illustration of these responses, a sample of student reports is drawn upon.

The excerpts below indicate how groups planned and implemented their approach, which included taking into consideration the task constraints, making informed decisions, and applying disciplinary knowledge.

Initial designing. The report from Sam's group demonstrates their thoroughness of task interpretation, a consideration of constraints in their design process, and cognizance of the contextual theme they had chosen.

> We had to consider our theme and criteria when we were designing our courses. The first criteria was that it had to be a square, rectangle or a combination of both for the shape of each and every course. Another criteria which we had to follow was that it must be significantly possible to get a hole in one. We lightly drew lines to indicate where to shoot for this. The last criteria was to have certain components in each course. We had to have one tee (where the ball is hit from), one hole (where it is hit into) and at least one obstacle. We had to stick to our theme which was National Monuments.

Design processes including decision-making. Student groups demonstrated future-oriented planning and decision-making as they were developing their design, taking into account the increasing complexity of their shapes and their subsequent placement on their golf course. For example, Sam's group explained:

> When designing a mini golf course, you roughly plan the obstacles beforehand. We did this, so we already know obstacles for most of our shapes. This means we knew the level of difficulty for our shapes. This is how we got the idea of the order of shapes. We made our first few shapes (1,2,3) quite simple to score and our last two (7,8) hard. Our most complex shape was most likely shape no. 6 because you had to make many rebounds around obstacles to get the ball in the hole.

Display of empathy and awareness of players' requirements. Both Tasha's and Sam's group displayed empathy for designing a golf course that would take into consideration accessibility and would cater for a range of players. Sam's group reported that their "…innovative design proved to us very suitable anywhere."

> We knew we had to create a variety of shape sizes to build up an experience that would be found enjoyable by many ages. The first thing we did was make three criteria for us to follow when we make the course. After doing a draft of the shapes then placing them on. Our first try for the course was a mistake because we put them all together with no space for the people playing to walk and the other problem with that was that if you miss the shot and hit someone else on another course [shape] which could mess up their shot or hit the person who's playing and they could get injured from the hit.

Susanna's group explained that they "didn't meet the criteria of it being fun for all ages because the layout and the designs look more for younger kids."

Other groups that displayed awareness of players' requirements and empathy in catering for these referred to safety aspects, ease of player movement, player enjoyment, and overall appeal. For example, Toby's group explained, "We designed our mini golf course by trying to make it suitable for all ages, and easy to walk around the course. We placed our shapes so that they could fit in easily and people could walk in between them." Comments from other groups included, "We wanted to make it challenging, cool, and creative;" "It's simple, fun, and unique;" and [we wanted] "to entertain the players."

Application of disciplinary knowledge. Sam's group applied their knowledge of perimeter and area as they commented on these measures and noted the "surprising" lack of relationship between the two:

> It [golf course] consisted of two rectangles. One had an area of 21m2 and the other had an area of 24m2. Our largest shape was shape 2 with an area of 85 m2. Although it had the biggest area, it had the same perimeter as shape 4 with 40 m. Shape 4 had an area of 72 m2, making it our second largest shape.

In extending further students' observations on perimeter and area, one class teacher asked, "Because one shape has the largest area, does that mean that it is automatically going to have the largest perimeter?" Franco responded that he did not think so, explaining why some shapes have large perimeters: "I think the reason this [shape] would have the largest perimeter is because it has a lot of bits, like this bit, sticking out" (Fig. 11.1).

Student groups also referred to the use of scale and spatial aspects, such as Sam's group, who reported that "*The scale for this was each cm2 would be 1* m^2" and "*Another reason they* [shapes] *were not the same size was because that could take up too much space.*" Tasha's group reported that they placed their shapes in order from easiest to hardest, taking advantage of the corners of their golf course:

> The square had an area of 100 square centimeters. It was the largest shape. We placed it in the corner because the corner is square shaped. It is the most difficult hole in the whole game, so that's why we put it in last. The cross/plus sign (one of their shapes) has an area of 20 square centimeters. It's a smaller shape. We placed it above the square on the left. It is

Fig. 11.1 Franco pointing out the perimeter of a shape on their golf course

number 4, so it means it's not too hard, and not too easy… We put the shapes easy to hard so people can access them easily.

In their group reports students commented on how they were using their disciplinary knowledge in planning their mini-golf course and also indicated this application when specifically asked in the workbook questions. They typically noted measurement of area and perimeter of their shapes, as well as their use of spatial awareness (e.g., "We measured the area and perimeter so we could fit all the shapes."). Standard procedures such as, "We used formula $(L \times W) = A$, $(L + W) \times 2 = P$", "We used multiplication to find the area of the shapes", and "We used length times width to calculate the area", were mentioned, along with reference to the use of measurement units, cm^2 and m^2.

Students' general reference to how and why they used measurement included "to find how big the obstacles were" and "to find out the perfect size and difficulty of the course". Other responses included, "The maths we used was the area and perimeter. To calculate their size we also used shapes to fit them. We used problem solving for where the obstacles go and to determine hardness," and "We calculated the distance between the holes." Reference to the use of angles was mentioned quite frequently (10 students), such as, a "knowledge of angles to develop an interesting mini-golf course" and "we have used our math knowledge of angles to design the course to make a whole [hole] in one possible." Unfortunately, there was limited opportunity for follow-up discussion on how the students applied their knowledge of angles in their designs.

Members of Stella's group reported how they used their perimeter and area measurements to create prototypes before finalising their model. They noted that the largest shape had both the largest perimeter and area, and that they trialled the positioning of their shapes before finalising their placement. As Stella's group explained in their report:

Firstly, we used our theme to help match with our criteria by making the mini golf course fun and sometimes challenging. We numbered the shapes from the easiest to the most challenging. Our most complex shape was our last shape or shape number 8. Why do we consider this shape challenging you ask? Well, this shape has many obstacles such as seashells, rocks and sand. This shape also has the largest perimeter of 26 cm and the largest area of 48 cm squared. We determined the position of the shapes by carefully placing them on the cardboard before gluing them down.

Display of empathy and awareness of players' requirements: Both Tasha's and Sam's group displayed empathy for designing a golf course that would take into consideration accessibility and would cater for a range of players. Sam's group reported that their "…innovative design proved to us very suitable anywhere."

We knew we had to create a variety of shape sizes to build up an experience that would be found enjoyable by many ages. The first thing we did was make three criteria for us to follow when we make the course. After doing a draft of the shapes then placing them on. Our first try for the course was a mistake because we put them all together with no space for the people playing to walk and the other problem with that was that if you miss the shot and hit someone else on another course [shape] which could mess up their shot or hit the person who's playing and they could get injured from the hit.

Susanna's group explained that they "didn't meet the criteria of it being fun for all ages because the layout and the designs look more for younger kids."

Other groups that displayed awareness of players' requirements and empathy in catering for these referred to safety aspects, ease of player movement, player enjoyment, and overall appeal. For example, Toby's group explained, "We designed our mini golf course by trying to make it suitable for all ages, and easy to walk around the course. We placed our shapes so that they could fit in easily and people could walk in between them." Comments from other groups included, "We wanted to make it challenging, cool, and creative;" "It's simple, fun, and unique;" and [we wanted] "to entertain the players."

Designing Models of Sustainable Towns

In the second example, sixth-grade students completed an integrated STEM activity where they designed and created 2-D representations of sustainable towns (see also English, 2023). Applying design and modelling processes, together with disciplinary knowledge, student groups created towns comprising systems of interconnected environmental and infrastructural components. The open-ended nature of the activity facilitated students' creation of a range of towns.

Task Requirements and Implementation Conducted across two half-days, with 3.5 h each day, the problem commenced with students documenting their knowledge of renewable energy and its sources, together with their advantages and disadvantages. This was followed by viewing video clips of their state's renewable energy sites and the government's plans, namely, to reach a goal of 50% renewable energy within 10–15 years. The problem required students to adopt the role of town planners and develop design proposals for sustainable town models. A panel of experts would then judge their overall designs, with the best chosen for construction.

Students were required to take into account a number of constraints including population limits (no more than 600 people, with 2–5 persons per household), the number of desired housing blocks, and the selection and placement of essential services (a list of essential services such as schools, hospitals, police stations etc. was provided). Each installed solar panel was expected to generate around 1,200 units of electrical energy per day, at a cost of $1000 per panel, and each installed wind turbine to generate around 300,000 units of electrical energy per month at $20,000 per turbine. Students were to calculate the quantity of electricity required to meet the minimum 50% renewable energy demand. A budget of $2.5 million was stipulated. Each student group was provided with a large cardboard sheet displaying a grid for documenting and identifying their town components. In addition to working with a budget and applying appropriate estimation and computational skills, students had to apply their mathematical knowledge of grids, their purpose, and how to use them appropriately.

To encourage critical reflection on their creations, students were to identify any changes they made from their initial designs through to their final town creation, the main ideas applied in their development, and the mathematics and science knowledge/processes used. Student groups were also required to rate their towns, justify their rating, and identify any changes they would make to improve the overall design of their town. Finally, each group was to critically evaluate another group's model indicating positive and negative features and recommendations for improvement.

Initial Planning and Designing Analysis of the students' responses (see English, 2023 for further details of the analysis process) indicated that their initial planning included scoping the overall requirements of the problem, including the conditions and constraints on services and housing, budgetary considerations, and environmental factors. Students displayed iterative processes in designing and creating their model towns, revisiting frequently the problem requirements in the course of developing and refining their model towns. Students' use of problem scoping contrasted with some reported literature on students' failure to adequately scope a problem or a lack of time provided to do so (e.g., Strimel et al., 2019; Svihla & Reeve, 2016).

Application of STEM Disciplinary Knowledge and Processes Students' application of STEM disciplinary knowledge and processes was evident as they used their prior science and mathematics learning of sustainability and renewable energy sources, applied visualisation processes in positioning their town features, considered budgetary constraints in their designing, and made effective use of the grid to locate their town components. Students displayed considerable variation in what they regarded as important in designing their towns. For example, one group was initially preoccupied with estimating the population size, while another was more concerned with the costs of the renewable resources. Several assumptions and uncertainties also appeared in students' initial planning and designing including the desired number of residents per household, the necessity of some services, the number of people a service could accommodate, and whether store visitors should be counted as town residents. Some of the groups' assumptions were subsequently overturned as they cycled through design processes in considering the positioning of renewables, services, and residential blocks.

Ensuring an appropriate placement of the renewable energy sources appeared important to the groups. Applying their existing knowledge of sustainable practices, students debated the positioning of the renewable sources within an envisioned town. With a focus on the problem requirements and constraints, students made several design decisions, including balancing benefits and trade-offs such as the environmental requirements of the renewables competing with the location of tourist attractions. It would seem that, given all the media reports on sustainability, this community issue held significant meaning for the students in their design-based decision-making.

Students' Critiques of their Town Models Critically analysing their completed town models prompted students to reflect on their decisions and actions, and contemplate how their town designs might be improved. Critical analyses, referred to here as critiques, are underrepresented in STEM studies yet provide rich opportunities for students to enhance their design processes, their model creations, and their overall learning (McDonnell, 2016; Schut et al., 2022). In critically analysing and justifying their town designs, students were generally positive about their end-products.

Their justifications referred mainly to design aspects and how they coordinated these within a town that would meet citizens' needs while also caring for the environment. References were made to the strategic placement of services, the enhancement and aesthetics of the tourist facilities, and empathetic considerations of both the residents and the town's sustainable character. For example, one group member reported how they designed an appealing and well-functioning community while also caring for the community's environment:

> I rated my group's model there [close to excellent] because we had everything a town requires and more. We also calculated enough turbines and solar panels to get more electricity than we need [in] a year. Our town considers the needs of the place such as shops near the homes and the motels near the forest for the view and close to a walk for tourist attractions. I rated it close to excellent because the layout of the town was convenient for the residents of the town. Also, we created an efficient form of renewable electricity sources in an area of land that was going to be unused.

Students' recommendations for improving their towns largely pertained to their positioning of components such as housing and service sectors, the location of the renewable energy sources, and issues regarding capping or extending population growth and town services. Students' arguments for capping population growth included: "I would have less houses so that we could incorporate a produce farm, as well as moving the community pool so that it is an easier distance to travel." In contrast, another group recommended adding "more houses, motels etc.... because there [the town] is going to end up with more and more people eventually." Increasing the number of schools to cater for future population growth was also mentioned: "If I would change something about our town, I would make two or more schools so there could be one hundred and twenty kids that can go to school instead of only 40 kids." Other student groups recommended repositioning the wind turbines to a more suitable site: "I would not put the wind turbines near the forest as the trees block the wind." Concerns about the aesthetics of the town were expressed by a student who explained, "I would spread out the goods and services buildings a bit more for the sake of aesthetics. Because if a town is more appealing to the eye, then more tourists will come which is better for the economy."

Students' Critiques of their Peers' Models Students' critiques of their peer groups' town models revealed the application of STEM-based knowledge (mathematics and science), together with a knowledge of citizens and communities (a "curriculum connections" component of the Australian Curriculum [ACARA, 2022]). The peer critiques highlighted the overall design of the town, that is, how the various features combined to form an effective town model, including access to services, positioning of the renewable resources and associated safety issues, and the placement of the housing sectors including aesthetic features. The use of mathematical procedures and spatial reasoning also appeared in some of the peer group critiques. For example, one group identified problems with the positioning and size of community components, as follows:

> They have many small shops rather than one big block of shops. The community pool is right next to the fire station, and forest and middle of nowhere. The schools and park are huge. The path is not clear what it is. The fumes from petrol will go to the houses. 2 of the

3 bridges are in the wind farm. The only option for accommodation for tourists is all in one spot, not spread out.

They recommended the following actions to the group:

Some things they could have improved to make their town design better is: Put all of the wind turbines together out of the way so everything would be safer and nothing would be blocking the wind. And lots of houses would have more access and everything would be more accessible for all the people. And people would be closer to hospitals, and they don't have roads so they wouldn't be able to use a car. And those are the things they could make better and improve on.

Other weaknesses identified in peer critiques included problems in mathematics application, such as omission of the number of solar panels for each house and in total, not drawing to scale, and a failure to meet the renewable energy target. For example: "… they did not provide any maths and they just got the total of everything. So we cannot check if they met the criteria 50% sustainable. They also did no working out whatsoever."

Recommendations for peer group improvement included a better distribution of buildings and a more even energy distribution, for example: "I would say to put the buildings more spread out to create the illusion of a bigger space. Also, they should make the house designs so each one can have an even amount of energy." Other recommendations included how unused budgetary amounts could be spent or how the budget could be better used for the renewable energy sources, for example: "They would improve this design by using all their extra money and space to get to a higher energy amount and put some wind turbines in, or solar panels to make it more energy efficient."

Student groups had an opportunity to defend their town models. One group responded to the criticism of scale by explaining that distances were larger than they appeared in their representation, which was "on a very small scale." Claims regarding fuel leaking from the petrol stations were also negated:

First of all, I would like to say about the petrol station being near the river. That is on a very small scale. In reality, that's a very large distance from the river. Another thing about the petrol station, when it rains on the petrol station, it doesn't leak oil all the way down the road into the river.

Another student explained further:

We meant if there was like a storm, if the petrol station is damaged, is set alight. If the wind is coming in from this angle, if it's coming in from this angle…the petrol station sets alight, then the petrol will leak into the creek, the river, from what we can see. Also, we haven't said that the roads, normally roads in real life, the roads have a one main road and then they branch out.

The above explanation did not satisfy a member of the critiquing group: "But this (design) does not include that which means there's a very unnecessary maze-like confusion."

Concluding Points

With the current world-wide concerns, such as global emissions, long-term Covid effects, and subsequent economic problems, the power of mathematics and STEM has been displayed on the international stage. The STEM disciplines played a crucial role in public and personal discourse, in describing and modelling current and potential situations, and in explaining and justifying societal regulations and restrictions—not all of which drew effectively on these disciplines, with many incorrect assumptions drawn. With the global upheavals and continued disruptions we face today, the need for greater consumer awareness of the role of mathematics and STEM has never been greater. We need to foster this awareness in our students and encourage more innovative and adaptive approaches to mathematical and STEM-based problem solving. The foundations for this development begin in the primary school. It is argued that developing students' ways of thinking as they undertake modelling problems can foster important mathematical learning. This chapter has presented one approach to commencing these foundations, namely, through design-based mathematical modelling.

The two problems presented in this chapter enabled multiple entry and exit points, various approaches to solution, and opportunities for the innovative application of existing knowledge—the problems thus provided agency for advancing STEM learning for all students. Furthermore, the problems required the application of mathematics knowledge and processes, as seen in the students' different ways of applying their measurement understandings. The problem contexts were also meaningful and appealing to the students, drawing on both their disciplinary knowledge and community awareness.

The integration of modelling and design in STEM contexts has been undervalued, indeed, neglected in the primary grades. As Stillman et al. (2023) noted, the links between mathematical modelling and the STEM domains seem so obvious yet there has not been much uptake in the classroom. In particular, the various ways in which design can enhance modelling such as designing for new directions appears underrepresented. The latter was featured within the second problem as students had to contemplate design features of their town in the process of creating it (cf., Öberg & Verganti, 2014). As an increasingly important form of design thinking in the broader community (Meyer & Norman, 2020), designing for new directions is worthy of inclusion in problem tasks in the curriculum.

At the same time, we cannot neglect the ever-present challenge of increasing the level of mathematics within integrated STEM problems. As Forde et al. (2023) stressed, mathematical tasks need to be intentionally included in integrated STEM problems at a higher level of cognitive demand than currently features in the literature—in stark contrast to science, which is "noticeably represented and emphasised in integrated STEM" (English, 2016; Forde et al., 2023, p. 15; Li & Schoenfeld, 2019). Rather than viewing mathematics as static and merely a tool in STEM-based problem solving, seeing it in terms of "sense making" and development of ideas and thinking can help promote conceptual changes in mathematical practice (Li &

Schoenfeld, 2019, p. 7). In the second problem, for example, mathematical considerations were required in students' decision-making on energy consumption, their placement of town features on their grid, and the balancing of environmental concerns with populations needs.

Modelling is inherently interdisciplinary, with real-world contexts underpinning these interdisciplinary aspects. More resources that feature this modelling and incorporate a range of examples in which design plays a key role could contribute to a greater uptake in the classroom. Various media sources that contain controversial claims and questionable data regarding new developments (e.g., in product design or new community developments) can provide rich bases for design-based modelling problems. Suh et al.'s (2023) principles for creating "justice-oriented community-based mathematical modeling" (p. 210) activities can assist educators in developing problems that emanate from meaningful community issues.

Drawing further on the community, STEM professionals can provide authentic input into mathematical modelling and design, and can thus enhance classroom implementation. In one project for example (Fry & English, 2023), the expertise of an agricultural scientist assisted in the creation of a mathematical modelling task in which fourth-grade students explored ways to predict the surface area of a leaf within an agricultural science context involving fruit production. Students learned how leaf area was important in determining how much sunlight a plant might absorb through photosynthesis, with the process being part of a mathematical model used by the agricultural scientist. A key component of crop production models is the relationship between the length or width of a leaf and its area. The authentic STEM context and professional audience provided a valuable opportunity for students to share their models and gain validation that mathematical modelling was in fact a process used by agricultural scientists.

Acknowledgements The studies reported in this chapter were supported by an Australian Research Council Discovery Grant. The views expressed in this chapter are attributed to the author, not the Research Council. I wish to express my sincere appreciation of all the work undertaken by my research assistants, Maryam and Satwant Sandhu and the willing participation of the students and teachers.

References

Abassian, A., Safi, F., Bush, S., & Bostic, J. (2020). Five different perspectives on mathematical modeling in mathematics education. *Investigations in Mathematics Learning, 12*(1), 53–65. https://doi.org/10.1080/19477503.2019.1595360

Ärlebäck, J., & Doerr, H. M. (2018). Students' interpretations and reasoning about phenomena with negative rates of change throughout a model development sequence. *ZDM-Mathematics Education, 50*, 187–200. https://doi.org/10.1007/s11858-017-0881-5

Ateyo, P. (2022). *Fuelling conflict? The impact of the green energy transition on peace and security.* Research paper. International Alert. https://www.international-alert.org/publications/fuelling-conflict-the-impact-of-the-green-energy-transition-on-peace-and-security/

Bakker, A., & Wagner, D. (2020). Pandemic: Lessons for today and tomorrow? *Educational Studies in Mathematics Education, 104*, 1–4. https://doi.org/10.1007/s10649-020-09946-3

Barbosa, J. C. (2006). Mathematical modelling in classroom: A socio-critical and discursive perspective. *ZDM, 38*(3), 293–301. https://doi.org/10.1007/BF02652812

Bhandari, A. (2023). Design thinking: From bibliometric analysis to content analysis, current research trends, and future research directions. *Journal of the Knowledge Economy, 14*(3), 3097–3152. https://doi.org/10.1007/s13132-022-00920-3

Burkhardt, H. (2018). Ways to teach modelling—A 50-year study. *ZDM Mathematics, 50*, 61–75. https://doi.org/10.1007/s11858-017-0899-8

Bybee, R. W. (2013). *The case for STEM education: Challenges and opportunities*. NSTA Press.

Carlgren, L., Rauth, I., & Elmquist, M. (2016). Framing design thinking: The concept in idea and enactment. *Creativity and Innovation Management, 25*(1), 38–57. https://doi.org/10.1111/caim.12153

Carreira, S., Baioa, A. M., & de Almeida, L. M. W. (2020). Mathematical models and meaning by school and university students in a modelling task. *Avances De Investigación Matemática, 17*, 67–83.

Chin, D. B., Blair, K. P., Wolf, R. C., Conlin, L. D., Cutumisu, M., Pfaffman, J., & Schwartz, D. L. (2019). Educating and measuring choice: A test of the transfer of design thinking in problem solving and learning. *Journal of the Learning Sciences, 28*(3), 337–380. https://doi.org/10.1080/10508406.2019.1570933

Couso, D., Grimalt-Álvaro, C., & Simarro, C. (2022). Problematizing STEM integration from an epistemological and identity perspective. In D. Ortega-Sánchez (Ed.), *Controversial issues and social problems for an integrated disciplinary teaching* (pp. 183–196). Springer. https://doi.org/10.1007/978-3-031-08697-7_13

Australian Curriculum, Assessment and Reporting Authority (ACARA). (2022). *Australian Curriculum: Learning areas*. https://v9.australiancurriculum.edu.au/f-10-curriculum/f-10-curriculum-overview/learning-areas

English, L. D. (2023). Multidisciplinary modelling in a sixth-grade tsunami investigation. *International Journal of Science and Mathematics Education: Special Issue on STEM Integration in Primary Schools: Theory, Implementation, and Impact, 21*, 41–65. https://doi.org/10.1007/s10763-022-10303-4

English, L. D., & Watters, J. J. (2005). Mathematical modelling in the early school years. *Mathematics Education Research Journal, 16*(3), 59–80. https://doi.org/10.1007/BF03217401

English, L. D., & Mousoulides, N. (2011). Engineering-based modelling experiences in the elementary classroom. In M. S. Khine & I. M. Saleh (Eds.), *Models and modeling: Cognitive tools for scientific enquiry*. Models and modeling in science education series (pp. 173–194). https://doi.org/10.1007/978-94-007-0449-7_8

English, L. D., Adams, R., & King, D. (2020). Design learning in STEM education. In C. C. Johnson, M. J. Mohr-Schroeder, T. J. Moore, & L. D. English (Eds.), *Handbook of research on STEM education* (pp. 76–86). Routledge. https://doi.org/10.4324/9780429021381-9

English, L. D. (2013). Complex modelling in the primary and middle school years: An interdisciplinary approach. In G. Stillman, G. Kaiser, W. Blum, & J. Brown (Eds.), *Mathematical modelling: Connecting to practice—Teaching practice and the practice of applied mathematicians* (pp. 491–505). https://doi.org/10.1007/978-94-007-6540-5_42

English, L. D. (2016). Advancing mathematics education within a STEM environment. In K. Makar, S. Dole, J. Visnovska, M. Goos, A. Bennison, & K. Fry (Eds.), *Research in mathematics education in Australasia* 2012–2015 (pp. 353–371). https://doi.org/10.1007/978-981-10-1419-2_17

English, L. D. (2021). Mathematical and interdisciplinary modeling in optimizing young children's learning. In J. M. Suh, M. H. Wickstrom, & L. D. English (Eds.). *Exploring mathematical modeling with young learners* (pp. 3–24). Springer. https://doi.org/10.1007/978-3-030-63900-6_1

English, L. D. (2022). Interdisciplinary mathematical modeling. In C. Michelsen, A. Beckmann, V. Freiman, U. T. Jankvist, & A. Savard (Eds.), *Mathematics and its connections to the arts and sciences (MACAS): 15 years of interdisciplinary mathematics education. Mathematics education in the digital era* (series vol. 19) (pp. 163–177). Springer. https://doi.org/10.1007/978-3-031-10518-0_9

Ericson, J. D. (2022). Mapping the relationship between critical thinking and design thinking. *Journal of the Knowledge Economy, 13*, 406–429. https://doi.org/10.1007/s13132-021-00733-w

Forde, E. N., Robinson, L., Ellis, J. A., & Dare, E. A. (2023). Investigating the presence of mathematics and the levels of cognitively demanding mathematical tasks in integrated STEM units. *Disciplinary and Interdisciplinary Science Education Research, 5*(3), 1–18. https://doi.org/10.1186/s43031-022-00070-1

Fry, K., & English, L. D. (2023). How big is a leaf? Mathematical modeling through STEM Inquiry. *Mathematics Teacher: Learning and Teaching PK-12, 116*(2), 99–107.

Gibbs, A. M., & Young Park, J. (2022). Unboxing mathematics: Creating a culture of modeling as critic. *Educational Studies in Mathematics, 110*, 167–192. https://doi.org/10.1007/s10649-021-10119-z

Gruber, M., de Leon, N., George, G., & Thompson, P. (2015). From the editors: Managing by design. *Academy of Management Journal, 58*, 1–7.

Hestenes, D. (2010). Modeling theory for math and science education. In: Lesh, R., Galbraith, P., Haines, C., Hurford, A. (Eds.), *Modeling students' mathematical modeling competencies*. Springer. https://doi.org/10.1007/978-1-4419-0561-1_3

Hjalmarson, M. A., Holincheck, N., Baker, C. K., & Galanti, T. M. (2020). Learning models and modelling across the STEM disciplines. In C. C. Johnson, M. J. Mohr-Schroeder, T. J. Moore, & L. D. English (Eds.), *Handbook of research on STEM education* (pp. 223–233). Taylor & Francis.

Kangas, K., Seitamaa-Hakkarainen, P. (2018). Collaborative design work in technology education. In M. de Vries (Ed.), *Handbook of technology education* (pp. 597–609). Springer. https://doi.org/10.1007/978-3-319-44687-5_44

Kazemi, E. (2021). Mathematical modeling with young learners: A commentary. In: Suh, J.M., Wickstrom, M.H., English, L.D. (Eds.), *Exploring mathematical modeling with young learners* (pp. 337–342). Springer. https://doi.org/10.1007/978-3-030-63900-6_14

Krajcik, J. & Delen, I. (2017). How to support learners in developing usable and lasting knowledge of STEM. *International Journal of Education in Mathematics, Science and Technology, 5*(1), 21–28. https://doi.org/10.18404/ijemst.16863

Larsen, B. (2006). In *Math handbook for water system operators: Math fundamentals and problem solving*. Outskirts Press.

Lesh, R., & Doerr, H. M. (2003). Foundations of a models and modeling perspective on mathematics teaching, learning, and problem solving. In R. Lesh & H. M. Doerr (Eds.), *Beyond constructivism: Models and modeling perspectives on mathematics problem solving, learning, and teaching* (pp. 3–33). Lawrence Erlbaum.

Lesh, R. A., & Harel, G. (2003). Problem solving, modeling, and local conceptual development. *Mathematical Thinking and Learning, 5*(2/3), 157–189. https://doi.org/10.1080/10986065.2003.9679998

Lesh, R. A., English, L. D., Riggs, C., & Sevis, S. (2013). Problem solving in the primary school (K-2). *Mathematics Enthusiast, 10*(1 &2). https://doi.org/10.54870/1551-3440.1259

Lester, F. K., & Cai, J. (2016). Can mathematical problem solving be taught? Preliminary answers from 30 years of research. In P. Felmer , E. Pehkonen, & J. Kilpatrick (Eds.), *Posing and solving mathematical problems: Research in mathematics education* (pp. 117–135). https://doi.org/10.1007/978-3-319-28023-3_8

Li, Y., & Schoenfeld, A. H. (2019). Problematizing teaching and learning mathematics as 'given' in STEM education. *International Journal of STEM Education, 6*, 44. https://doi.org/10.1186/s40594-019-0197-9

Maass, K., Geiger, V., Romero Ariza, M., & Goos, M. (2019). The role of mathematics in interdisciplinary STEM education. *ZDM Mathematics Education, 51*(6), 869–884. https://doi.org/10.1007/s11858-019-01100-5
Maass, K., Artigue, M., Burkhardt, H., Doorman, M., English, L. D., Geiger, V., & Schoenfeld, A. (2022). Mathematical modelling–A key to citizenship education. In N. Buckholtz, B. Schwarz, & K. Vorhöter (Eds.), *Initiationen mathematikdidaktischer Forschung* (pp. 31–50). Springer.
Maass, K., Zehetmeier, S., Weihberger, A., & Flößer, K. (2023). Analysing mathematical modelling tasks in light of citizenship education using the COVID-19 pandemic as a case study. *ZDM—Mathematics Education, 55*, 133–145. https://doi.org/10.1007/s11858-022-01440-9
Makar, K., Fry, K., & English, L. D. (2023). Primary students' learning about citizenship through data science. *ZDM—Mathematics Education.* https://doi.org/10.1007/s11858-022-01450-7
Martinovic, D., & Milner-Bolotin, M. (2022). Problematizing STEM: What it is, what it is not, and why it matters. In C. Michelsen, A. Beckmann, V. Freiman, U. T. Jankvist, & A. Savard (Eds.), *Mathematics and its connections to the arts and sciences (MACAS). Mathematics education in the digital era* (Vol 19, pp. 135–162). Springer. https://doi.org/10.1007/978-3-031-10518-0_8
McDonnell, J. (2016). Scaffolding practices: A study of design practitioner engagement in design education. *Design Studies, 45*, 9–29. https://doi.org/10.1016/j.destud.2015.12.006
Meyer, J. F. C. A., & Lima, M. (2023). Relevant mathematical modelling efforts for understanding COVID-19 dynamics: An educational challenge. *ZDM Mathematics Education, 55*, 49–63. https://doi.org/10.1007/s11858-022-01447-2
Meyer, M. W., & Norman, D. (2020). Changing design education for the 21st century. *The Journal of Design, Economics, and Innovation, 6*(1), 13–49. https://doi.org/10.1016/j.sheji.2019.12.002
Micheli, P., Wilner, S. J. S., Bhatti, S. H., Mura, M., & Beverland, M. B. (2018). Doing design thinking: Conceptual review, synthesis, and research agenda. *Journal of Product Innovation Development, 36*(2), 124–128. https://doi.org/10.1111/jpim.12466
Micheli, P., Wilner, S., Bhatti, S., et al. (2019). Doing design thinking: Conceptual review, synthesis and research agenda. *Innovation Management, 36*, 124–148. ISSN 0737-6782.
Moyer, P. S. (2001). Using representations to explore perimeter and area. *Teaching Children Mathematics, 8*, 1.
National Council of Supervisors of Mathematics and National Council of Teachers of Mathematics (NCSM-NCTM). (2018). *Building STEM education on a sound mathematical foundation.* Author. https://www.nctm.org/uploadedFiles/Standards_and_Positions/Position_Statements/Building%20STEM%20Education%20on%20a%20Sound%20Mathematical%20Foundation%20(NCSM-NCTM%202018).pdf
Öberg, A., & Verganti, R. (2014). An unexplored path of innovation. *International Journal of Innovation in Management, 2*(2), 77–92.
Rhodes, T., Lancaster, K., & Rosengarten, M. (2020). A model society: Maths, models and expertise in viral outbreaks. *Critical Public Health, 30*(3), 253–256.
Sanchez Aguilar, M., & Castaneda, A. (2021). What mathematical competencies does a citizen need to interpret Mexico's official information about the COVID-19 pandemic? *Educational Studies in Mathematics, 108*, 227–248. https://doi.org/10.1007/s10649-021-10082-9
Sarama, J., Clements, D. H., Barrett, J. E., Cullen, C. J., Hydyma, A., & Vanegas, Y. (2021). Length measurement in the early years: Teaching and learning with learning trajectories. *Mathematical Thinking and Learning, 24*(4), 267–290. https://doi.org/10.1080/10986065.2020.1858245
Schut, A., van Mechelen, M., Klapwijk, R. M., Gielen, M., & de Vries, M. J. (2022). Towards constructive design feedback dialogues: Guiding peer and client feedback to stimulate children's creative thinking. *International Journal of Technology and Design Education, 32*, 99–127. https://doi.org/10.1007/s10798-020-09612-y
Skovsmose, O., Moura, A. Q., & Carrijo, M. (2023). Inclusive citizenship through mathematics education: a conceptual investigation. *ZDM—Mathematics Education.* https://doi.org/10.1007/s11858-023-01470-x
Stillman, G. A., Ikeda, T., Schukajlow, S., de Loiola, Araújo, J., & Bergman Ärlebäck, J. (2023). Survey of interdisciplinary aspects of the teaching and learning of mathematical modelling in

mathematics education. In Greefrath, G., Stillman, G. A., & Carreira, S. (Eds.), *International perspectives on the teaching and learning of mathematical modelling*: *Research from ICME 14* (pp. 3–20). Springer.

Stillman, G. A. (2019). State of the art on modelling in mathematics education—Lines of inquiry. In G. A. Stillman and J .P. Brown (Eds.), *Lines of inquiry in mathematical modelling research in education, ICME-13 Monographs* (pp. 1–20). Springer Open. https://doi.org/10.1007/978-3-030-14931-4_1

Strimel, G. J., Kim, E., Grubbs, M. E., & Huffman, T. J. (2019). A meta-synthesis of primary and secondary student design cognition research. *International Journal of Technology and Design Education, 30*(4), 243–274. https://doi.org/10.1007/s10798-019-09505-9

Suh, J. M., Tate, H., Seshaiyer, Rossbach, M., Green, S., Matson, K., Aguire, J., Seshaiyer, P., & Steen, S. (2023). Dilemmas and design principles in planning for justice-oriented community-based mathematical modeling lessons. *Mathematics Teacher Educator, 11*(3), 210–229.

Suh, J. M. & Seshaiyer, P. (2017). In *Modeling mathematical ideas across the learning progressions.* Rowman & Littlefield Publishing.

Svihla, V., & Reeve, R., (2016). Facilitating problem framing in project-based learning. *The Interdisciplinary Journal of Problem-based Learning, 10*(2). https://doi.org/10.7771/1541-5015.1603

Tan-Sisman, G., & Aksu, M. (2016). A study on sixth grade students' misconceptions and errors in spatial measurement: length, area, and volume. *International Journal of Science and Mathematics Education, 14*(7), 1293–1319. https://doi.org/10.1007/s10763-015-9642-5

The Australia Curriculum, Assessment and Reporting Authority (ACARA 2022). Australian Curriculum 9.0.

van Kuijk, J., Daalhuizen, J., & Christiaans, H. (2019). Drivers of usability in product design practice: Induction of a framework through a case study of three product development projects. *Design Studies, 60*, 139–179.

Chapter 12
Lessons Learned from Research on Fermi Problems and Mathematical Modelling: Theoretical and Practical Implications for School STEM Education

Jonas Bergman Ärlebäck and Lluís Albarracín

Abstract This chapter aims to demonstrate how adopting a mathematics education research perspective on Fermi problems can contribute to and enhance the practice and understanding of STEM education. To achieve this, we first discuss mathematical modelling from a mathematics education perspective and then briefly elaborate on modelling as discussed in STEM education. Our attention is then directed towards the inherent coupling of Fermi problems to mathematical activities and contents of the mathematical curriculum, in particular, estimates, estimation and mathematical modelling in mathematics education research. We present a framework for connecting recent advances in the research on Fermi problems to teaching and learning in STEM education and discuss the example of *The tin foil boat challenge.*

Introduction

The STEM subjects—Science, Technology, Engineering and Mathematics—which have been brought together as an integrated educational research discipline (Kelley & Knowles, 2016), have gained both interest and prominence in educational research over the last 20 years (Li et al., 2020). Although there are different interpretations in the literature of what the acronym *STEM* and *STEM education* entail (Honey et al., 2014), Li et al. (2020) argue, based on their review of literature published between 2000 and 2018, that there is increasing clarity of terminology and a converging understanding of the identity of STEM education in the research field. However, as noted by many researchers (e.g. Fitzallen, 2015; Maass et al., 2019), the role and

J. B. Ärlebäck (✉)
Department of Mathematics, Linköping University, 58183 Linköping, Sweden
e-mail: jonas.bergman.arleback@liu.se

L. Albarracín
Department de Didàctica de la Matemàtica Clències, Universitat Autònoma de Barcelona, Barcelona, Spain
e-mail: lluis.albarracin@uab.cat

J. Anderson and K. Makar (eds.), *The Contribution of Mathematics to School STEM Education*, https://doi.org/10.1007/978-981-97-2728-5_12

function of mathematics in STEM education is often invisible. In many cases, instead of STEM activities promoting the exploration of the uses of mathematics in rich contexts, the use of mathematics is reduced to serve as a tool in the service of other disciplines. Therefore, from the perspective of teaching and learning mathematics in the context of STEM education, a major concern is whether mathematics is taught and learned only incidentally, when mathematical knowledge and competence are needed in the application of mathematics in situ, or whether it is considered more fundamental in integrated STEM education and requires explicit and systematic attention (Anderson et al., 2020).

Hallström and Schönborn (2019) as well as Hjalmarson et al. (2020) have discussed how *models* and *modelling* are key concepts in STEM education. The use of these concepts shows some similarities (as well as some differences) across STEM disciplines. These authors argue that models and modelling could serve as a bridge between STEM disciplines in integrated STEM education practices. This argument is consistent with the findings of a literature review by Ärlebäck and Albarracín (2019) on the use of Fermi problems (FPs) in the STEM disciplines. Two examples of FPs are *How many liters of water does one person use in a year?* (Okamoto et al., 2023) and *How many cars can fit in the faculty parking?* (Segura & Ferrando, 2021). FPs have been characterised as contextualised miniature modelling problems (Robinson, 2008; see below for further elaboration), and the findings of the literature review by Ärlebäck and Albarracín (2019) illustrate how the use of FPs in educational settings can act as integrators of STEM disciplines as well as facilitators of learning content within one or more STEM disciplines. In addition, recent research on the use of FPs in mathematics education highlights how this problem type can connect to and make mathematical curricula content explicit to teachers and learners (Albarracín & Ärlebäck, 2019; Ärlebäck & Albarracín, 2022). Taken together, and supported by the established field of research on modelling in mathematics education (Blum, 2015; Leung et al., 2021) that focuses on modelling both as a mathematical curriculum goal in its own right and as a vehicle for learning other mathematical curriculum content (Julie & Mudaly, 2007), this suggests that a viable strategy for making the role, function and importance of mathematics in STEM education more visible and explicit may be to use FPs and adopt a modelling perspective on STEM education.

This chapter aims to demonstrate how adopting a mathematics education research perspective on FPs can contribute to and enhance the practice and understanding of STEM education. To achieve this, we first discuss mathematical modelling from a mathematics education perspective and then briefly elaborate on modelling as discussed in STEM education. Our attention is then directed towards the inherent coupling of FPs to mathematical activities and contents of the mathematical curriculum, in particular, estimates, estimation and mathematical modelling in mathematics education research. We connect recent advances in the research on FPs to teaching and learning in STEM education and discuss the example of the FP *The tin foil boat challenge*.

Mathematical Modelling

Just as the words models and modelling are used with different meanings and interpretations in everyday life, many different approaches and perspectives to mathematical modelling can be found in the mathematics education research literature (Kaiser & Sriraman, 2006). In this chapter, we want to use a broad and inclusive understanding of mathematical modelling in order to connect the research in mathematics education on modelling to the general discussion of modelling in the other STEM subjects. Hence, we draw on Niss et al. (2007) and consider

- That a *mathematical model* can be seen as consisting of a triplet including an extra-mathematical situation or domain, a mathematical domain, and a mapping connecting the extra-mathematical and mathematical domains; and
- That *mathematical modelling* can be seen as "the whole process of structuring, generating real-world facts and data, mathematising, working mathematically, and interpreting/validating (perhaps several times around the loop)" (9–10).

From this broad perspective, mathematical modelling therefore essentially involves identifying, delimiting and selecting both the non-mathematical and mathematical domains relevant to a problem situation, and specifying the mapping between the two.

As noted by Niss (2015), a modelling process that results in a mathematical model can serve both descriptive and prescriptive purposes. In descriptive modelling, mathematics is used to describe and represent an existing situation or phenomenon in order to gain insight, provide explanations or predictions of real-world aspects. Prescriptive modelling, on the other hand, concerns informing decision-making or action in order to influence or organize (future) real-world scenarios. During both descriptive and prescriptive modelling, finding a suitable mathematical domain that can map and connect to the extra-mathematical domain of the situation or phenomena is crucial to achieve useful results. The proficiency and comprehension of the modeller in mathematics are key factors for success. In other words, in mathematical modelling the extra-mathematical situation or problem encourages and motivates the exploration, teaching and learning of useful and applicable mathematical content and topics. For the mathematical models, both descriptive and prescriptive, to effectively and efficiently connect to the extra-mathematical domain, it is crucial that the modeller has a deep understanding of both the problem situation at hand and the extra-mathematical domain involved. In other words, mathematical modelling also encourages, supports and motivates the teaching and learning of the knowledge and skills in the extra-mathematical domain. Regarding models and modelling in mathematics education research, Julie and Mudaly (2007) have discussed a similar duality in the teaching of mathematical models and modelling. According to Julie and Mudaly (2007), mathematical models and modelling as a mathematics curriculum content can be regarded both as a goal in its own right and as a vehicle for learning other mathematical content. In this chapter, we use and expand on Julie and Mudaly's (2007) concept of dual use to encompass not only mathematics but also other STEM subjects as the extra-mathematical domain within the modelling process.

The Models and Modeling Perspective

Ärlebäck and Albarracín (2023a, 2023b) argue that *the models and modeling perspective* (Lesh & Doerr, 2003) is one perspective in mathematics education research that effectively encompasses the teaching and learning of mathematical modelling as both content and vehicle. According to this perspective, models are considered to be human constructs and conceptual systems that are inherently social in nature. They aid in making sense of problems, situations and phenomena. Models are systems composed of elements, relationships, operations and rules that can predict, explain or describe the behaviour of another system. From this perspective, learning fundamentally involves creating and developing useful and generalized models. Therefore, in teaching and learning, the role of modelling activities is to enable students to develop models by involving them in the creation, development, understanding, modification, and use of their own models, in order to make sense of various situations and contexts (Lesh & Doerr, 2003).

Compared to the above more general description of mathematical models and modelling based on Niss et al. (2007) and Niss (2015), the models and modeling perspective (Lesh & Doerr, 2003) provides a more practice-oriented framework for working with models and modelling in classrooms. This is evident in mathematics education research (see for example Lesh and Doerr (2003)), but it has also been used and explored outside of mathematics education in science and engineering education (e.g. Diefes-Dux et al., 2006; Hjalmarson et al., 2006, 2008). In this chapter, we build on and contribute to this line of research and argument by focusing on the use of FPs as miniature modelling problems (Robinson, 2008), as the characteristics of FPs bear a strong resemblance to the characteristics the three types of modelling activities used in the models and modeling perspective (Ärlebäck, 2009). Considering that we want to broadly situate and elaborate on FPs as modelling activities in STEM education, we next briefly discuss how modelling is used and understood in the STEM education context.

Mathematical Modelling in a STEM Education Context

According to Hallström and Schönborn (2019), there is a growing consensus in STEM education research that it is important for STEM teaching and learning to be authentic in order to be successfully implemented. Authentic in this context is understood to mean that activities and problems should "reflect the natures of the parent disciplines as much as possible" (Gilbert et al., 2000, p. 3). With regard to the notions of models and modelling in STEM education, Hallström and

Schönborn (2019) compared and reviewed how models and modelling are defined, what epistemic roles and functions models and modelling have, and how models and modelling are used and come to the fore in educational practices in the STEM disciplines. Their findings show that models and modelling: (1) bridge the gap between the STEM disciplines and facilitate a more integrated STEM education; (2) promote STEM literacy and facilitate the transfer of skills and knowledge between the STEM disciplines; (3) serve as a route to a more authentic STEM education; and emphasise that (4) models and modelling need to be taught rooted in the STEM disciplines. Hjalmarson et al. (2020) argue along the same lines, and additionally discuss the notions of models and modelling as typically used in computational thinking, and the common theme of general problem-solving competencies.

In Science, models and modelling are fundamentally "always conceptual in nature" (Hjalmarson et al., 2020, p. 226) and are often used to develop descriptions and explanations of natural phenomena (Hallström & Schönborn, 2019). In terms of Niss' (2015) terminology, this emphasis on models and modelling in Science is primarily in line with the mathematical aspects of descriptive models and modelling. In Engineering, on the other hand, models and modelling are used in design processes to generate prototypes or test initial designs (Hjalmarson et al., 2020) or to analyse existing solutions to engineering problems (Hallström & Schönborn, 2019) in order to improve the solutions. Thus, the mathematical aspects of models and modelling in engineering can be said to be primarily prescriptive in nature (cf. Niss, 2015). Although a simplified generalisation, the main difference between the use and function of models and modelling in Science and Engineering is summarised by Hjalmarson et al. (2020): "Where science models the real world, engineering uses models to design for the world" (p. 228).

As a STEM discipline, Technology displays a considerable amount of variety in what it encompasses and how it is comprehended (Power, 2018). Drawing on Honey et al. (2014), Ellis et al. (2020) recognized the following four different perspectives on Technology in STEM: (1) as a product of engineering; (2) as educational or instructional technology; (3) as a tool used by practitioners of Science, Engineering and Mathematics; and (4) as coding or computational thinking. The wide variation in how Technology is used in and related to mathematics (cf. Hansson, 2020) and the other STEM disciplines (Ellis et al., 2020) makes it difficult to characterise the general role and function of the mathematical aspects of Technology in terms of Niss' (2015) notions of descriptive and prescriptive modelling. However, in Hjalmarson et al.'s (2020) discussion of models and modelling across STEM disciplines, Technology is concretised in terms of computational thinking. Based on Shute et al.'s (2017) literature review and the resulting six facets of computational thinking, Hjalmarson et al. (2020) suggest that "[s]tudents' processes of designing, creating, evaluating, and using models as CT [computational thinking] practices can provide a common language for connecting problem-solving across the STEM disciplines" (p. 229).

In this chapter, we take a mathematical perspective on STEM education, understanding STEM education as a general integrated curriculum approach in which mathematics is linked to one or more of the other STEM subjects. We in particular focus on mathematical modelling as a process that connects mathematical knowledge with the real world, and as such integrate mathematical activities and well defined mathematical learning objectives with one or more other STEM disciplines. Given the broad characterizations of modelling and the mathematical aspects of Science, Engineering and Technology discussed above, we next discuss research on FPs to argue how the role of estimation and FPs can be used as a way of transferring research advances on FPs as mathematical modelling activities in mathematics education to STEM education activities. We provide a framework that facilitating making explicit the connections needed to enable this transfer of knowledge across disciplines.

Estimates, Estimation and Fermi Problems

An *estimate* is an approximate numerical value that quantifies a given quantity and is often used in everyday life as part of on-the-spot decision-making. For example: *Do I need to buy more flour for tomorrow's baking according to this recipe, or will the amount of flour left in the packet be enough?* or, *Given that the news has reported a traffic jam nearby, do I need to leave 10 min earlier than usual to get to work on time?* The general process and reasoning behind an individual making an estimate is called estimation. Estimation is used not only in everyday life, but also in various forms in many scientific disciplines when faced with the need for numerical values of quantities for which no specific methods or reliable data are available, when an accurate calculation is not needed, or is too costly to perform, such as estimating the global carbon budget (Friedlingstein et al., 2019).

Traditionally, estimation in mathematics education has been discussed in terms of computational estimation, numerosity and measurement estimation (Hogan & Brezinski, 2003). However, mathematics curricula in various countries, including the UK, Greece, and Turkey (Andrews et al., 2022), include a more general type of estimation activity that brings the different types of estimation mentioned above to the forefront. These estimation activities involve situations in which the estimators estimate quantities. The estimators, who are the individuals doing the estimation, do this without necessarily having prior developed estimation referents or extensive knowledge of the context for these estimations. It is worth noting that these activities have not been widely discussed in the literature. The type of activity requiring this type of estimation is known as *Fermi problems*. Two examples of Fermi problems (FPs) used in research are:

How many semitrailer loads of bottled water [16.9 fluid ounces] were needed by New Orleans' refugees in the week following Hurricane Katrina? (Taggert et al., 2007).

How many rolls of toilet paper are used in the schools every year? (Ärlebäck & Albarracín, 2023a, 2023b).

Sriraman and Knott (2009) describe an FP as an estimation problem used for the didactic purpose of identifying the initial conditions of the problem and making conjectures based on the various quantities or variables that arise when solving the problem. Characterised as open-ended questions, FPs typically provide little or no specific information to guide the problem-solving process (Efthimiou & Llewellyn, 2006). This emphasises the need to analyse and think carefully about the problem situation at hand. In an FP, the development of a solution method can be equated with the creation of a model that describes the situation under investigation (Robinson, 2008). An important aspect of FPs from a mathematical modelling perspective is to perform a detailed analysis of the situation presented in the problem statement. The aim is to break down the initial problem into simpler sub-problems, so that a solution can be obtained for the original question through reasoned estimates. Instead of relying solely on reasoned estimates to arrive at a solution, these reasoned estimates can be supported and arrived at by engaging in class activities or problem-solving processes. If these activities and processes are linked to reality, they in addition facilitate validation and critical reflection about the estimates (Albarracín et al., 2022; Sriraman & Knott, 2009). Robinson (2008) characterised FPs as miniature modelling problems partly due to the similarities between the processes involved in solving FPs and more general modelling tasks. Additionally reasons for this characterisation are that FPs as classroom activities typically are more modest in scope, time-requirement, and prior content knowledge specific requirements.

The Fermi Estimation Method

Fermi problems (FPs) are named after Enrico Fermi, a Nobel Prize winning physicist. Fermi used this type of problem in his own research and also included them in his teaching. In one well-documented case, Fermi witnessed the detonation of the Trinity test in the Alamogordo desert in New Mexico and, at the moment of the explosion, scattered small pieces of paper before and during the passage of the shock wave. By measuring the distance travelled by the pieces of paper, he estimated the power of the explosion to be about 10,000 tons of TNT, which turned out to be of the same order of magnitude as the estimate obtained from calculations based on the more robust data provided by the sensors placed in the blast zone (Allison et al., 1955). Fermi utilised this type of problem in his physics classes to highlight the capabilities of inductive reasoning and to equip his students with the knowledge required to undertake laboratory experiments (Robinson, 2008). Even though the use of this type of problem, which was pioneered by Fermi, has spread to classrooms of all levels and even other technological fields, research on the educational applications of FPs has been relatively limited until recent decades when it began to receive more attention. Interestingly, the majority of research on the use of FPs has been conducted in the field of mathematics education (Ärlebäck & Albarracín, 2019).

Several authors have pointed out links between FPs and mathematical modelling (Ärlebäck, 2009; Czocher, 2016; Albarracín et al., 2022; Peter-Koop, 2009; Sriraman & Lesh, 2006). The literature shows that FPs promote mathematical modelling in primary education (Albarracín & Gorgorió, 2014; Henze & Fritzlar, 2010; Peter-Koop, 2009), secondary education (Ärlebäck, 2009; Keune & Henning, 2003) and tertiary education (Czocher, 2016; Ferrando & Segura, 2020). This research frequently centres on decomposing and breaking down the initial problem into simpler sub-problems, which are solved independently using estimates. This approach is commonly referred to as the *Fermi method* (or *Fermi estimation method*; see Fig. 12.2a and below for further elaboration). Research suggests that students can be successfully taught to use this method appropriately to solve a wide range of FPs through direct instruction (Barahmeh et al., 2017; Raviv et al., 2016). While the majority of research on FPs has been conducted in mathematics education (Ärlebäck & Albarracín, 2019), the Fermi estimation method has also been applied in other disciplines such as economics to identify potential new markets (Anderson & Sherman, 2010), astrobiology to estimate the amount of energy the Earth receives from the Sun (Cordry, 2010), and engineering projects to assess the feasibility of proposed solutions (Shakerin, 2006). Taken collectively, this indicates that utilizing FPs could be a viable strategy for engaging students in integrated STEM educational settings.

One line of recent research, represented for instance by Albarracín and Gorgorió (2018), has focused on the use of sequences of FPs rather than as stand-alone tasks. These authors focused on initial ideas related to density and used FPs with the contexts of the tasks becoming progressively more general and removed from the students' everyday experience:

FP1: How many people can we fit in the schoolyard?
FP2.1: How many people could we fit into the Palau St. Jordi sports pavilion for a concert?
FP2.2: How many people could we fit in the town hall square during a public protest?
FP2.3: How many people could we fit in Plaça Catalunya square during a public protest?
FP3: How many trees are there in Central Park?

The use of sequences of FPs illustrates the feasibility of using FPs to focus on specific mathematical content and concepts. In addition, it also illustrates that FPs can naturally have various functions in a sequence of tasks, making them adaptable for flexible implementation in integrated STEM tasks that typically involve sequences of tasks belonging to different STEM disciplines.

To illustrate the Fermi estimation method and the way in which a problem is decomposed into related sub-problems, we will use the problem *How many coins fit into a 1 m3 cubic safe*? discussed and analysed in Albarracín and Gorgorió (2014). One of the students in their study developed the following proposal to solve this FP (Fig. 12.1).

By using the cylindrical shape of the coins, the student proposed solving the problem by assuming that the coins could be arranged in a three-dimensional grid distribution. Consequently, the student aimed to solve the following two sub-problems: *How many coins can fit into one of the dimensions of the base of the safe?* and *How many coins can be stacked on top of each other to cover the height of the safe?* Fig. 12.2b illustrates how these two sub-problems are intended to answer the original problem.

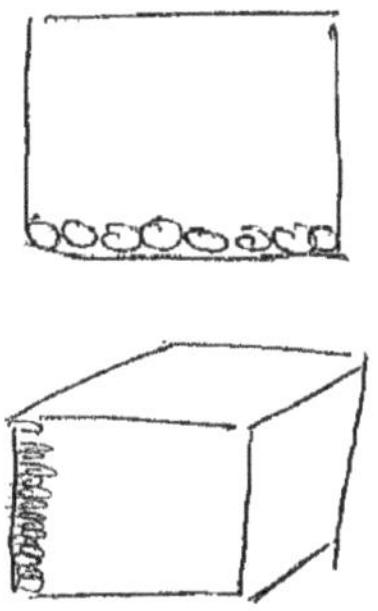

I would calculate the diameter of the coin, so I would know how many coins fit in the base by dividing a meter by this diameter (call the result of this calculation X).

I would then find the thickness of the coin and divide a meter by the thickness (call the result of this calculation Y).

Then what I would calculate next would be: $X^2 \cdot Y$

Then you would have the number of coins that fit in a cubic meter.

Fig. 12.1 A student's proposed solution to the FP *How many coins fit in a 1 m³ cubical safe?*

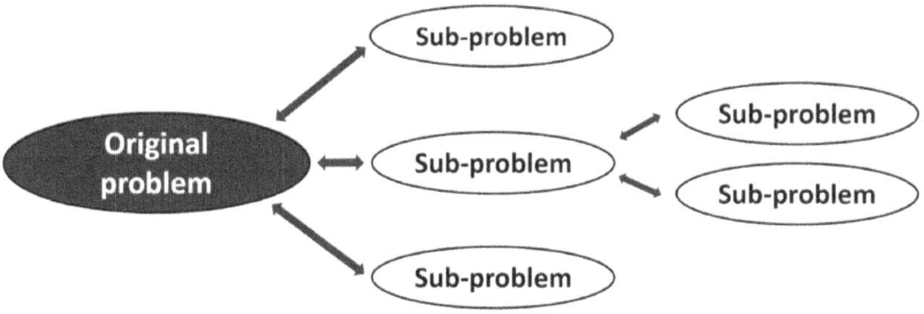

(a)

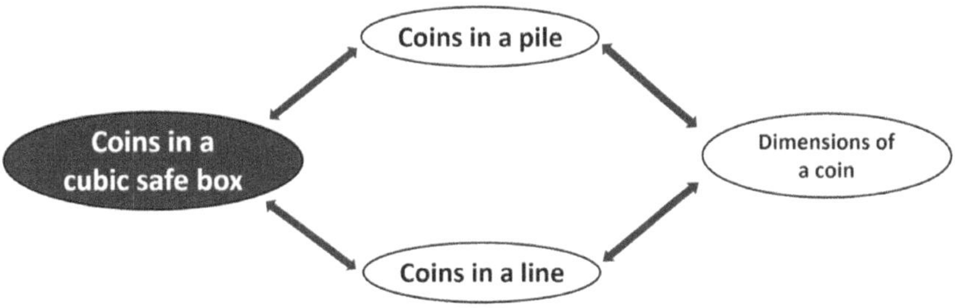

(b)

Fig. 12.2 **a** A generic illustration of the Fermi estimate method; **b** The Fermi estimate method illustrated on the FP *How many coins fit in a 1 m³ cubical safe?*

Fermi Problem Activity Templates

To support teachers in getting their students to work on FPs in the mathematics classroom Albarracín and Ärlebäck (2019) have introduced *Fermi Problem Activity Templates* (FPATs) to explain and facilitate the use of FPs and the Fermi method. Similar to the Fermi estimation method illustrated in Fig. 12.2, the FPATS are graphical representations that show the structure of the problem broken down into sub-problems. Furthermore, a FPAT also shows the mathematical activities that are used to, or suggested to be used to, determine the necessary quantities required to obtain an answer to the problem. An FPAT focuses on the structure of the solution to the problem and the model developed to solve the problem, and can be used as both a didactical tool or as a tool for analysis. As a didactic tool to it can represent a problem's solution, derived from an a priori analysis, in terms of the sub-problems that need to be solved. As an analytical tool FPATs depict the structure of students' activities when solving a problem. It's worth noting that each FP can be approached differently, depending on the students' extra-mathematical knowledge at each educational level (Ärlebäck & Albarracín, 2023a, 2023b). Thus, an FPAT only characterizes a single approach to the problem's solution, and should not be thought of as a static recipe or algorithm to blindly follow to solve the problem. One significant characteristic of an FPAT is that it broadens the conventional emphasis of FPs on estimation, with other mathematical activities available for obtaining the necessary numerical values of quantities to address a problem. As a result, work on FPs in the classroom can be linked and honed towards specific mathematics curriculum content and topics with greater ease.

In order to explicitly visualise the structure of a solution to a FP and the different types of mathematical activities that can be used to solve it, the FPATs employ different graphical representations for four mathematical activities associated with different contents and parts of the mathematics curriculum. The four mathematical activities, which were identified in an exhaustive review of the literature on the use of FPs in the STEM disciplines (Albarracín & Ärlebäck, 2019), are the following:

(a) *guesstimation* (providing a fast answer as in traditional FP solving);
(b) *measurement and experimentation* (of relevant values, in the lab or the field and using tools);
(c) *looking up data* (in official databases or online newspapers); and
(d) *polling or statistical data collection* (using surveys and questionaries, etc.).

To provide an instant overview of the solution, each activity in an FPAT is represented by a specific symbol or shape, according to the following scheme: guessing (ellipse), experimenting (trapezoid), looking up data (rectangle) and polling or collecting statistical data (hexagon). In Fig. 12.3, an FPAT depicts the student's solution to the earlier problem of determining the number of coins that can fit into a cubic safe measuring 1 m^3. In Fig. 12.3, the FPAT illustrates: (1) the two sub-problems identified by the student (denoted by the two inner square brackets); (2) what activities are proposed to obtain the required numerical values (represented by the trapezoids—measurement in both sub-problems); and (3) how the sub-problems and their

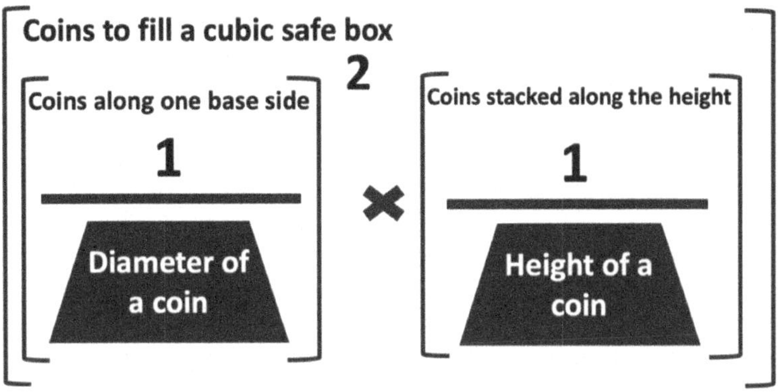

Fig. 12.3 A FPAT for the coins in safe box problem

respective results are related to the original problem in order to provide an answer to the posed question expressed as a formula (here capturing the three-dimensional lattice distribution arrangement of the coins in the safe model).

FPATs are a versatile tool. They have been used for research purposes to identify and characterise diverse approaches to the same problem, thereby facilitating comparison of students' work (Albarracín & Ärlebäck, 2019). In addition, FPATs can be useful in classroom activities. On the one hand, research indicate that FPATs can support students in their individual work in classrooms or during group discussions by serving as a tool for clarifying complex ideas to their peers and teachers (Ärlebäck & Albarracín, 2023a, 2023b). On the other hand, by conducting an a priori analysis of an FP and FPAT has been shown to assist pre-service teachers in designing and preparing classroom work (Ärlebäck & Albarracín, accepted). FPATs can aid teachers at various educational levels in creating classroom tasks related to predetermined learning goals or specific curriculum content, comparing methods, and engaging with the progressively intricate facets of problem-solving and mathematical modelling, which are crucial challenges in the instruction and acquisition of mathematic.

We now pull together the issues as aspects discussed about by situating and elaborating on FPs and FPATs from a modelling perspective in a STEM education context.

Fermi Problems: Integrators and Facilitators in Learning STEM Disciplines

The literature review on the use of FPs in STEM by Ärlebäck and Albarracín (2019) focused both on the arguments invoked for using FPs in educational settings, and on the categorisation of research findings involving FPs. In terms of the arguments used for using FPs, four main reasons were identified: (i) *estimation as an important skill and tool*; (ii) *promoting (critical) understanding and importance of quantities within the disciplines*; (iii) *facilitating and practising general problem-solving skills*; and (iv) *promoting, learning and engaging in modelling*. Here, (iii) in particular is consistent with Hjalmarson et al.'s (2020) argument that the use and development of general problem-solving skills is a unifying and common theme across the STEM disciplines. In educational settings, both (i) and (ii) are rooted in the nature of the parent disciplines and in this sense resonate with Hallström and Schönborn's (2019) emphasis on authenticity. Reason (iv) on the other hand supports the idea that FPs are emerging as a novel tool and a pathway towards more integrated STEM disciplines, as suggested by both Hallström and Schönborn (2019) and Hjalmarson et al. (2020). Indeed, the latter point is further emphasised in the conclusions of the review by Ärlebäck and Albarracín (2019), namely that FPs—in their capacity as miniature modelling problems—and FPATs have both the practical and theoretical prerequisites to act as integrators between STEM disciplines. A second result of the literature review was the function of FPs as facilitators of learning in the STEM disciplines. In this context a FPAT has the capacity to provide a map over the sub-problems involved and explicitly show the interplay and relationship of the different relevant STEM disciplines coming to the fore as well as are integrated in the problem at hand. In particular, the FPAT facilitates (1) emphasising the role and function of Mathematics in solving the problem; and (2) explicitly linking the mathematics used to the mathematics curriculum content. Therefore, a well-designed FP can serve as a starting point for teaching and learning in a STEM educational setting focused on different disciplines, as demonstrated by Hansson (2020) in relation to technology, Robinson (2008) in relation to science, and Shakerin (2006) in relation to engineering.

Overall, the preceding discussion supports the view that FPs and FPATs are capable of promoting interdisciplinary integration within STEM education, which confers benefits both to teaching and learning both in the individual STEM disciplines and in integrated STEM education. In particular, a FPAT has the potential to explicitly highlight the important roles and functions of Mathematics in the solution of a FP in relation to mathematical curricula content. Next, we explore and elaborate on this capacity of FPs and FPATs by discussing the activity called *The Tin Foil Boat Challenge.*

The Tin Foil Boat Challenge

Building on the discussion in the previous sections, we now demonstrate the potential of FPs and FPATs in facilitating the integration of STEM disciplines as well as highlight the mathematics coming to the fore using a modified version of *The tin foil boat challenge* (Lee, 2014). In this challenge, students design and construct boats from aluminium foil to investigate how their size and shape affect the weight they can carry. The question posed to the students is the following *How many coins does it take to sink a tin foil boat made from a sheet of tin foil measuring* 30×40 *cm?*

This challenge aims to engage students with STEM concepts such as Archimedes' principle (which states that any object immersed in a fluid exerts an upward or buoyant force equal to the weight of the fluid displaced by the object), buoyancy (the ability of an object to float in water or another fluid), material density (the mass of an object per unit volume) and weight distribution (the arrangement or positioning of weight within an object). These concepts are all under the umbrella of Physics (i.e. Science); therefore, this challenge involves some science content. Furthermore, this challenge necessitates the design and construction of a boat that optimises the usage of a tin foil sheet of a given dimension. The boat should be capable of carrying as many coins as possible while considering the arrangement of coins in the boat. This is an engineering problem that entails several variables and calls for a substantial amount of trial and error, which can benefit from estimation (i.e. Mathematics) as a form of anticipation and skill development in the laboratory (Robinson, 2008). Engineers routinely use estimation in their professional practice to provide answers to ill-defined problems or when a detailed solution would be too costly in terms of resources such as money or time (Shakerin, 2006). From a Mathematics point of view, and depending on the grade level in which the tin foil boat challenge is used, different areas and topics related to geometry, optimisations and differentiation can be used and explored.

Since intuitively a tin foil boat full of coins is likely to sink, one concrete way of thinking about this problem is to estimate an upper bound for the number of coins needed to perform the experiment. That is, to ask the FP: *How many coins do we need to fill a boat made of 30 × 40 cm tin foil?*

The solution to this introductory FP can be approached in a number of ways, each represented by a different FPAT. One of the decisions to be made is the shape of the boat, given the constraints of the tin foil provided. From an engineering point of view, the boat must be able to float, transport objects and navigate, which will influence the design of the boat as it must be able to cut through the water (Fig. 12.4, left). From a mathematical point of view, however, it makes sense to first try a design of the boat as a rectangular box, as this simplifies calculations and estimations (Fig. 12.4, right). Thinking about how to find the maximum volume of a rectangular box-shaped boat, for example, offers different opportunities for teachers to link and explore different mathematics curriculum contents and activities, depending on the grade level in which the challenge is implemented. For example, trial and error, building your own designs and then taking measurements, or finding an algebraic function of volume and then either using the method of completing the squares or differentiation. Technology can come into play here in the tools used to take measurements or, for example, as a spreadsheet used to explore the algebraic function of volume to approximate the dimensions that give the largest boat volume. In this example presented here however, we will opt for the following proposal to solve the introductory FP:

- First, the volume of the boat is obtained by approximating it to a rectangular box and measuring its height, width and length.

Fig. 12.4 Two different tin foil boat designs

- Secondly, to obtain the total weight of the coins tin boat can carry, the available volume is multiplied by the effective density of the space occupied by a coin, which in turn is calculated by multiplying
  - o the density of a coin which can be found by consulting official sources of the bank issuing the coins; and
  - o the "occupancy factor" derived from observing how much of the area of a square with the same side length as the diameter of the coin is occupied by a single coin.
- Thirdly and finally to get the number of coins that the tin foil boat can hold, dividing the total weight obtained by the weight of a coin. The weight of a coin is obtained by weighing a set of coins and dividing by the number of coins, since we are unlikely to have a scale accurate and precise enough to weigh a single coin.

This process is shown in Figure 12.5 in the form of an FPAT and can be interpreted as a chain of experiments and investigations leading to an answer to the introductory FP: *How many coins do we need to fill a boat made of 30x40 cm tin foil*?

The solution of this introductory FP involves elements from different STEM disciplines. The key elements that affect the total weight of the coins in the boat are the mathematical and structural considerations that come to the fore in the way the coins (cylinders) occupy the design volume of the boat, given the tin foil provided. The FPATs explicitly illustrate how the (mathematical and/or scientific) concept of density emerges naturally, which, as it can be complex for students to grasp, can then be discussed and developed. Furthermore, the FPAT in Fig. 12.5 explicitly provides an accessible and qualitative quasi-algebraic structure of how the quantity of coins necessary to fill a tin foil boat is dependent on the various quantities involved. This representation can be used as a starting point for introducing more formal variables and a continued mathematisation to derive a mathematical formulae. Alternatively, if the aim is to maximize the number of coins one can fit in a thin foil boat, it can be used to engage in a discussion with the students about mathematical strategies

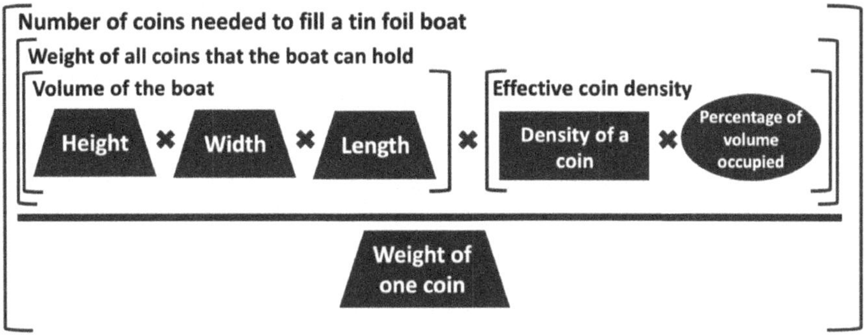

Fig. 12.5 A FPAT for the coins and tin foil boats problem

and principles of designing the boat (its height, width and length) to maximize its volume.

The actual estimate obtained from the introductory Fermi can be used to prepare the materials needed to tackle and experiment with the original problem: *How many coins does it take to sink a tin-foil boat made from a tin sheet of foil measuring 30 × 40 cm?* The introductory FP introduce and prepare the students for this more complex and open-ended question, and to begin to address and understand the scientific concepts of buoyancy and Archimedes' principle. In a classroom setting, the tin foil boat challenge has the potential to generate many different boat designs and results that can be collected for collaborative investigation and analysis. Here, the use of mathematical and statistical tools and representations, created and explored using appropriate technology, facilitates critical investigation and exploration of the scientific concepts and principles involved, as well as drawing conclusions about the 'best boat design' from an engineering perspective.

Brief Summarising Remarks

This chapter discusses how research on FPs can contribute to and enhance the practice and understanding of STEM education. We positioned FPs in relation to modelling in both mathematics and STEM education and contended that FPs can function as integrators of STEM disciplines and facilitators of learning content across one or more STEM disciplines. FPs have proven to be effective as introductory activities to mathematical modelling as well as suitable components of sequences of tasks aiming at developing specific mathematical ideas and content. However, for teachers to productively support their students in this work they require training and support. In this context, we illustrated how FPATs can function as both didactic tools and tools for more systematic analysis. In particular, we argued that FPATs can be used to make the important role and function of Mathematics in STEM education more visible and explicit. In the example of the introductory FP for the tin foil boat challenge, the FPAT in Fig. 12.5 highlights the mathematical structure of the solution to the problem as well as the importance of understanding how the dimension of the boat affects the available volume.

In our opinion, FPs are potential candidates aligned with the call to use models and modelling to attain more integrated STEM education, as suggested by Hallström and Schönborn (2019) and Hjalmarson et al. (2020). Nonetheless, further empirical studies are required to validate and develop the ideas and framework presented in this chapter. As shown in the literature presented, FPs can play different roles in STEM activities due to their various ties to the STEM disciplines and their accessible nature. A particularly interesting line of future research would be to use the FPAT framework to design FPs that focus on integrating the STEM disciplines. This this is analogous to what is outlined in the tin foil boat challenge, where engineering design meets physical and mathematical concepts, and technology is employed in data collection and analysis. These components intertwine to fulfil the requirement of producing an estimate, which is a key decision-making activity when starting to engage in a STEM project. In activities such as these, the use of FPs and FPATs can assist learners at all levels of education in decision making and understanding complex concepts.

Acknowledgements This work has been done with the support of grant PID2021-126707NB-I00 funded by MCIN/AEI/https://doi.org/10.13039/501100011033 and by "ERDF—A way of making Europe". Lluís Albarracín is Serra Húnter Fellow in Universitat Autònoma de Barcelona and member of the research group GIPEAM—2021SGR00497, funded by AGAUR (Generalitat de Catalunya).

References

Albarracín, L., & Ärlebäck, J. B. (2019). Characterizing mathematical activities promoted by Fermi problems. *For the Learning of Mathematics, 39*(3), 10–13.

Albarracín, L., & Gorgorió, N. (2014). Devising a plan to solve Fermi problems involving large numbers. *Educational Studies in Mathematics, 86*(1), 79–96.

Albarracín, L., & Gorgorió, N. (2018). Students estimating large quantities: From simple strategies to the population density model. *Eurasia Journal of Mathematics, Science and Technology Education, 14*(10), em1579. https://doi.org/10.29333/ejmste/92285

Albarracín, L., Segura, C., Ferrando, I., & Gorgorió, N. (2022). Supporting mathematical modelling by upscaling real context in a sequence of tasks. *Teaching Mathematics and Its Applications: An International Journal of the IMA, 41*(3), 183–197. https://doi.org/10.1093/teamat/hrab027

Allison, S. K., Segrè, E., & Anderson, H. L. (1955). Enrico Fermi 1901–1954. *Physics Today, 8*, 9–13.

Anderson, P. M., & Sherman, C. A. (2010). Applying the Fermi estimation technique to business problems. *Journal of Applied Business and Economics, 10*(5), 33–42.

Anderson, J., English, L., Fitzallen, N., & Symons, D. (2020). The contribution of mathematics education researchers to the current STEM education agenda. In J. Way, J. Anderson, C. Attard, J. Bobis, K. Cartwright, & H. McMaster (Eds.), *Research in mathematics education in Australasia, 2016–2019* (pp. 27–57). Springer.

Andrews, P., Xenofontos, C., & Sayers, J. (2022). Estimation in the primary mathematics curricula of the United Kingdom: Ambivalent expectations of an essential competence. *International Journal of Mathematical Education in Science and Technology, 53*(8), 2199–2225.

Ärlebäck, J. B. (2009). On the use of realistic Fermi problems for introducing mathematical modelling in school. *The Mathematics Enthusiast, 6*(3), 331–364.
Ärlebäck, J. B., & Albarracín, L. (2022). FPAT—A framework for facilitating the teaching and learning of using Fermi problems originating in mathematics education research. *The Physics Teacher, 60*(7), 569–571. https://doi.org/10.1119/5.0040017
Ärlebäck, J. B., & Albarracín, L. (2023a). Fermi problems as a hub for task design in mathematics and STEM education. *Teaching Mathematics and Its Applications: An International Journal of the IMA, 60*(7), 1–13. https://doi.org/10.1093/teamat/hrad002
Ärlebäck, J. B., & Albarracín, L. (accepted). Characterising pre-service primary teachers' posed Fermi problems. In *Proceedings of CERME13, Budapest*, 10–14 July 2023.
Barahmeh, H. M., Hamad, A. M. B., & Barahmeh, N. M. (2017). The effect of Fermi questions in the development of science processes skills in physics among Jordanian ninth graders. *Journal of Education and Practice, 8*(3), 186–194.
Blum, W. (2015). Quality teaching of mathematical modelling: What do we know, what can we do?. In Cho, S. (eds), *The Proceedings of the 12th International Congress on Mathematical Education*. Springer.
Cordry, S. M. (2010). Thermodynamics and human population. *The Physics Teacher, 48*(6), 403–407.
Czocher, J. A. (2016). Introducing modeling transition diagrams as a tool to connect mathematical modeling to mathematical thinking. *Mathematical Thinking and Learning, 18*(2), 77–106.
Diefes-Dux, H. A., Hjalmarson, M. A., Zawojewski, J. S., & Bowman, K. J. (2006). Quantifying aluminum crystal size part 1: The model-eliciting activity. *Journal of STEM Education: Innovations and Research, 7*(1&2), 51–63.
Efthimiou, C. J., & Llewellyn, R. A. (2006). Avatars of Hollywood in physical science. *The Physics Teacher, 44*, 28–33.
Ellis, J., Wieselmann, J., Sivaraj, R., Roehrig, G., Dare, E., & Ring-Whalen, E. (2020). Toward a productive definition of technology in science and STEM education. *Contemporary Issues in Technology and Teacher Education, 20*(3), 472–496.
Ferrando, I., & Segura, C. (2020). Fomento de la flexibilidad matemática a través de una secuencia de tareas de modelización. *Avances De Investigación En Educación Matemática, 17*, 84–97.
Fitzallen, N. (2015). STEM education: What does mathematics have to offer?. In M. Marshman, V. Geiger & A. Bennison (Eds.), *Proceedings of the 38th Annual Conference of the Mathematics Education Research Group of Australasia* (pp. 237–244). MERGA.
Friedlingstein, P., et al. (2019). Global carbon budget 2019. *Earth System Science Data, 11*, 1783–1838.
Gilbert, J. K., Boulter, C. J., & Elmer, R. (2000). Positioning models in science education and in design and technology education. In J. K. Gilbert &, C. J. Boulter (Eds.), *Developing models in science education*. Kluwer Academic Publishers.
Hallström, J., & Schönborn, K. J. (2019). Models and modelling for authentic STEM education: Reinforcing the argument. *International Journal of STEM Education, 6*, 22.
Hansson, S. O. (2020). Technology and mathematics. *Philosophy and Technology, 33*(1), 117–139.
Henze, J. & Fritzlar, T. (2010). Primary school children's model building processes by the example of Fermi questions. Problem solving in mathematics education. In A. Ambrus & E. Vásárhelyi (Eds.), *Proceedings of the 11th ProMath conference* (pp. 60–75). Eötvös Loránd University.
Hjalmarson, M., Diefes-Dux, H. A., Bowman, K., & Zawojewski, J. S. (2006). Quantifying aluminum crystal size part 2: The model development sequence. *Journal of STEM Education: Innovations and Research, 7*(1/2), 64–73.
Hjalmarson, M. A., Diefes-Dux, H. A., & Moore, T. J. (2008). Designing modeling activities for engineering. In J. S. Zawojewski, H. A. Diefes-Dux, & K. J. Bowman (Eds.), *Models and modeling in engineering education: Designing experiences for all students* (pp. 37–54). Sense Publishers.

Hjalmarson, M. A., Holincheck, N., Baker, C. K., & Galanti, T. M. (2020). Learning models and modelling across the STEM disciplines. In C. C. Johnson, M. J. Mohr-Schroeder, T. J. Moore, & L. D. English (Eds.), *Handbook of research on STEM education* (pp. 223–233). Taylor & Francis.

Hogan, T. P., & Brezinski, K. L. (2003). Quantitative estimation: One, two, or three abilities? *Mathematical Thinking and Learning, 5*(4), 259–280.

Honey, M., Pearson, G. & Schweingruber, H. (Eds.). (2014). *STEM integration in K-12 education: Status, prospects, and an agenda for research.* National Academies Press.

Julie, C., & Mudaly, V. (2007). Mathematical modelling of social issues in school mathematics in South Africa. In W. Blum, P. Galbraith, H.-W. Henn & M. Niss (Eds.), *Modelling and applications in mathematics education. The 14th ICMI study* (pp. 503–510). Springer.

Kaiser, G., & Sriraman, B. (2006). A global survey of international perspectives on modelling in mathematics education. *ZDM, 38*(3), 302–310.

Kelley, T. R., & Knowles, J. G. (2016). A conceptual framework for integrated STEM education. *International Journal of STEM Education, 3*(1).

Keune, M., & Henning, H. (2003). Modelling and spreadsheet calculation. In Q. X. Ye, W. Blu, S.-K. Houston, & Q. Y. Yiab (Eds.), *Mathematical modelling in education and culture* (pp. 101–110). Horwood.

Lee, M. Y. (2014). Tinkering with buoyancy. *Teaching Children Mathematics, 20*(9), 574–578.

Lesh, R., Hoover, M., Hole, B., Kelly, A., & Post, T. (2000). Principles for developing thought-revealing activities for students and teachers. In A. E. Kelly & R. A. Lesh (Eds.), *Handbook of research design in mathematics and science education* (pp. 591–645). Lawrence Erlbaum.

Lesh, R. A. & Doerr, H. M. (Eds.). (2003). *Beyond constructivism: Models and modeling perspectives on mathematics problem solving, learning, and teaching.* Lawrence Erlbaum Associates.

Leung, F., Stillman, G. A., Kaiser, G., & Wong, K. L. (Eds.). (2021). *Mathematical modelling education in East and West.* Springer.

Li, Y., Wang, K., Xiao, Y., & Froyd, J. E. (2020). Research and trends in STEM education: A systematic review of journal publications. *International Journal of STEM Education, 7*(11).

Maass, K., Geiger, V., Romero Ariza, M., & Goos, M. (2019). The role of mathematics in interdisciplinary STEM education. *ZDM Mathematics Education, 51*(6), 869–884.

Niss, M., Blum, W., & Galbraith, P. (2007). Introduction. In W. Blum, P. L. Galbraith, H.-W. Henn, & M. Niss (Eds.), *Modelling and applications in mathematics education: The 14th ICMI study* (pp. 3–32). Springer.

Niss, M. (2015). Prescriptive modelling. Challenges and opportunities. In G. A. Stillman, W. Blum & M. S. Biembengut (Eds.), *Mathematical modelling in education research and practice* (pp. 67–80). Springer.

Okamoto, H., Hartmann, M., & Kawasaki, T. (2023). Analysis of the relationship between creativity in Fermi problems measured by applying information theory, creativity in psychology, and mathematical creativity. *Education Sciences, 13*(3), 315. https://doi.org/10.3390/educsci13030315

Peter-Koop, A. (2009). Teaching and understanding mathematical modelling through Fermi-Problems. In B. Clarke, B. Grevholm & R. Millman (Eds.), *Tasks in primary mathematics teacher education* (pp. 131–146). Springer.

Power, R. (Ed.). (2018). *Technology and the curriculum: Summer 2018.* [eBook]. Power Learning Solutions. Available from https://techandcurriculum.pressbooks.com/

Raviv, D., Harris, A., & Dezotti, T. (2016). Estimation as an essential skill in entrepreneurial thinking. In *Proceedings 123rd ASEE annual conference and exposition.* American Society for Engineering Education.

Robinson, A. W. (2008). Don't just stand there—Teach Fermi problems! *Physics Education, 43*(1), 83–87.

Segura, C., & Ferrando, I. (2021). Classification and analysis of pre-service teachers' errors in solving Fermi problems. *Education Sciences, 11*(8), 451. https://doi.org/10.3390/educsci11080451

Shakerin, S. (2006). The art of estimation. *International Journal of Engineering Education, 22*(2), 273–278.

Shute, V. J., Sun, C., & Asbell-Clarke, J. (2017). Demystifying computational thinking. *Educational Research Review, 22*, 142–158.

Sriraman, B., & Knott, L. (2009). The mathematics of estimation: Possibilities for interdisciplinary pedagogy and social consciousness. *Interchange, 40*(2), 205–223.

Sriraman, B., & Lesh, R. (2006). Modeling conceptions revisited. *ZDM, 38*(3), 247–254.

Taggart, G. L., Adams, P. E., Eltze, E., Heinrichs, J., Hohman, J., & Hickman, K. (2007). Fermi questions. *Mathematics Teaching in the Middle School, 13*(3), 164–167.

Chapter 13
Extending Data-Driven Modelling from School Mathematics to School STEM Education

Takashi Kawakami and Akihiko Saeki

Abstract This study focuses on 'modelling' and 'data' to ensure the effective implementation of STEM education, specifically emphasising the 'M' (Mathematics) at its core. While the modelling process is used to develop real-world interdisciplinary problem-solving and concepts in STEM subjects, data are inherently interdisciplinary. In this chapter, we illustrate the potential of extending *data-driven modelling* (DDM), which involves the construction of mathematical and statistical models that describe and explain variability in data to make better predictions about real-world events, from school mathematics to school STEM education. First, we review a framework for describing DDM in school mathematics. Second, given the interdisciplinary nature of data and modelling, we extend the DDM framework to school STEM education and explore a framework of *interdisciplinary* DDM that can connect knowledge and practice across STEM subjects. Third, we explore the paper helicopter and seed dispersal tasks within the interdisciplinary DDM framework. Finally, we present an example of interdisciplinary DDM practice for Year 4 (9–10-year-olds) students using the seed dispersal task. The extended interdisciplinary DDM framework provides research implications for school STEM education and practical implications for developing STEM tasks and units that highlight the role of the 'M'.

Keywords Real-world problem solving · Mathematical models and modelling · Statistical models and modelling · Variability and prediction · Data-driven modelling

---

T. Kawakami (✉)
Utsunomiya University, Tochigi, Japan
e-mail: t-kawakami@cc.utsunomiya-u.ac.jp

A. Saeki
Kanazawa Institute of Technology, Ishikawa, Japan

J. Anderson and K. Makar (eds.), *The Contribution of Mathematics to School STEM Education*, https://doi.org/10.1007/978-981-97-2728-5_13

Introduction

The involvement of the 'M' (Mathematics)—as a blend of mathematics and statistics in K-12 mathematics (Zbiek, 2016)[1]—in STEM education is often limited to computation and symbolisation, which limits the role of mathematics (e.g. Anderson et al., 2020; English, 2017; Maass et al., 2019). To address this issue, this study focuses on the emergent interaction between 'modelling' and 'data' (Kawakami, 2017; Kawakami & Mineno, 2021; Kawakami & Saeki, 2022). Modelling can be described as an approach to promote the role of mathematics in integrated STEM education (English, 2017; Maass et al., 2019; Zbiek, 2016), while data and statistics are recognised as the foundation of integrated STEM education education (Watson et al., 2020). Both modelling and data or statistical investigation have become perspectives for linking STEM subjects in national mathematics curricula (e.g. Australian Curriculum, Assessment and Reporting Authority, 2023).

STEM teaching practices based on modelling and data have recently been introduced at the primary and secondary school levels. Studies have assessed their potential to improve students' mathematics learning and STEM competencies (e.g. English, 2017, 2023; Mulligan et al., 2023). However, the expected emergent interaction between models, modelling, and data while designing STEM tasks and practices with mathematics at their core remains poorly understood. Additionally, how this interaction can be optimally reflected in STEM tasks and practices remains ambiguous.

Research on mathematics and statistics education has explored the role and use of modelling, which puts data at the heart of the activities in educational contexts (e.g. Ärlebäck & Kawakami, 2023). These studies include two main perspectives: *empirical modelling* (EM; Berry & Houston, 1995; Stillman & Brown, 2023), which is a type of mathematical modelling with data, and *data modelling* (DM; Lehrer & English, 2018; Pfannkuch et al., 2018), which is a type of statistical modelling. EM and DM emphasise real-world context, establish the purpose of data-based predictions, and focus on tensions between data/context and models. They differ in whether these tensions relate to data and mathematical models or data and statistical models (Kawakami, 2023).

In contrast, drawing on English and Watson's (2018) idea of modelling with data, we have explored modelling in school mathematics that generates, validates, and modifies both mathematical and statistical models based on data/context to make better predictions (Kawakami, 2023; Kawakami & Saeki, 2022). In the current study, such modelling is referred to as *data-driven modelling* (DDM). DDM, comprising data/context, a mathematical model, and a statistical model, can conceptualise the dynamic aspects of transitions between mathematical (i.e. deterministic reasoning) and statistical reasoning (i.e. non-deterministic/stochastic reasoning), using data as the starting point or mediator. The essence of stochastic reasoning involves taking time to examine data and considering the imperfections in human thinking (Silver, 2012). In addition, variability can be intrinsically important for understanding the

[1] In this chapter, 'mathematics' is used in this sense unless a distinction is required.

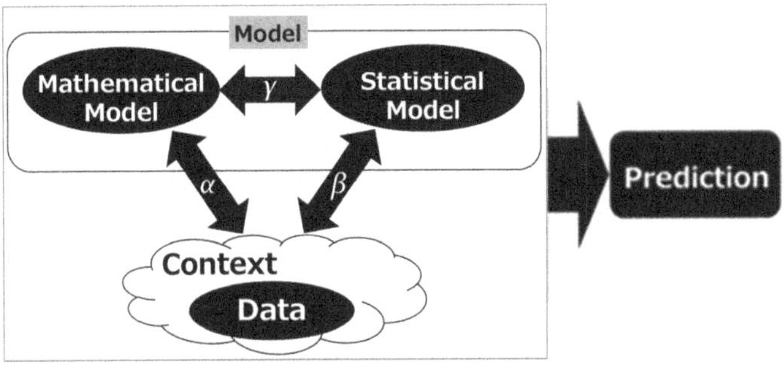

Fig. 13.1 DDM framework. Adapted from Kawakami and Saeki (2022)[2]

random mechanisms of complex phenomena in scientific inquiry (Wilensky & Jacobson, 2014). It is important to use mathematical and statistical reasoning when dealing with data. In addition to combining mathematical and statistical reasoning, and given the interdisciplinary nature of modelling and data (e.g. Develaki, 2020; Hjalmarson et al., 2020; Watson et al., 2020), DDM can connect knowledge and practice across STEM subjects.

In this chapter, we illustrate the potential of extending DDM from school mathematics to school STEM education. Accordingly, we also aim to highlight the importance of mathematics in school STEM education.

Conceptual Framework

Data-Driven Modelling (DDM) in School Mathematics

We have developed a framework that describes DDM in the context of school mathematics (Fig. 13.1; adapted from Kawakami & Saeki, 2022). It comprises transitions between data/context and mathematical models (α), data/context and statistical models (β), and mathematical and statistical models (γ). This framework, which integrates mathematical and statistical modelling characteristics, can generate perspectives that capture learners' DDM more precisely and design appropriate DDM activities and teacher interventions (Kawakami, 2023; Kawakami & Saeki, 2022).

It is important to understand and interpret data in the the real-world context (Cobb & Moore, 1997). The data structure comprises signal and noise (Konold & Pollatsek, 2002). Signal refers to regular patterns or central tendencies in the data, whereas noise refers to irregular random patterns or random errors due to several sources of variability (e.g. natural, measurement, sampling, and accident; Wild & Pfannkuch, 1999).

[2] Transitions α, β, and γ refer to the generation, validation, and revision of models, respectively.

A *model* is the representation of the structure of a given system (Hestenes, 2010), and it reflects the modeller's series of interpretations of the object (Lesh & Doerr, 2003; Piaget, 1968). Based on a constructivist view of models and the perception that they are built on the modeller's deterministic or stochastic interpretation (e.g. Dvir & Ben-Zvi, 2023; Ikeda & Stephens, 2015), this study did not consider the same representation as 'absolutely', but as relatively, i.e. as a mathematical or statistical model which depends on the learner's intention and interpretation of the model. By *mathematical model*, we mean a representation of the signal inherent in the data, reflecting the modeller's deterministic interpretation of the data and context. A typical example is the linear model $y = ax + b$ (a and b are parameters), where the value of the variable y can be determined if the value of the variable x is determined. By *statistical model*, we mean a representation of the noise inherent in the data, reflecting the modeller's non-deterministic/stochastic interpretation of the data and context. A typical example is the linear model $y = ax + b + \varepsilon$ (a and b are parameters), where the value of the variable y cannot be determined even if the value of the variable x is determined and distributed by a random error ε.

Transition α refers to generating and validating a mathematical model based on the data/context and, if necessary, revising it. Mathematical modelling with data (i.e. EM) focuses on transition α. *Transition* β refers to generating and validating a statistical model based on the data/context and, if necessary, revising it. Statistical modelling (i.e. DM) focuses on transition β. *Transition* γ refers to generating and validating a mathematical model based on a statistical model and vice versa. For example, generating a statistical model based on a mathematical model involves the non-deterministic/stochastic representation and interpretation of the mathematical model or generating a new statistical model through representation and interpretation. Note that no sequentially is assumed for transitions α, β, and γ. These transitions may coincide.

Extended Interdisciplinary DDM for School STEM Education

We aim to extend the DDM framework (Fig. 13.1) from school mathematics to school STEM education with mathematics at its core. While the nuances and emphases of models and modelling differ among STEM subjects, the model as an idealised and perspectival representation of the real world and the modelling process (generation, validation, and revision) as an attempt to explain and make predictions about phenomena are both universal to STEM disciplines and subjects (Develaki, 2020; Hjalmarson et al., 2020). Data are also interdisciplinary in nature and are not limited to mathematics but are utilised in the other STEM disciplines in a general way (Watson et al., 2020). Indeed, the variability associated with data is inherent in nature and many artificial systems, and is essential to scientific inquiry (Lehrer et al., 2020; Wilensky & Jacobson, 2014).

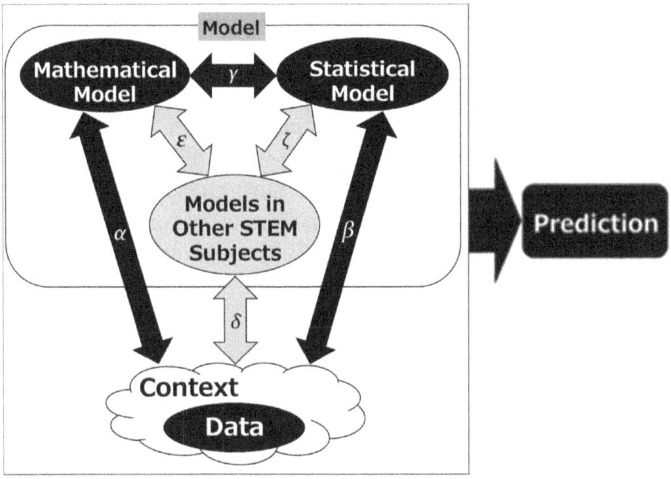

Fig. 13.2 Interdisciplinary DDM framework (an extension of the DDM framework)[3]

Given the interdisciplinary nature of modelling and data, DDM activities have an inherent potential to be extended to interdisciplinary activities connecting knowledge and practices among STEM subjects. In this study, DDM, which focuses on interdisciplinary activities, is called *interdisciplinary* DDM. Interdisciplinary DDM generates, validates, and revises mathematical and statistical models and models in other STEM subjects based on data/context to make better predictions.

A framework that describes interdisciplinary DDM is shown in Fig. 13.2. This framework is an extension of the DDM framework (Fig. 13.1) for school STEM education; it adds transitions between data/context and models in other STEM subjects (δ), mathematical models and models in other STEM subjects (ε), and statistical models and models in other STEM subjects (ζ). Elements and interactions added from Fig. 13.1 are highlighted in grey.

Building on the models in Fig. 13.1, *models in other STEM subjects* include modellers' interpretations of data and their representations in terms of big ideas about STEM, such as scientific and engineering models (e.g. scale model/prototypes) (Chalmers et al., 2017; English et al., 2017; Gilbert & Justi, 2016). Following transitions α and β, *transition δ* refers to generating models in other STEM subjects from the data/context and validating or revising those models based on the data/context. Following transition γ, *transitions ε and ζ* refer to generating and validating a mathematical or statistical model based on the models in other subjects and vice versa. For example, generating a mathematical or statistical model based on other STEM subjects' models involves the deterministic or non-deterministic/stochastic representation and interpretation of those models or the generation of a new mathematical or statistical model through representation and interpretation. Note that no sequentially is assumed for transitions δ, ε, and ζ. These transitions may coincide.

[3] Transitions α, β, γ, δ, ε, and ζ refer to generation, validation, and revision of models, respectively.

Paper Helicopter and Seed Dispersal Tasks

One of the barriers for teachers seeking to implement STEM education with mathematics at its core is related to the development of rich STEM tasks (Anderson et al., 2020; Tytler et al., 2019). As a first step towards STEM task design, we focus on two statistical tasks that can be used to realise DDM in an embedded STEM context: the paper helicopter and seed dispersal tasks. We explore both tasks within the interdisciplinary DDM framework (Fig. 13.2) and an integrated STEM context.

Paper Helicopter Task

A paper helicopter (Fig. 13.3) is a simple helicopter-shaped object made of paper and a paper clip (Box, 1992; Takahashi, 2018). Paper helicopter-based material has been adopted in mathematics and statistics education (e.g. Ainley et al., 2000; Doerr et al., 2017; Kawakami, 2017; Mulligan et al., 2023). When the paper helicopter is dropped, the paper clip acts as a weight, and the helicopter should fall vertically while rotating. It is possible to measure the time it takes for the helicopter to reach the ground after being dropped from a certain height. Measuring such flight times was originally developed as a teaching material for quality engineering (Box, 1992), making it inherently interdisciplinary. The independent variables that constitute a paper helicopter include wingspan, wing length, wing elevation angle, body length, and body width. In contrast, the dependent variables include flight time, fall speed, and number of revolutions (e.g. Takahashi, 2018).

We refer to the paper helicopter task (Fig. 13.4) in Kawakami's (2017) lessons for Year 5 (10–11-year-olds) students. The main task is to consider ways to reduce the variability of experimental data to compare the flight times of paper helicopters with different wing lengths. Sub-tasks involve predicting the distribution of flight times and validating these predictions with experimental results. In sub-tasks 1 and 2,

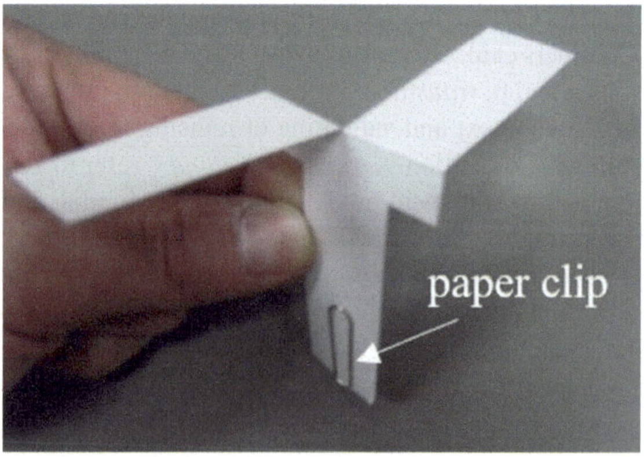

Fig. 13.3 The paper helicopter. Photo taken by the first author

An experiment was conducted to measure the flight times of paper helicopters with a wingspan of 5 cm and 10 cm, respectively; however, the results varied too much to facilitate a good comparison. Consequently, what can be done to reduce the variability?

Sub-task 1: What do you think caused the different results in the experiment? Also, predict the distribution of flight times if the causes were removed. After making a prediction, validate it through a second experiment with improved sources of variability.

Sub-task 2: Predict the distribution of flight times if we conducted a third experiment with further improved sources of variability.

Fig. 13.4 Paper helicopter task (Kawakami, 2017)

students predict, in their own way, the theoretical distribution behind the phenomena of the paper helicopter's flight time.

If students interpret the phenomena deterministically, they may predict the distribution using a mathematical model such as mean value. On the contrary, if they interpret the phenomena stochastically, they may make predictions using a statistical model such as mean value with range and a dot plot with the central range. Kawakami (2023) illustrated the dynamic DDM activities of a student working on the paper helicopter task and moving between data/context and the mathematical and statistical models for better prediction using the DDM framework.

The paper helicopter task can also be associated with models in other STEM subjects, such as engineering prototypes and scientific models of falling motion. Consequently, this task can potentially realise/formulate big ideas about STEM (Chalmers et al., 2017). However, it does not sufficiently emphasise the need to use and redesign prototypes (i.e. paper helicopters) in the context of the real world and the connection to natural systems in science.

Seed Dispersal Task

In the context of the real world, we notice that the spinning and falling of the paper helicopter is similar to the shape and movement of rotary wind seed dispersal, such as those of the acer palmatum and buckleya lanceolata seeds (Fig. 13.5). Accordingly, seed dispersal-based material has been adapted in STEM education, keeping data and statistics at its core (e.g. Fitzallen et al., 2019). Fitzallen et al. (2019) developed and implemented seed dispersal material for STEM education. The lessons for Year 5 (10–11-year-olds) students were developed by combining statistical problem-solving with the design process to determine the type of seed that flies furthest in the wind. The STEM elements inherent in the seed dispersal material are summarised in Table 13.1 (Smith et al., 2019). Therefore, the seed dispersal material also has the potential to realise/formulate big ideas about STEM (Chalmers et al., 2017).

The main difference between the seed dispersal material and the paper helicopter task is the emphasis on 'the the real-world context of seed dispersal and its diversity'. The seed dispersal task is more closely connected to the real-world context (i.e.

Fig. 13.5 Buckleya lanceolata seeds. Photo taken by Shohei Chiba

Table 13.1 STEM elements inherent in the seed dispersal material.

Science	• Flowers, fruit, and seeds: structure and function • Diversity of seed dispersal mechanisms: wind, water, animal, fire (smoke/heat), gravity, and explosion • Comparison of the effectiveness of different designs
Technology	• Use of TinkerPlots to represent and interpret the collected data
Engineering	• Engineering design brief for building a wind dispersal device: 'Which design can disperse a seed the farthest using a fan as wind?' • Diversity of designs for dispersal. Focus on one of the following three wind-based mechanisms: 1. Parachute (e.g. dandelion clock), 2. Sail (e.g. elm seed), 3. Helicopter (e.g. maple seed)
Mathematics	• Measurement parameters: height for drop; distance from drop point; number of trials • Representation of data in tables and graphs before loading into TinkerPlots to explore variation within data • Statistics: e.g. the issue of number of trials

Adapted from Smith et al. (2019, p. 45)

natural systems in science), in spite of the fact that the specific concepts explored through the data are fundamentally similar to the paper helicopter task.

We refer to the seed dispersal task (Fig. 13.6) in Chiba and Kawakami's (2022) lessons—which incorporates predictions into the seed dispersal material—for Year 4 (9–10-year-olds) students. The main task involves redesigning the shape of the buckleya lanceolata seed to maximise its flight time. After several experiments that measure the flight time of seed prototypes (Fig. 13.7), sub-tasks 1 and 2 involve predicting trends in flight times with prototype shape as a variable and validating the predictions with real data. The four wings of buckleya lanceolata seeds are wiry and spread out when they fall due to the centrifugal force of rotation. Thus, Chiba

The buckleya lanceolata seeds come in various sizes, as shown in Fig. 5. The speed at which they fall seems to vary depending on the seeds. Therefore, how can the flight time of the seeds be increased?

Sub-task 1: A prototype of the seed (Fig. 7a) is used to investigate changes in the flight time. After several experiments with the prototype to measure the flight time, predict what will happen to it if the vertical length of the prototype is varied, as shown in Fig. 7b. After making a prediction, validate it through experiments.

Sub-task 2: After several experiments with the prototype to measure the flight time, predict what will happen to it if the slit length of the prototype is varied, as shown in Fig. 7c. After making a prediction, validate it through experiments.

Fig. 13.6 Seed dispersal task (Chiba & Kawakami, 2022)

a b c

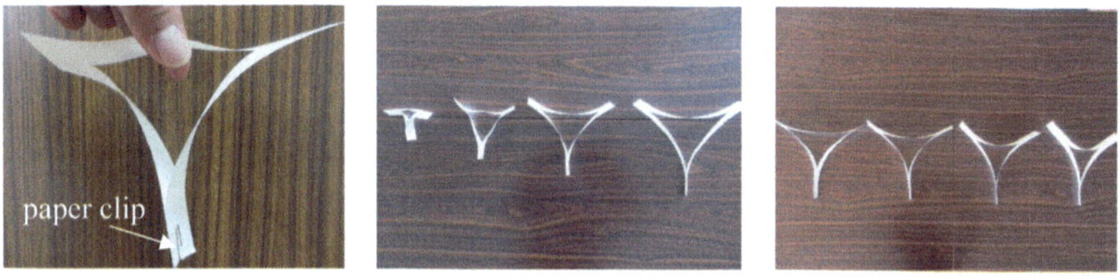

Fig. 13.7 **a–c** Prototypes of the buckleya lanceolata seed. Photos taken by Shohei Chiba

and Kawakami (2022) prepared a prototype that behaved similarly to the seeds when they fell. The paper clip on the prototype represents the fruit.

In sub-tasks 1 and 2, like in the paper helicopter task (Fig. 13.4), students may generate mathematical and statistical models (e.g. line graphs, dot plots, statistics) based on the data/context. In addition, students may also generate models in other STEM subjects. For example, with regards to engineering, students may interpret trends deterministically or stochastically in terms of the size, weight, and shape of the seed prototypes shown in Fig. 13.7. Conversely, the prototypes can be redesigned to maximise the flight time from graphs that organise the actual data.

Scientific models of the experimental process, including conditional control, can also be prepared to conduct experiments with different variables that constitute the seed prototype. Given the nature of the activity—creating seed prototypes and observing their falling motion—students can also generate scientific models about the motion of a falling body under air resistance or about the structure and function of seeds, as shown in Table 13.1. These scientific models can also provide insights to generate line graphs and dot plots, and make predictions or interpret graphs that organise actual data. Thus, the seed dispersal task allows students to experience interdisciplinary DDM, which moves between data/context, mathematical models, statistical models, and models in other STEM subjects, to make better predictions.

Example of Interdisciplinary DDM Using the Seed Dispersal Task

Outline of Classroom Practice

We will begin this section by presenting an outline of the interdisciplinary DDM practice for the seed dispersal task described in Chiba and Kawakami (2022). The practice was designed in collaboration with a researcher and a teacher (i.e. a classroom practitioner). The researcher explained the concepts and aims of the activities to the teacher, who then prepared a teaching plan and worksheets. At the end of each lesson, the teacher briefly reported the results and discussed the next lesson with the researcher. The plan was adjusted based on the results and discussions.

The participants were students ($n = 30$) from one class of Year 4 in a public elementary school in Japan. They had previously learned about bar graphs, line graphs, and two-dimensional tables. However, they did not know about representative values, dot plots, histograms, and proportional graphs.[4]

The teaching experiment comprised nine 45-min lessons in the mathematics classroom and 'the period for integrated studies'.[5] The first five lessons were spent on sub-task 1, and the remaining four on sub-task 2.[6] Before working on sub-task 1, the teacher presented a picture of a dandelion seed and asked the students what they thought the fluff was for. The students realised that plants try to spread their seeds in different ways to increase their range. Then, they dropped the buckleya lanceolata seeds and realised individual differences in the seeds and their flight times. When the teacher asked how the flight time of the seeds could be increased, they started focusing on the structure of the seed (e.g. 'reduce the size of the seed and increase the size of the leaf'). When asked what could be done to check the validity of these hypotheses, the students started thinking about transforming the real seeds into prototypes. Here, the teacher introduced the seed prototype shown in Fig. 13.7a. The students made a prototype from a paper pattern and dropped it several times. The whole class then confirmed the experimental method of dropping the prototype (e.g. 'drop the prototype from the same height', 'on cue, drop the prototype and start the stopwatch simultaneously').

After this, the students were divided into groups experimenting with dropping prototypes with vertical lengths of 15 and 20 cm (Fig. 13.7b) and recorded the median, as they had not yet learned about mean values. Due to time constraints, the teacher presented the experiment results with the 10 cm prototype and used the median results for all groups to generate the data. After the students plotted these

---

[4] See also Kawakami and Aoyama (2018) for details on the elementary-level statistics curriculum in Japan.

[5] It is a cross-curricular inquiry class in the national curriculum standards in Japan. Its content is left at the discretion of each school (Ministry of Education, Culture, Sports, Science and Technology, 2008).

[6] The worksheets for sub-tasks 1 and 2 are attached in Appendices A and B, respectively.

data on a line graph, the teacher introduced sub-task 1 by asking, 'How does the flight time change as the vertical length increases?' The students presented their predictions by drawing line graphs on worksheets and calculating differences in the data. After making predictions, they collected data on the flight times of the prototype with longer vertical lengths, plotted the data sets on a line graph, and compared the predicted graph with the actual data to validate their predictions.

Sub-task 2 was established based on the students' validations. Based on their observation that 'the flight time increases as the vertical length increases, but decreases after a certain length', the teacher asked, 'What can we do to further increase the flight time of the prototype?'. The students commented, 'change the length of the sides', 'change to lighter paper', or 'change the length of the slits'. Of these, the teacher focused on the length of the slits. Data collected for sub-task 1 showed that the prototype with a vertical length of 30 cm had the maximum flight time. Thus, the students collected data by experimenting with the flight time of a prototype with a fixed vertical length and slit lengths of 3, 6, and 9 cm (Fig. 13.7c), as in sub-task 1.

The students then plotted this data on line graphs and presented their predictions regarding the change in the flight time if the slit lengths were increased future. They did this by drawing line graphs on the worksheet and calculating the differences in the data. Once the predictions were made, they collected data on the flight time of the prototype with a further increase in slit length, plotted the data on a line graph, and validated their predictions by comparing the predicted graph with the graph of the actual data. Finally, based on the experiment results, the students drew conclusions about the optimal shape of the seed prototype (e.g. 'the vertical and slit lengths are 30 cm and 9 cm, respectively').

A Student's Interdisciplinary DDM Activities: Miyu's Case

We now present an exemplary case demonstrated by a student, Miyu, who had traced all the transitions in tackling the seed dispersal task using the interdisciplinary DDM framework (Fig. 13.2). Table 13.2 includes Miyu's worksheet excerpts and descriptions of her activities. We identified mathematical models, statistical models, and models in other STEM subjects based on the framework, descriptions of the student's worksheets, and the post-class interview. Miyu's prediction for sub-task 1 was based on only a mathematical model, but for sub-task 2, she made predictions using a statistical model and models in other STEM subjects.

In her prediction for sub-task 1 (Table 13.2a), Miyu interpreted the vertical length of the prototype and the flight time as having a regular proportional relationship. Referring to the regular pattern of steep and gradual slopes of the line graph, from 10 to 25 cm in 5 cm increments, she predicted that the slope of the line graph between 25 and 40 cm would also be regular, steep, and gradually become steeper in 5 cm increments (i.e. 'Between the first place [of the vertical length] 0 and 5 [e.g. 10 cm versus 15 cm], the flight time does not increase much, and from 5 to 0 [e.g. 15 cm

Table 13.2 Descriptions of Miyu's activities using the interdisciplinary DDM framework

<table>
<tr><th>Miyu's worksheet excerpts</th><th>Descriptions using the framework</th></tr>
<tr><td>

(**a**) Prediction in sub-task 1

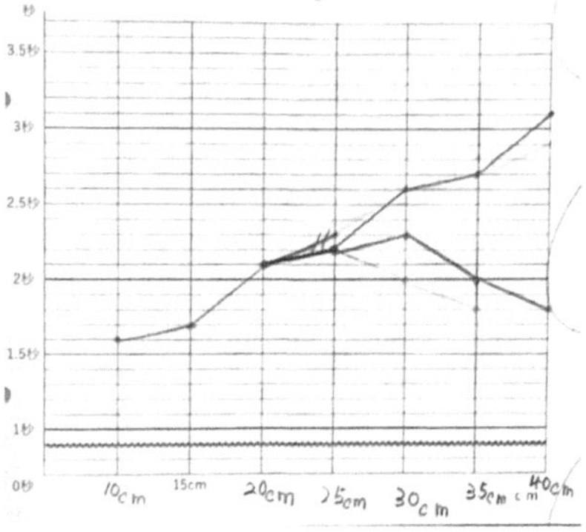

Note: Graph of hypothetical data (black); Graph of actual data (red); Horizontal axis: Vertical length of the prototype; Vertical axis: Flight time

'The longer the vertical length, the longer the flight time. I thought the flight time would be longer if I increased the vertical length. Between the first place [of the vertical length] 0 and 5 [e.g. 10 cm versus 15 cm], the flight time does not increase much, and from 5 to 0 [e.g. 15 cm versus 20 cm], the flight time increases a lot'. [Square brackets indicate author's notes.]

</td><td>

(**a**) Prediction with a mathematical model generated from data/context [transition α]

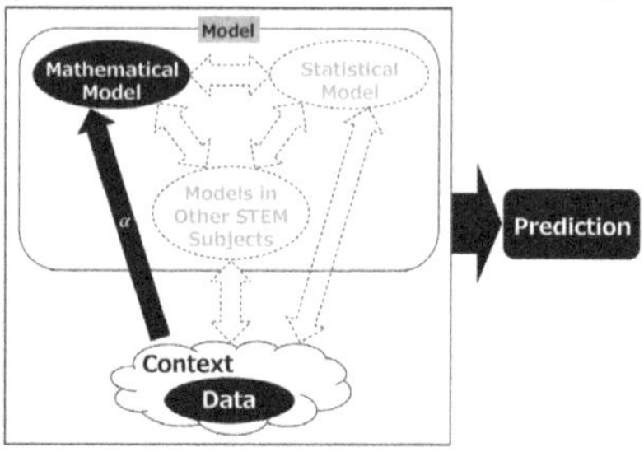

</td></tr>
<tr><td>

(**b**) Validation of the prediction in sub-task 1

'The prediction was that [the slope of the graph] would go up, but the actual graph [slope] went up to the middle, then down from the middle. I think the flight time was shortened because the prototype got longer and heavier. Next time I make a prediction, I want to consider the prototype's length, weight, and size. I also found [from the data and graphs] that the flight time got longer up to the halfway point, but as the prototype got longer, the flight time got shorter, depending on the weight, length, size and other factors of the prototype'. [Square brackets indicate author's notes.]

</td><td>

(**b**) Validation of a mathematical model based on both data/context [transition α] and models in other STEM subjects generated from context [transitions δ and ε]

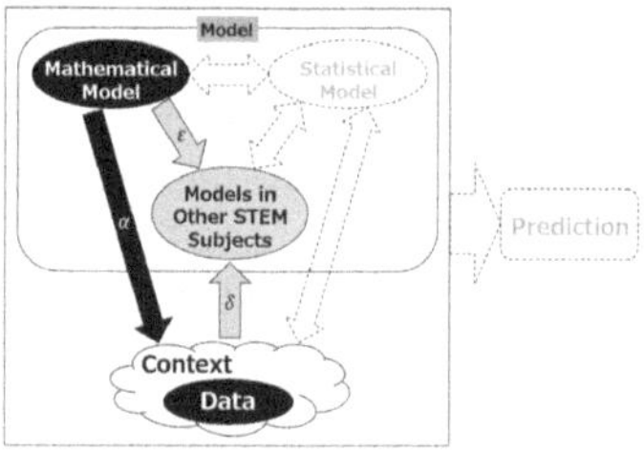

</td></tr>
</table>

versus 20 cm], the flight time increases a lot'). Here, the line graph was interpreted deterministically, focusing on the signal (i.e. the regular pattern) using what can be considered a mathematical model. A mathematical model was generated based on the data collected through the experiment (transition α).

When validating her sub-task 1 predictions (Table 13.2b), Miyu believed that the line graph depicting the actual data decreased in the middle of the graph due to the relationship between the vertical length and weight of the prototype (i.e. 'I think the flight time was shortened because the prototype got longer and heavier'). Thus, she validated a mathematical model (i.e. line graph) based on real data and the experiment context (transition α). She also validated the mathematical model using models in other STEM subjects, such as prototype and scientific models (transition ε). The expression of the relationship—the longer the vertical length of the prototype, the heavier it becomes, and the faster it falls—is an informal scientific model of falling motion and gravity. This scientific model was generated from the experiment's context by observing the prototypes' fall (transition δ).

In her prediction for sub-task 2 (Table 13.3a), Miyu interpreted the slit length of the prototype and the flight time as almost a curve. In the post-classroom interview, Miyu was asked why she did not predict a regular change in slope as in sub-task 1; she replied, 'it would not always change in the same way'. Thus, Miyu determined the slope of the line graph in Table 13.3a as partially irregular. The line graph in Table 13.2a (as a mathematical model) was interpreted stochastically to generate the line graph in Table 13.3a (as a statistical model; transition γ). This statistical model was also generated based on the data at hand (i.e. flight times for slit lengths of 3, 6, and 9 cm), the context of the experiment, and models in other STEM subjects such as prototypes and an informal scientific model of aerodynamic forces (transitions β and ζ). Miyu built and dropped a prototype and observed the results (i.e. 'When I made the 3–6 cm prototypes, I thought the gap would gradually get narrower') and believed that the smaller the 'gap' in the prototype by a certain amount, the shorter the flight time (i.e. 'I thought the gap would shorten the flight time again').

For validating the prediction in sub-task 2 (Table 13.3b), Miyu wrote that 'my prediction that the graph would not be a straight line as the slope of the graph was correct because the gap in the prototypes would get smaller' and 'I thought the flight time also went up and down because the gaps were too big, too small, or just right'. Here, Miyu verified her predicted line graph in Table 13.3a (as a statistical model) based on real data, the context of the experiment (transition β), and models in other STEM subjects (i.e. a prototype and an informal scientific model of aerodynamic forces; transition ζ). Thus, Miyu's interdisciplinary DDM activities involved data-driven transitions between a mathematical model, a statistical model, and models in other STEM subjects.

Discussion and Implications for School STEM Education

This chapter explored the potential of extending DDM from school mathematics to school STEM education. In this section, we will discuss the characteristics of interdisciplinary DDM activities by reflecting on Miyu's case and derive the implications of using the interdisciplinary DDM framework (Fig. 13.2).

We have characterised interdisciplinary DDM activities as taking data as a starting point and making predictions using deterministic (based on mathematical models) and non-deterministic/stochastic reasoning (based on statistical models)

Table 13.3 Descriptions of Miyu's activities using the interdisciplinary DDM framework

Miyu's worksheet excerpts	Descriptions using the framework
(**a**) Prediction in sub-task 2 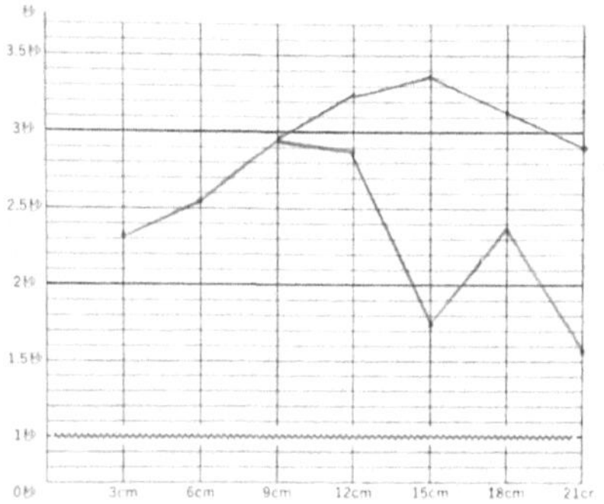 *Note*: Graph of hypothetical data (black); Graph of actual data (red); Horizontal axis: Slit length of the prototype; Vertical axis: Flight time *'I increased [the slope of the graph] because I thought the prototype gap was neither too big nor too small until halfway through, but then the gap became too small, and I believed the prototypes would not fly very well'. [Square brackets indicate author's notes.]*	(**a**) Prediction with a statistical model generated from a mathematical model [transition γ], data/context [transition β], and models in other STEM subjects [transition ζ] 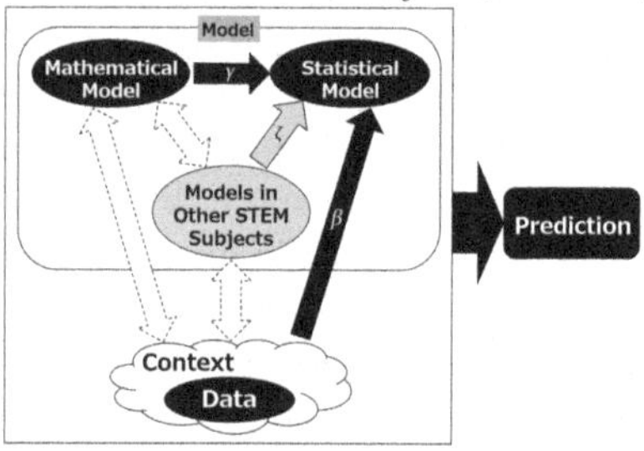
(**b**) Validation of the prediction in sub-task 2 *'I think my prediction that the graph would not be a straight line as the slope of the graph would increase was correct because the gap in the prototypes would get smaller, but the reason the flight time increased between 15 and 18 cm [in the actual graph] is that the gap was neither large nor small, but just right. I also thought that the gap in the prototype had something to do with the flight time of the prototypes going up and down. I thought the flight time also went up and down because the gaps were too big, too small, or just right'. [Square brackets indicate author's notes.]*	(**b**) Validation of a statistical model based on both data/context [transition β] and models in other STEM subjects [transition ζ] 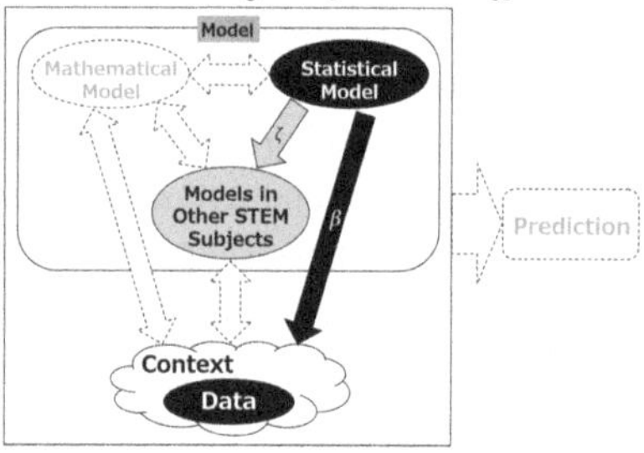

while supported by models from other STEM subjects such as prototypes and scientific models. This aspect was not illuminated in DDM activities using the paper helicopter task (Fig. 13.4) (cf. Kawakami, 2023). In Miyu's case, models in other STEM subjects played the role of referents, providing a different perspective to interpret mathematical and statistical models (Tables 13.2b and 13.3b) and generate statistical models from mathematical ones (Table 13.3a). In addition, the case also indicated the interaction between models from other STEM subjects (prototypes versus scientific models). In transitions ε and ζ, the use of within-subject big ideas in the context of other STEM subjects, the comparison of dissimilar big ideas, and the encompassing of big ideas can occur (Chalmers et al., 2017). Therefore, a detailed process analysis will be necessary for the future research.

The interdisciplinary DDM framework provides research implications for school STEM education and practical implications for developing STEM tasks and lessons that highlight the role of the 'M'. In terms of research implications, the framework comprises the back-and-forth between data and models (i.e. transitions α, β, and δ) and between models (i.e. transitions γ, ε, and ζ). This facilitates the generation, evaluation, and revision of each model to make better predictions. By unravelling the complex and dynamic process of interdisciplinary DDM, we can construct STEM education that places the 'M' and interdisciplinary DDM at its core. In addition, as DDM provides the platform and the authentic need for mathematical and statistical modelling, the framework can be used as a research tool for linking research in mathematics education, statistics education, and STEM education. If the concept of data is taken in a broader sense, or if the focus is on the format, sources, and ethics of data, interdisciplinary DDM can also be seen as a potential aid to research in data science education (e.g. Biehler, 2022; Fry et al., 2023; Makar et al., 2023).

Regarding practical implications, the framework can be used as a pedagogical tool for extending mathematics teaching practices to STEM classroom units by considering the relevance of the mathematical and statistical tasks that deal with data to models in other STEM subjects. In the case of tasks such as the paper helicopter and seed dispersal tasks (Figs. 13.4 and 13.6), teachers will inevitably realise that statistical tasks and practices in the mathematics curriculum are connected to various STEM subjects. This realisation will, in turn, serve as a stepping stone to developing integrated STEM curricula and units (Drake & Burnes, 2004). Broadly speaking, the present study's approach is to implement the 'STEMisation of everyday learning' by using modelling and data as core perspectives. The framework's description of transitions α–ζ can also provide a lens for understanding students' interdisciplinary activities and highlighting appropriate opportunities for teacher intervention in interdisciplinary DDM.

We hope that this framework offers one solution to teachers' challenges in developing quality STEM tasks and units that foreground the 'M' in school STEM education (Anderson et al., 2020; English, 2017; Maass et al., 2019; Tytler et al., 2019). In future research, this framework should be elaborated through further case analysis and development, focusing on informing the creation of rich STEM education practices.

Acknowledgements We thank Shohei Chiba for conducting the classroom practice and providing class details.

Funding This study was partially supported by the Japan Society for the Promotion of Science KAKENHI Grant Numbers JP21K02513 and JP21K02553.

Appendix 1: A Worksheet for Sub-Task 1

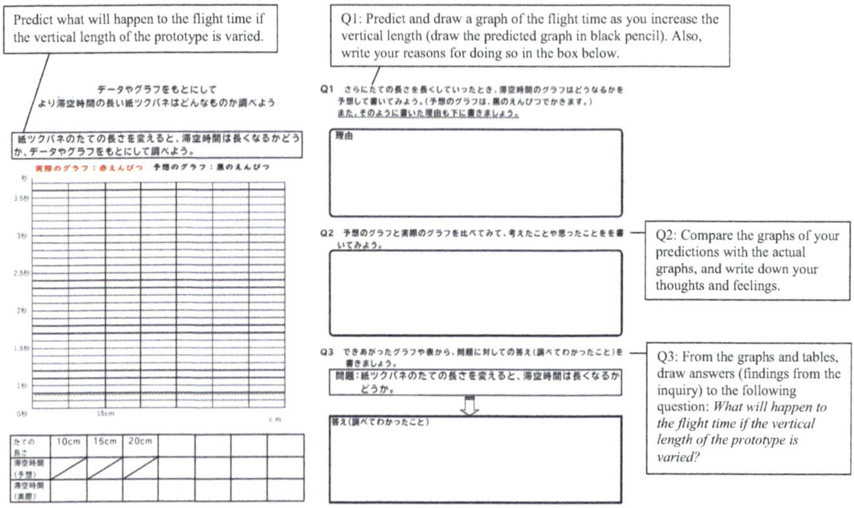

Appendix 2: A Worksheet for Sub-Task 2

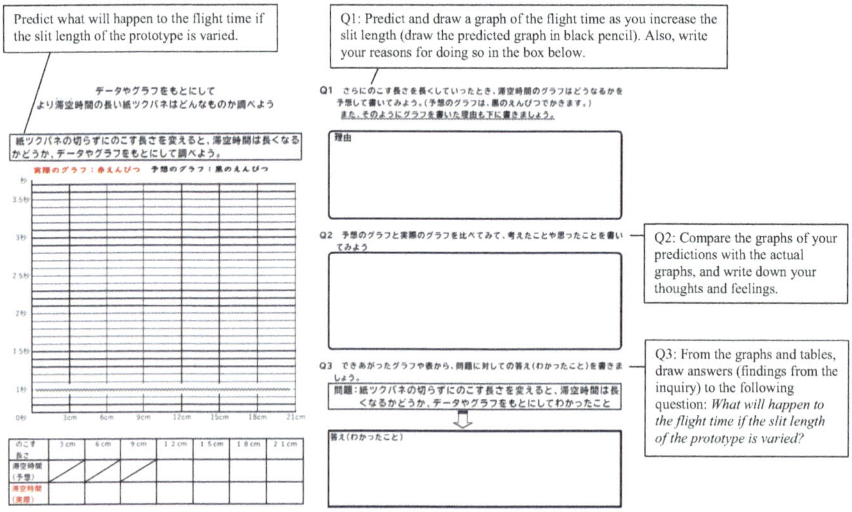

References

Ainley, J., Nardi, E., & Pratt, D. (2000). The construction of meanings for trend in active graphing. *International Journal of Computers for Mathematical Learning, 5*, 85–114. https://doi.org/10.1023/A:1009854103737

Anderson, J., English, L., Fitzallen, N., & Symons, D. (2020). The contribution of mathematics education researchers to the current STEM education agenda. In J. Way, C. Attard, J. Anderson, J. Bobis, H. McMaster, & K. Cartwright (Eds.), *Research in mathematics education in Australasia 2016–2019* (pp. 27–57). Springer. https://doi.org/10.1007/978-981-15-4269-5_3

Ärlebäck, J. B., & Kawakami, T. (2023). The relationship between statistics, statistical modelling and mathematical modelling. In G. Greefrath, S. Carreira, & G. Stillman (Eds.), *Advancing and consolidating mathematical modelling: Research from ICME-14* (pp. 293–309). Springer. https://doi.org/10.1007/978-3-031-27115-1_18

Australian Curriculum, Assessment and Reporting Authority. (2023). *The Australian Curriculum Version 9.0*. Retrieved from https://v9.australiancurriculum.edu.au

Berry, J., & Houston, K. (1995). *Mathematical modelling*. Edward Arnold.

Biehler, R. (2022). Revisiting fundamental ideas for statistics education from the perspective of machine learning and its applications. In S. A. Peters, L. Zapata-Cardona, F. Bonafini, & A. Fan (Eds.), *Proceedings of the 11th international conference on teaching statistics*. IASE, Rosario, Argentina. https://doi.org/10.52041/iase.icots11.T1A2

Box, G. E. P. (1992). George's column: Teaching engineers experimental design with a paper helicopter. *Quality Engineering, 4*, 453–459. https://doi.org/10.1080/08982119208918925

Chalmers, C., Cartler, M., Cooper, T., & Nason, R. (2017). Implementing "big ideas" to advance the teaching and learning of science, technology, engineering and mathematics (STEM). *International Journal of Science and Mathematics Education, 15*(Suppl 1), 25–43. https://doi.org/10.1007/s10763-017-9799-1

Chiba, S., & Kawakami, T. (2022). *Teaching for statistical thinking in the middle grades of primary school: The STEM approach*. [Oral presentation] 104th Annual Meeting of Japan Society of Mathematics Education, Online, JSME. (in Japanese) https://researchmap.jp/kenshin/presentations/39321640

Cobb, G. W., & Moore, D. S. (1997). Mathematics, statistics, and teaching. *The American Mathematical Monthly, 104*, 801–823. https://doi.org/10.2307/2975286

Develaki, M. (2020). Comparing crosscutting practices in STEM disciplines: Modeling and reasoning in mathematics, science, and engineering. *Science & Education, 29*, 949–979. https://doi.org/10.1007/s11191-020-00147-1

Doerr, H., delMas, R., & Makar, K. (2017). A modeling approach to the development of students' informal inferential reasoning. *Statistics Education Research Journal, 16*, 86–115. https://doi.org/10.52041/serj.v16i2.186

Drake, S. M., & Burns, R. C. (2004). *Meeting standards through integrated curriculum*. Association for Supervision and Curriculum Development.

Dvir, M., & Ben-Zvi, D. (2023). Informal statistical models and modeling. *Mathematical Thinking and Learning, 25*, 79–99. https://doi.org/10.1080/10986065.2021.1925842

English, L. (2017). Advancing elementary and middle school STEM education. *International Journal of Science and Mathematics Education, 15*, 5–24. https://doi.org/10.1007/s10763-017-9802-x

English, L. (2023). Multidisciplinary modelling in a sixth-grade tsunami investigation. *International Journal of Science and Mathematics Education, 21*, 41–65. https://doi.org/10.1007/s10763-022-10303-4

English, L., King, D., & Smeed, J. (2017). Advancing integrated STEM learning through engineering design: Sixth-grade students' design and construction of earth quake resistant buildings. *The Journal of Educational Research, 110*, 255–271. https://doi.org/10.1080/00220671.2016.1264053

English, L., & Watson, J. (2018). Modelling with authentic data in sixth grade. *ZDM, 50*, 103–115. https://doi.org/10.1007/s11858-017-0896-y

Fitzallen, N., Wright, S., & Watson, J. (2019). Focusing on data: Year 5 students making STEM connections. *Journal of Research in STEM Education, 5*, 1–19. https://doi.org/10.51355/jstem.2019.60

Fry, K., English, L., & Makar, K. (2023). Cognitive tuning in the STEM classroom: Communication processes supporting children's changing conceptions about data. *Mathematics Education Research Journal*. https://doi.org/10.1007/s13394-023-00465-x

Gilbert, J., & Justi, R. (2016). Modelling-based teaching in science education. *Springer*. https://doi.org/10.1007/978-3-319-29039-3

Hestenes, D. (2010). Modeling theory for math and science education. In R. Lesh, P. Galbraith, C. Haines, & A. Hurford (Eds.), *Modeling students' mathematical modeling competencies: ICTMA 13* (pp. 13–41). Springer. https://doi.org/10.1007/978-1-4419-0561-1_3

Hjalmarson, M., Holincheck, N., Baker, C. K., & Galanti, T. M. (2020). Learning models and modeling across the STEM disciplines. In C. C. Johnson, M. J. Mohr-Schroeder, T. J., Moore, & L. D. English (Eds.), *Handbook of research on STEM education* (pp. 223–233). Routledge.

Ikeda, T., & Stephens, M. (2015). Reconsidering the roles and characteristics of models in mathematics education. In G. A. Stillman, W. Blum, & M. Salett Biembengut (Eds.), *Mathematical modelling in education research and practice: Cultural, social and cognitive influences* (pp. 351–361). Springer. https://doi.org/10.1007/978-3-319-18272-8_29

Kawakami, T. (2017). Combining models related to data distribution through productive experimentation. In G. A. Stillman, W. Blum, & G. Kaiser (Eds.), *Mathematical modelling and applications: Crossing and researching boundaries in mathematics education* (pp. 95–105). Springer. https://doi.org/10.1007/978-3-319-62968-1_8

Kawakami, T. (2023). A triplet of data/context, mathematical model, and statistical model: Conceptualising data-driven modelling in school mathematics. In P. Drijvers, C. Csapodi, H. Palmér, K. Gosztonyi, & E. Kónya (Eds.), *Proceedings of the 13th Congress of the European Society for Research in Mathematics Education* (pp. 1243–1250). Budapest: Hungary, Alfréd Rényi Institute of Mathematics and ERME.

Kawakami, T., & Aoyama, K. (2018). Ingenuity and challenges to incorporate statistical-inquiry process into statistics lessons in primary schools: The case of Japan. In M. A. Sorto, A. White, & L. Guyot (Eds.), *Proceedings of the 10th international conference on teaching statistics*. IASE, Kyoto, Japan. https://iase-web.org/icots/10/proceedings/pdfs/ICOTS10_2F2.pdf?1531364243

Kawakami, T., & Mineno, K. (2021). Data-based modelling to combine mathematical, statistical, and contextual approaches: Focusing on ninth-grade students. In. F. K. S. Leung, G. A. Stillman, G. Kaiser, & K. L. Wong (Eds.), *Mathematical modelling education in East and West* (pp. 389–400). Springer. https://doi.org/10.1007/978-3-030-66996-6_32

Kawakami, T., & Saeki, A. (2022). A framework for describing and analysing data-driven modelling activities in school mathematics: From the perspectives of mathematical and statistical models. *Journal of Science Education in Japan, 46*, 421–437. (in Japanese) https://doi.org/10.14935/jssej.46.421

Konold, C., & Pollatsek, A. (2002). Data analysis as the search for signals in noisy processes. *Journal for Research in Mathematics Education, 33*, 259–289. https://doi.org/10.2307/749741

Lehrer, R., & English, L. (2018). Introducing children to modeling variability. In D. Ben-Zvi, K. Makar, & J. Garfield (Eds.), *International handbook of research in statistics education* (pp. 229–259). Springer. https://doi.org/10.1007/978-3-319-66195-7_7

Lehrer, R., Schauble, L., & Wisittanawat, P. (2020). Getting a grip on variability. *Bulletin of Mathematical Biology, 82*(8). https://doi.org/10.1007/s11538-020-00782-3

Lesh, R. A., & Doerr, H. M. (Eds.). (2003). *Beyond constructivism: Models and modeling perspectives on mathematics problem solving, learning, and teaching*. Routledge. https://doi.org/10.4324/9781410607713

Maass, K., Geiger, V., Ariza, M. R., & Goos, M. (2019). The role of mathematics in interdisciplinary STEM education. *ZDM, 51*, 869–884. https://doi.org/10.1007/s11858-019-01100-5

Makar, K., Fry, K., & English, L. (2023). Primary students' learning about citizenship through data science. *ZDM, 55*, 967–979. https://doi.org/10.1007/s11858-022-01450-7

Ministry of Education, Culture, Sports, Science and Technology. (2008). Chapter 5 the period for integrated studies. In Ministry of Education, Culture, Sports, Science and Technology (Ed.), *Elementary school curriculum guideline English version (Tentative translation)*. Retrieved from https://www.mext.go.jp/component/a_menu/education/micro_detail/__icsFiles/afieldfile/2011/10/12/1261037_13_1.pdf

Mulligan, J., Tytler, R., Prain, V., & Kirk, M. (2023). Implementing a pedagogical cycle to support data modelling and statistical reasoning in years 1 and 2 through the Interdisciplinary Mathematics and Science (IMS) project. *Mathematics Education Research Journal*. https://doi.org/10.1007/s13394-023-00454-0

Pfannkuch, M., Ben-Zvi, D., & Budgett, S. (2018). Innovation in statistical modeling to connect data, chance and context. *ZDM, 50*, 1113–1123. https://doi.org/10.1007/s11858-018-0989-2

Piaget, J. (1968). Structuralism. *Psychology Press*. https://doi.org/10.4324/9781315722368

Silver, N. (2012). *The signal and the noise: Why so many predictions fail—but some don't*. The Penguin Press.

Smith, C., Fitzallen, N., Watson, J., & Wright, S. (2019). The practice of statistics for stem: Primary students and pre-service primary teachers exploring variation in seed dispersal. *Teaching Science, 65*(1), 38–47.

Stillman, G., & Brown, J. (2023). Modeling the phenomenon versus modeling the data set. *Mathematical Thinking and Learning, 25*, 270–295. https://doi.org/10.1080/10986065.2021.2013144

Takahashi, T. (2018). Active learning of statistical quality control in practice with hope-add-inn. In M. A. Sorto, A. White, & L. Guyot (Eds.), *Proceedings of the 10th international conference on teaching statistics*. IASE, Kyoto, Japan. https://iase-web.org/icots/10/proceedings/pdfs/ICOTS10_5J2.pdf?1531364274

Tytler, R., Williams, G., Hobbs, L., & Anderson, J. (2019). Challenges and opportunities for a STEM interdisciplinary agenda. In B. Doig, J. Williams, D. Swanson, R. Borromeo Ferri, & P. Drake (Eds.), *Interdisciplinary mathematics education: The state of the art and beyond* (pp. 51–81). Springer. https://doi.org/10.1007/978-3-030-11066-6_5

Watson, J., Fitzallen, N., & Chick, H. (2020). What is the role of statistics in integrating STEM education? In J. Anderson, & Y. Li (Eds.), *Integrated approaches to STEM education* (pp. 91–115). Springer. https://doi.org/10.1007/978-3-030-52229-2_6

Wild, C. J., & Pfannkuch, M. (1999). Statistical thinking in empirical enquiry. *International Statistical Review, 67*(3), 223–248. https://doi.org/10.1111/j.1751-5823.1999.tb00442.x

Wilensky, U., & Jacobson, M. (2014). Complex systems and the learning sciences. In R. K. Sawyer (Ed.), *The Cambridge handbook of the learning sciences* (2nd ed., pp. 319–338). Cambridge University Press.

Zbiek, R. M. (2016). Response 2: Quantitative reasoning: Capturing a tension between structure and variability. In R. Duschl, & A. S. Bismack (Eds.), *Reconceptualizing STEM education* (pp. 189–193). Routledge.

Part V
Mathematics and the Development of Responsible Citizenship

Chapter 14
How Mathematics in STEM Can Contribute to Responsible Citizenship Education in Schools

Judy Anderson

Abstract Developing students' responsible citizenship is an essential outcome of a quality school education. Understanding one's rights and responsibilities as a citizen is critical to addressing many local, national and global issues such as sustainability and climate change. Through integrated STEM curriculum tasks and lessons, students can begin by investigating local issues connected to their lives and based on their experiences. Since place and context matter, these experiences will vary across schools, districts and regions. Early understandings should then lead to exploring more complex social issues such as food sustainability, biodiversity, human rights and social justice at the national and international levels. Through modelling scenarios, using data to explore trends, collecting and analysing information, or critically evaluating claims made in the media, students have opportunities to use mathematical understandings to develop the skills and dispositions to become responsible citizens.

Keywords Responsible citizenship · Sustainability · Place-based experiences · Social justice · STEM literacy

Introduction

Citizenship involves being a member of a given state or nation and following the rules and regulations governing membership. Hence "citizenship is both the condition of being a citizen and the reciprocal obligation of duties, rights and privileges" (Bybee, 2018, p. 57). A citizenship education involves developing students' understanding of the responsibilities inherent in membership of the state or country and developing knowledge to make decisions and take responsibility for themselves within the community. In a democratic society, citizens have the freedom to act, speak and think freely although they are still governed by the laws and rules of the

J. Anderson (✉)
The University of Sydney, Camperdown, NSW, Australia
e-mail: Judy.anderson@sydney.edu.au

J. Anderson and K. Makar (eds.), *The Contribution of Mathematics to School STEM Education*, https://doi.org/10.1007/978-981-97-2728-5_14

society. A democracy is based on the will of citizens to determine their own political, economic, social and cultural systems, and provides the environment for the protection of human rights (Bybee, 2018; Matthews, 2019). The role of STEM in citizenship education is to ensure citizens develop STEM literacies that enable them to understand and respond to problems and challenges, "make reasonable judgments based on evidence, and fulfil their civic duties at the local, national and global levels" (Bybee, 2018, p. 82).

This chapter at the beginning of this section of the book explores the development of responsible citizenship, and the importance of place and context including social justice perspectives. A discussion of STEM and citizenship in school education follows including the development of STEM literacy and the types of tasks that help promote responsible citizenship education with a focus on mathematics. Finally, there is an introduction to the other chapters in this section of the book.

Responsible Citizenship

With a focus on understanding and addressing global issues such as sustainability (Oxfam, 2019; United Nations, 2015), global citizenship education has been promoted within the STEM education community although it has been contested because of perceived Western cultural biases (Andreotti, 2015). However, the realities of climate change are more evident now than ever before (Barwell et al., 2022; Burkhardt et al., 2024; Bybee, 2018; Fuentes et al., 2020) and require a commitment to a school education that ensures students appreciate the challenges and develop appropriate responsible behaviours to advance social change for climate action (Nusche et al., 2024). As well as the term 'global citizenship', Bybee (2018) uses the term 'global interdependence' to highlight the "reality of living on earth—with one biosphere and the interrelatedness of individuals, environments and communities ..." (p. 61).

Understanding one's citizenship responsibilities and taking appropriate actions leads to the notion of responsible citizenship which is the third approach recommended by Maass et al. (2019) for promoting the role of mathematics within an interdisciplinary STEM curriculum. Advocating for the development of students' social, civic and intercultural understandings, education for responsible citizenship involves addressing issues such as ethics, equity and social justice. As scientific development is linked to social issues such as environmental sustainability and equity, and 'critical mathematics' (Skovsmose, 2021) promotes using the tools of mathematics for "effective socially-conscious decision-making" (Maass et al., 2019, p. 876), there is an opportunity to connect science and mathematics in examining ethical, moral and socially just issues (Zembylas, 2005). In addition, engineering and technology offer avenues for exploring issues of sustainability at the local level, particularly for school students. However, Nusche et al. (2024) highlight one approach to responsible citizenship education will not work in all school contexts because of the diversity of cultures, ethnic backgrounds, and local environmental challenges, thus raising the importance of place and context.

The Importance of Place and Context

Understanding the role of place and context is critical to improving student engagement and interest in social issues that impact their lives and the lives of others (Gallay et al., 2021; Nicol et al., 2023). Place-based education connects learning with the local context (Yemini et al., 2023) which can be particularly helpful in relation to a range of sustainability issues. Direct experiences of local places facilities learning as well as eliciting "positive emotions such as a sense of connectedness, belonging and care towards these places" (Nusche et al., 2024, p. 25). STEM and ethnomathematics education can build on students' personal, cultural and community knowledge (Rosa & Orey, 2018) or use ethnomodelling to identify the mathematical practices of social and cultural groups (Orey et al., 2020). A contrasting approach involves viewing *STEM as place* where Nicol and colleagues (2023) explore "core ideas inherent to Indigenous and ecological perspectives". Based on the understanding that place matters for all but particularly for Indigenous peoples, Nicol et al. advocate for an integrated approach to STEM education that addresses "ongoing needs, purposes and innovations in science and technology focused on STEM but also meets the needs of our ever-changing society" (p. 1236).

Context is crucial to addressing students' needs as well as engagement with local communities and their concerns (Hobbs & Kelly, 2020; Nusche et al, 2024). Referring to a contextualised curriculum that addressed a local environmental issue, Gallay et al. (2021) reported outcomes of a study into middle- and high-school science and mathematics classes in urban schools serving racial and ethnic minority students. Students' reflections yielded evidence of a developing understanding of the usefulness of science to change community beliefs, and an appreciation of the links between STEM and civic contributions. Using STEM initiatives that develop responsible citizenship through school-community partnerships aimed at addressing local issues ensures relevance and heightens student engagement, particularly if they have a social justice perspective (SJP). Clearly students also need to learn about global issues but the local community can be a useful starting point for integrated STEM curriculum investigations (see for example Anderson et al., 2019).

Taking a Social Justice Perspective

Acknowledging the importance of place and context raises issues about inclusivity and social justice. Social justice perspectives in schools raise awareness of local issues with students becoming empowered to develop a voice and sense of agency when they learn STEM concepts and practices that help to critique society (Leonard & Moore, 2014). There has been a movement for some time in the mathematics education community to teach mathematics through a SJP (Coles et al., 2013; Gutstein & Peterson, 2013; Wager & Stinson, 2012) as there has been in science education (Bybee, 2018; Finkel, 2018; Zembylas, 2005). More recently, integrated STEM through a social justice perspective has gained attention (Cowie & Mildenhall, 2021; Nicol et al., 2019). Taking an SJP also highlights the importance of creating equitable opportunities for all students (Gutstein, 2003; Wright, 2016).

Since citizens need to have an education that prepares them for understanding and making critical judgements about an overload of information and if schooling is not equitable, some disadvantaged communities may be more susceptible to the manipulation of opinions than others. Several researchers have explored how taking a mathematics and SJP impacts the lives of students from marginalised communities. For example, located in a working class low-income community, Gutstein's (2003) project focussed on changing students' orientation towards mathematics through real-world projects with a SJP. The use of critical pedagogies helped students develop both critical consciousness as well as mathematics competencies (Frankenstein, 1990; Gutstein, 2003) and to view mathematics as a tool to make sense of the world and their place in it. Rather than use longer projects, this study investigated the use of rich mathematics tasks with a SJP involving higher-order thinking, connections to community, discussion and communication, collaboration, and student agency (Gutstein & Peterson, 2013). Of importance was the need to choose tasks that were intellectually challenging but also potentially engaging so that students would persevere and be committed to working in a group to find solutions.

Teachers need to provide students with opportunities to use mathematics to analyse the world critically (Frankenstein, 1990; Gutstein, 2003; Wright, 2016). In a study by Anderson and Kreisler (2018), contextualising the mathematics with tasks that have a SJP was investigated in three grade 7 classrooms of schools in lower socio-economic areas of a large city. Students engaged with several of the tasks from Gutstein and Peterson (2013) during mathematics lessons observed by the researchers who noted their interest, enthusiasm and curiosity. Discussions allowed students to share their experiences, ask further questions and become emotionally invested in their learning. Student surveys and teacher interviews indicated students were more engaged when the mathematics was perceived as relevant with connections to real world issues, enabled longer and deeper conversations about SJPs, and was challenging with appropriate scaffolding when necessary.

Several studies have noted the importance of teacher professional learning when using such tasks (Nicol et al., 2019; Wright, 2016), particularly mathematics teachers who may be more accustomed to didactic teaching and less inclined to use contexts that raise ethical and moral dilemmas (Forgasz et al., 2015). Allowing students to discuss sensitive issues can be challenging and confronting for teachers, particularly if they are concerned about how the students might react to issues affecting their lives. Geiger et al. (2015) also note teachers need substantial support to design suitable tasks that require students to make decisions and judgments. This leads to a discussion of some of the approaches to the design and implementation of a STEM and citizenship education in schools.

STEM and Citizenship in School Education

The approach taken to citizenship education in schools will differ depending on location but it is an inherent responsibility of school education to develop each students' citizenship capability. For example, citizenship education may be a component of the social sciences curriculum with prescribed content evident in school subjects such as geography or history, it may be a separate subject, or it may involve both approaches. In the Australian curriculum, there is a *Civics and Citizenship 7–10* curriculum document, and citizenship is also included in three of the seven 'general capabilities'—Personal and Social Capability, Ethical Understanding and Intercultural Understanding—that are embedded across the curriculum where appropriate including in science, mathematics and technologies subjects (ACARA, 2020).

Regarding the role of the STEM subjects, particularly mathematics, in developing responsible citizenship, Maass et al. (2019) proposed two approaches—numeracy and socio-scientific issues. Based on a model of numeracy for the twenty-first century (Goos et al., 2014), Maass et al. (2019) argued knowledge of mathematics is insufficient to engage with and understand social issues since a critical orientation is required to engage with quantitative data, develop arguments and propose solutions to problems. Coupled with a need for numerate citizens, there is a need for the citizens to develop scientific literacy to engage with socio-scientific issues. Such issues might include the location of wind power plants and their impact on local fauna. Rather than just advocating for the development of the literacies of mathematics and science, Bybee (2018) proposed the need to develop a STEM literacy requiring teachers to design suitable tasks and problems.

Developing STEM Literacy

Policy documents typically describe the desired outcome of an integrated STEM curriculum as developing STEM literate citizens (McDonald, 2016). According to Balka (2011, p. 7), "STEM literacy is the ability to identify, apply and integrate concepts from science, technology, engineering and mathematics to understand complex problems and to innovate to solve them". Hence, STEM literacy requires the development of knowledge and understanding of each of the STEM subjects and their contribution to solving real-world problems, or developing literacies in all four of the STEM subjects (McDonald, 2016). Further, Bybee (2018) proposes STEM literacy refers to an individual's:

- knowledge, attitudes and skills to identify questions and problems in life situations, to explain the natural and designed world, and to draw evidence-based conclusions about STEM related-issues;
- understanding of the characteristic feature of STEM disciplines as forms of human knowledge, inquiry and design;
- awareness of how STEM disciplines shape our material, intellectual, and cultural environments; and
- willingness to engage in STEM-related issues, and with the ideas of science, technology, engineering, and mathematics as a constructive, concerned and reflective citizen (p. 10).

This means that to develop STEM literacy and become a STEM literate citizen, students need opportunities to learn how to solve STEM problems that require using knowledge from each of the STEM subjects, understand how each of these subjects contributes to problem solutions, and appreciate how such solutions help to address social and environmental issues (Roehrig et al., 2021). Using suitable integrated STEM problems in classrooms requires certain teaching approaches.

The teaching approaches recommended to develop STEM literate students typically involve inquiry-based learning (Kennedy & Odell, 2014; Makar et al., 2023), argumentation and reasoning (Erduran et al., 2015; Fielding-Wells et al., 2014), and digital learning such as game-based learning and computer simulations through modelling as well as coding and robotics (McDonald, 2016). To this list of teaching approaches, Larkin and Lowrie (2022) add project-based learning and problem-based learning. In early years education, Larkin and Lowrie (2022) also argue for 'intentional teaching' in STEM education where play-based environments are advocated. Intentional teaching is "purposeful, thoughtful and deliberate" (p. 34) based on careful teacher planning to achieve specific student learning outcomes. While I understand the notion of 'play' might suggest random experiences that the child selects as they feel inclined, I would argue *all* teaching should be intentional. This notion of intentional teaching also addresses criticisms of inquiry-based learning where some suggest it involves students discovering knowledge without teacher support (e.g., Kirschner et al., 2006).

On the contrary, teachers play a key role in the design and implementation of an integrated STEM education through their knowledge and understanding as well as their attitudes and the school context. Thibaut et al. (2018) investigated the variation in teachers' instructional practices in integrated STEM. In their study, the instructional practices were classified as integration, problem-centered learning, inquiry-based learning, design-based learning, and cooperative learning. By measuring teachers' attitudes, Thibaut et al. (2018) found positive correlations between teachers' attitudes and all five teaching practices. However, they were more positive for inquiry-based learning, design-based learning and cooperative learning, and much lower for integration of STEM content and problem-centred learning. The authors proposed that both practices required significant collaboration with colleagues, and not just a positive attitude. Designing suitable integrated STEM tasks that promote responsible citizenship education requires further examination.

Designing Integrated STEM Problems that Promote Responsible Citizenship

As noted in Anderson (2024) in this volume, designing STEM problems or tasks is a challenge because teachers need to have knowledge and understanding of all the STEM subjects or they need to work collaboratively with colleagues from different STEM subject backgrounds. Adding a responsible citizenship perspective creates further challenges, particularly if an SJP is adopted. According to Wright (2016) SJP tasks that focus on mathematics:

- employ collaborative, discursive, problem-solving, and problem-posing pedagogies, which promote the engagement of learners with mathematics;
- recognize and draw upon learners' real-life experiences in order to emphasize the cultural relevance of mathematics;
- promote mathematical inquiries that enable learners to develop greater understanding of their social, cultural, political, and economic situations;
- facilitate mathematical investigations that develop learners' agency, enabling them to take part in social action and realize their foregrounds; and
- develop a critical understanding of the nature of mathematics and its position and status within education and society (p.107)

This set of criteria would be equally suitable for designing integrated STEM tasks with an SJP. Another approach to help the design and evaluation of STEM tasks with an SJP was developed by Nicol et al. (2019). A matrix was used to record subject content and context, the type of social justice issue addressed (personal, community, global, social, ecological), and the twenty-first century skills needed (critical thinking, communication, cross-cultural understanding, creativity, collaboration).

Working with colleagues from different countries and cultural backgrounds, Nicol et al. (2019) sought to identify the challenges in designing SJP STEM problems that foregrounded mathematics. Many of the problems they created were focused on middle elementary to early secondary school curriculum, where the mathematics content included concepts related to number, functions and relationships, and statistics. They challenged themselves to design problems for lower elementary and upper secondary schooling, that used a broader range of mathematical concepts. For example, using data from the *Gapminder* website (Gapminder.com), they designed a task for lower elementary students but when trialled in a classroom, students did not engage with the social issue involved. While students readily participated in the task, they were distracted by creating graphs rather than engaging with, and talking about, the social justice issue. This experience raises questions about when younger children are ready to engage with issues such as conservation and sustainability.

Other researchers have explored the use of a range of problems and activities to promote responsible citizenship with a focus on mathematics (Burkhardt et al., 2024; Bybee, 2018; Coles et al., 2013; Cowie & Mildenhall, 2021; English, 2023; Gutstein & Peterson, 2013; Wager & Stinson, 2012). These problems can be used from primary to secondary school settings and offer a diverse set of resources and ideas for teachers and curriculum developers. Some examples are presented here.

The *Long Walk* module (Cowie & Mildenhall, 2021) for primary school students was designed to help students connect with the refugee experience by designing shoes for children using resources likely to be available in a refugee camp. Using a four-phase problem-based approach of research, investigate, design, and evaluate and communicate, teachers used a range of stimulus material to engage the students with the experience of refugees and their plight.

By exploring the 12 sets of data presented in *If World were a Village of 100 People* (Smith & Armstrong, 2020), primary school students can discover that life in other nations is often very different from their own. By reducing the world population to 100 people, they can more easily compare data about nationalities, schooling and literacy, languages spoken in the home, availability of electricity and running water, and age distributions, among others. Raising awareness of poverty and availability of resources helps to promote students' questions about equitable access to resources and poverty levels in other nations. Changing the parameters of the situation by considering the population of Australia or of the school as 100 people, further helps to develop a sense of community and an understanding of difference. By providing relevant data, students are encouraged to ask their own questions. See also a further suggestion from Cotton (2013) who proposes teachers could "imagine the world as represented by the class you teach" (p. 102).

In her classroom-based research, English has used a range of design challenges with primary school students to develop their knowledge and understanding of using data to address real-world issues of climate change and renewable energy. For example, English (2022) worked with grade 5 students who used a quantitative modelling activity to determine the evacuation times of villagers living 10 kms downhill from an erupting volcano. Further, English (2023) worked with grade 6 students who were tasked with designing a sustainable town with at least 50% renewable energy sources—see further information about this challenge in Chap. 11 of this book.

The *Population Education* website (populationeducation.org) contains many rich tasks and teaching ideas for mathematics and climate change at various grade levels. For example at grades 6–8, *People and Climate Change* (https://populationeducation.org/resource/people-and-climate-change-the-data-is-in/) allows students to examine a range of representations of data that enable them to identify relationships between population growth, greenhouse gas emissions, temperature rise, land ice melt and sea level rise.

The *Sustainable Development Goals* (SDGs) (UN, 2015, https://sdgs.un.org/goals) outline critical issues to be considered in classrooms addressing responsible

Fig. 14.1 The 17 global goals for sustainable development promoted by the OECD (https://www.oecd.org/dac/sustainable-development-goals.htm)

citizenship (see Fig. 14.1). Oxfam (2019) discusses the importance of developing global citizenship and recommends using the SDGs to develop knowledge and understanding of "social justice and equity, identity and diversity, globalisation and interdependence, sustainable development, peace and conflict, human rights, and power and governance" (p. 8). Advice is offered about using many of the school subject areas in tasks including mathematics, science and technology. One recommended task is to collect data on daily water consumption or energy use and consider ways to reduce consumption, set targets and monitor progress. I worked with teachers in the STEM Academy (Anderson et al., 2019) who encouraged their students to investigate the sustainability of their school. Grade 8 students used the school's invoices to determine the average daily water consumption and energy use of the school, followed by a recommendation to the principal of strategies to reduce consumption.

In their recently released book, Burkhardt et al. (2024) define mathematical literacy as the ability to "use one's mathematical capabilities to better understand the world, its structures and events, and with that understanding to make better-informed decisions" (p. 5). They present a range of scenarios for students to investigate using data including risk taking, climate change, equity, and wealth. While some focus on personal decision making, others relate to more national and global issues suitable for secondary school students. For example, they describe the use of a task from the Mathematics Assessment Project (UK) called *Muddying the Waters* (http://map.mathshell.org) to support the development of students' critical thinking by modelling changes to the pattern of pollution when a dam is built near a factory. This Shell Centre Resource from the University of Nottingham is a rich source of mathematical problem solving tasks and lessons suitable for investigating a range of responsible citizenship issues.

New Contributions to Our Understanding of Responsible Citizenship Education Through Mathematics in STEM

In this section of the book, the editors have included four chapters by authors who have recently investigated approaches to developing responsible citizenship education through integrated STEM approaches. Different perspectives are presented with Hilton and Hilton (2024) focusing on citizenship education in democratic societies, Rosa and Orey (2024) elaborating a more theoretical stance of ethnomodelling as a pedagogical action in responsible citizenship education, Steffensen (2024) using the refugee crisis to demonstrate the power of engaging students with real-world issues for responsible citizenship and Maass and Straser (2024) focusing on supporting teachers as they design interdisciplinary STEM tasks for responsible citizenship education.

Positioning their paper within the context of an ever-changing world of geopolitical instability, Hilton and Hilton (2024) argue the availability of information and vast amounts of data means students need to develop STEM literacy to be able to critically assess claims for active, engaged citizenship. By defining STEM education in terms of both disciplinary and cross-disciplinary approaches, they highlight the importance of subject-matter knowledge as well as developing the capacity to solve problems involving knowledge from across the STEM subjects, particularly to solve problems associated with ethical, moral, cultural and political issues. The authors outline the importance of students developing each of the separate STEM literacies as well as STEM literacy, particularly for participation in a democratic, equitable society promoting social activism and individual and collective agency. An activity from the Population Education website is used to demonstrate how to engage prospective and practising teachers with many social and environmental issues. The final section of the chapter provides the authors reflections on the challenges of implementing such approaches for teachers, school leaders, policy makers, and teacher educators.

A more theoretical stance is adopted by Rosa and Orey (2024) as they describe ethnomodelling through STEM education to help students develop ways of thinking, reasoning, inferring, modelling and developing mathematical knowledge in socio-cultural contexts. By developing as critical and reflective citizens, students learn about local cultural practices and links to systems from multiple worldviews. The challenge for teachers is to balance the traditional curriculum with making connections to local communities and students' cultural backgrounds. Describing STEM as a meta-discipline, the authors draw on the literature to propose STEM education can become a cohesive and holistic approach to provide students with new opportunities to solve daily problems using data and evidence-based reasoning. Rosa and Orey present two examples of ethnomodelling for students that highlight the importance of mathematics in the design and construction of cultural artefacts.

In the chapter by Steffensen (2024), the author works with her prospective teachers to develop lessons where grade 7 students examine stimulus material and model data to learn about the refugee crisis in Europe as well as worldwide. The approach emphasises the use of mathematics to analyse and interpret data so that students become critical, responsible citizens as they develop their understanding about a significant social and global issue. From a critical mathematics education perspective, Steffensen argues the task helps to empower students socially and politically, thus supporting the development of responsible citizens. The students demonstrated empathy for the refugees' plight advocating in their final presentations for all governments to take responsibility and accept refugees.

Finally Maass and Straser (2024) describe a project where they worked with researchers and teachers in six countries to design STEM tasks for responsible citizenship education. By using a combination of mathematical modelling, socio-scientific issues, and inquiry-based learning, the authors designed a professional development program for teachers that was adapted to accommodate differences in each participating country. The aim was to use the program at scale by involving more stakeholders from across education systems. Participating teachers implemented tasks in their schools and completed questionnaires at the beginning and end of the program. While there was evidence that teachers' self-efficacy beliefs improved, longer professional development programs with greater support were more successful.

The four chapters in this section provide evidence of successful implementation of integrated STEM curriculum problems and activities that can have a positive impact on students' responsible citizenship development. All tasks used mathematics in significant ways to demonstrate the power and potential of using mathematics as a focus in integrated STEM education.

References

Anderson, J., & Kriesler, A. (2018). Making maths matter: Engaging students from low SES schools through social justice contexts. In E. Bergqvist, M. Osterholm, C. Granberg, & L. Sumpter (Eds.), *Proceedings of the 42nd conference of the International Group for the Psychology of Mathematics Education* (Vol. 2, pp. 35–42). PME.

Anderson, J., Wilson, K., Tully, D., & Way, J. (2019). Can we build the wind powered car again? Students' and teachers' responses to a new integrated STEM curriculum. *Journal of Research in STEM Education, 5*(1), 20–39. https://doi.org/10.51355/jstem.2019.61

Anderson, J. (2024). Approaches to analysing the mathematics in integrated STEM tasks. In J. Anderson, & K. Makar (Eds.), *The contribution of mathematics to school STEM education: Current Understandings*. Springer.

Andreotti, V. (2015). Global citizenship education otherwise: Pedagogical and theoretical insights. In A. Abdi, L. Shultz, & T. Pillay (Eds.), *Decolonizing global citizenship education* (pp. 221–230). Sense Publishers.

Australian Curriculum Assessment and reporting Authority (ACARA). (2020). *The shape of the Australian curriculum, Version 5.0.* ACARA.

Balka, D. (2011). *Standards of mathematical practice and STEM*. School Science and Mathematics Association.

Barwell, R., Boylan, M., & Coles, A. (2022). Mathematics education and the living world: A dialogic response to a global crisis. *The Journal of Mathematical Behaviour*, 68, 1–10. https://doi.org/10.1016/j.jmathb.2022.101013

Burkhardt, H., Pead, D., & Stacey, K. (2024). *Learning and teaching for mathematical literacy: Making mathematics useful for everyone*. Routledge.

Bybee, R. W. (2018). *STEM education: Now more than ever*. National Science Teachers Association.

Coles, A., Barwell, R., Cotton, T., Winter, J., & Brown, L. (2013). *Teaching secondary mathematics as if the planet matters*. Routledge.

Cotton, T. (2013). Starting from handling data. In A. Coles, R. Barwell, T. Cotton, J. Winter, & L. Brown, *Teaching secondary mathematics as if the planet matters* (pp. 98–107). Routledge.

Cowie, B., & Mildenhall, P. (2021). Exploring STEM learning in primary classrooms: In support of social justice agendas. In A. Berry, C. Buntting, D. Corrigan, R. Gunstone, & A. Jones (Eds.), *Education in the 21st century: STEM creativity and critical thinking* (pp. 49–62). Springer.

English, L. D. (2022). Fifth-grade students' quantitative modeling in a STEM investigation. *Journal for STEM Education Research, 5*(2), 134–162. https://doi.org/10.1007/s41979-022-00066-6

English, L. D. (2023). Sixth-grade students as sustainable town designers. *Research in Integrated STEM Education, 1*(2), 283–315. https://doi.org/10.1163/27726673-bja00009

Erduran, S., Oxdem, Y., & Park, J. Y. (2015). Research trend on arguementation in science education: A journal content analysis from 1998–2014. *International Journal of STEM Education, 2*(5), 1–12. https://doi.org/10.1186/s40594-015-0020-1

Fielding-Wells, J., Dole, S., & Makar, K. (2014). Inquiry pedagogy to promote emerging proportional reasoning in primary students. *Mathematics Education Research Journal, 26*(1), 47–77.

Finkel, L. (2018). Infusing social justice into the science classroom: Building a social justice movement in science education. *The Journal of Educational Foundations, 31*(1&2), 38–556.

Forgaz, H., Bleazby, J., & Sawatzki, C. (2015). Ethics and the challenges for inclusive mathematics teaching. In A. Bishop, H. Tan, & T. N. Barkatsas (Eds.), *Diversity in mathematics education: Towards inclusive practices* (pp. 147–165). Springer.

Frankenstein, M. (1990). Incorporating race, gender, and class issues into a critical mathematics literacy curriculum. *Journal of Negro Education, 59*(3), 336–347.

Fuentes, R., Galeotti, M., Lanza, A., & Manzano, B. (2020). COVID-19 and climate change: A tale of two global problems. *Sustainability, 12*(8560), 1–14. https://doi.org/10.3390/su12208560

Gallay, E., Flanagan, C., & Parker, B. (2021). Place-based environmental civic science: Urban students using STEM for public good. *Frontiers in Education, 6*, 693455. https://doi.org/10.3389/feduc.2021.693455

Geiger, V., Forgasz, H., & Goos, M. (2015). A critical orientation to numeracy across the curriculum. *ZDM, 47*(4), 611–624.

Goos, M., Geiger, V., & Dole, S. (2014). Transforming professional practice in numeracy teaching. In Y. Li, E. Silver, & S. Li (Eds.), *Transforming mathematics instruction: Multiple approaches and practices* (pp. 81–102). Springer International Publishing. https://doi.org/10.1007/978-3-319-04993-9_6

Gutstein, E. (2003). Teaching and learning mathematics for social justice in an urban, Latino school. *Journal for Research in Mathematics Education, 34*, 37–73.

Gutstein, E., & Peterson, B. (Eds.). (2013). *Rethinking mathematics: Teaching social justice by the numbers*. Rethinking Schools Ltd.

Hilton, A., & Hilton, G. (2024). The importance of STEM education to responsible citizenship within a democracy. In J. Anderson & K. Makar (Eds.), *The contribution of mathematics to school STEM education: Current understandings*. Springer.

Hobbs, L., & Kelly, H. (2020). STEM into industry: Brokering partnerships between schools and local industry. *Curriculum Perspectives, 40*(2), 247–255.

Kennedy, T. K., & Odell, M. R. L. (2014). Engaging students in STEM education. *Science Education International, 25*(3), 246–258.

Kirschner, P. A., Sweller, J., & Clark, R. E. (2006). Why minimal guidance during instruction does not work: An analysis of the failure of constructivist, discovery, problem-based, experiential, and inquiry-based teaching. *Educational Psychologist, 41*(2), 75–86.

Larkin, K., & Lowrie, T. (2022). *STEM Education in the early years: Thinking about tomorrow.* Springer Monographs.

Leonard, J., & Moore, C. M. (2014). Learning to enact social justice pedagogy in mathematics classrooms. *Action in Teacher Education, 36*, 76–95.

Maass, K., & Straser, O. (2024). Developing teachers' capacity to implement interdisciplinary STEM tasks specifically targeted to responsible citizenship education. In J. Anderson, & K. Makar (Eds.), *The contribution of mathematics to school STEM education: Current Understandings.* Springer.

Maass, K., Geiger, V., Romero Ariza, M., & Goos, M. (2019). The role of mathematics in interdisciplinary STEM education. *ZDM, 51*(6), 869–884.

Makar, K., Fry, K., & English, L. (2023). Primary students' learning about citizenship through data science. *ZDM–Mathematics Education, 55*(5), 967–979.

Matthews, T. (2019). *To be equal and free: The nexus between human rights and democracy.* Heinrich Boll Foundation.

McDonald, C. V. (2016). STEM education: A review of the contribution of the disciplines of science, technology, engineering and mathematics. *Science Education International, 27*(4), 530–569.

Nicol, C., Bragg, L. A., Radzimski, V., Yaro, K., Chen, A., & Amoah, E. (2019). Learning to teach the M in/for STEM for social justice. *ZDM, 51*, 1005–1016. https://doi.org/10.1007/s11858-019-01065-5

Nicol, C., Thom, J. S., Doolittle, E., Glafield, F., & Ghostkeeper, E. (2023). Mathematics education for STEM as place. *ZDM, 55*, 1231–1242. https://doi.org/10.1007/s11858-023-01498-z

Nusche, D., Rabella, M. F., & Lauterback, S. (2024). *Rethinking education in the context of climate change: Leverage points for transformative change. OECD Education Working Paper No. 307.* OECD.

Orey, D. C., Rosa, M., & Soares, R. B. (2020). Mathematical modelling as a learning environment to transform a street activity into a sport practice. In G. A. Stillman, G. Kaiser, & C. E. Lampen (Eds.), *Mathematical modelling education and sense-making* (pp. 199–207). Springer International Publishing. https://doi.org/10.1007/978-3-030-37673-4_18

Oxfam. (2019). *The sustainable development goals: A guide for teachers.* Oxfam.

Roehrig, G. H., Dare, E. A., Ellis, J. A., & Ring-Whalen, E. (2021). Beyond the basics: A detailed conceptual framework of integrated STEM. *Disciplinary and Interdisciplinary Science Education Research, 3*(1), 1–18. https://doi.org/10.1186/s43031-021-00041-y

Rosa, M., & Orey, D. (2018). STEM education in the Brazilian context: An ethnomathematical perspective. In R. Jorgensen, & K. Larkin (Eds.), *STEM education in the junior secondary* (pp. 221–248). https://doi.org/10.1007/978-981-10-5448-8_11

Rosa, M., & Orey, D. (2024). Contributions of the pedagogical action of ethnomodelling to STEM education. In J. Anderson & K. Makar (Eds.), *The contribution of mathematics to school STEM education: Current Understandings.* Springer.

Skovsmose, O. (2021). Mathematics and crises. *Educational Studies in Mathematics, 108*, 1–15. https://doi.org/10.1007/s10649-021-10037-0

Smith, D. J., & Armstrong, S. (2020). *If the world were a village of 100 people: A book about the world's people* (2nd ed.). Kids Can Press.

Steffensen, L. (2024). Developing responsible citizenship through integrated STEM education activities. In J. Anderson, & K. Makar (Eds.), *The contribution of mathematics to school STEM education: Current understandings.* Springer.

Thibaut, L., Knipprath, H., Dehaene, W., & Depaepe, F. (2018). The influence of teachers' attitudes and school context on instructional practices in integrated STEM education. *Teaching and Teacher Education, 71*, 190–205. https://doi.org/10.1016/j.tate.2017.12.014

United Nations. (2015). *Transforming our world: The 2030 agenda for sustainable development.* UN

Wager, A. A., & Stinson, D. W. (Eds.). (2012). *Teaching mathematics for social justice: Conversations with educators.* National Council of Teachers of Mathematics.

Wright, P. (2016). Social justice in the mathematics classroom. *London Review of Education, 14*(2), 104–118.

Yemini, M., Engel, L., & Ben Simon, A. (2023). Place-based education: A systematic review of literature. *Educational Review*, pp. 1–21, https://doi.org/10.1080/00131911.2023.2177260

Zembylas, M. (2005). Science education: For citizenship and/or for social justice? *Journal of Curriculum Studies, 37*(6), 709–722.

Chapter 15
The Importance of STEM Education to Responsible Citizenship Within a Democracy

Annette Hilton and Geoff Hilton

Abstract The world currently faces challenging problems, such as climate change, food and water insecurity, and mass displacement of people. All such challenges are influenced by ethical, moral, cultural, and political factors. To engage with these challenges as both individuals and members of society, it is important that we are equipped with the skills and knowledge needed to make sense of them and to make decisions related to them. In today's digital world, it can be difficult to wade through ever-increasing volumes of information and data and to be able to do this in a critical and informed way, and yet, as citizens in a democratic society, this is essential if we are to evaluate policies and hold our elected representatives to account. The important roles played by schools and STEM educators in developing future citizens with the capabilities needed to engage in these ways cannot be overstated. In this chapter, we focus on the importance of STEM education and the centrality of mathematics education within STEM to develop and strengthen our capacities to participate in democratic society. We consider the challenges and opportunities for STEM educators and describe how mathematics might be foregrounded with STEM education.

Keywords Democracy · Citizenship education · STEM literacy · Critical and creative thinking

A. Hilton (✉) · G. Hilton
University of Technology Sydney, Ultimo, NSW, Australia
e-mail: annette.hilton@uts.edu.au

The University of Queensland, Saint Lucia, QLD, Australia

G. Hilton
e-mail: g.hilton@uq.edu.au

J. Anderson and K. Makar (eds.), *The Contribution of Mathematics to School STEM Education*, https://doi.org/10.1007/978-981-97-2728-5_15

Introduction

The world faces significant challenges, including climate change, food and water security, habitat loss, geopolitical instability, mass displacement of people, and disease management. These issues are influenced by ethical, moral, economic, political, and cultural factors and impact on social justice, human rights, and stewardship of the planet (Maass et al., 2022; McNamara et al., 2023). Society needs the skills and knowledges to understand the information and claims related to such challenges to be able to address them, while also respecting fundamental values of freedom and democracy (Maass et al., 2019). Understanding the information citizens encounter is a prerequisite for their capacity for active citizenship and engagement in political issues, which are a "trademark of a healthy democracy" (Bredberg, 2020, p. 517). In today's digital world, the volume and availability of information is increasing and much of it involves claims based on mathematics. This has implications for mathematics education and STEM education more broadly. In this chapter we focus on the importance of STEM education, and mathematics in particular, to build future citizens' capacity to participate actively and critically in democratic society.

Policy agendas across the world emphasise the need for education to develop young people's capacity for responsible citizenship. For example, the Paris declaration in 2015 identified the need to.

- ensure young people acquire social, civic, and intercultural competences, by promoting democratic values and fundamental rights, social inclusion, and non-discrimination, as well as active citizenship; and
- enhance critical thinking and media literacy, particularly in the use of the Internet and social media, to develop resistance to discrimination and indoctrination. (European Commission, 2016).

These goals align with the key arguments in this chapter. First, there is a need to develop young people's capacity for active citizenship within a democracy and second, it is important to ensure that young people have the skills and literacies to make sense of the vast quantities of information with which they are confronted to make informed and critical decisions.

Schools play an essential role in educating young people to build their critical and creative thinking skills and their capacity for engaging as active citizens (Riddle et al., 2022). We argue that the development of students' STEM literacies makes an essential contribution to the skill set needed by students to function in a democratic society as active and informed citizens who can engage critically with political and societal issues. The chapter begins with a discussion of STEM education, arguments for integrated STEM versus stand-alone STEM disciplines, and the importance of STEM literacies. Later in the chapter, we discuss the importance of mathematics within STEM and the contribution of STEM to the strength of democracy before discussing challenges and opportunities for STEM educators.

What is STEM Education?

In this section we define STEM and STEM education. The acronym STEM refers to the disciplines of science, technology, engineering, and mathematics and there are many definitions of *STEM education*. For example, the *National STEM School Education Strategy 2016–2026* (Education Council, 2015) defines STEM education as.

- the teaching of the disciplines of science, technology, engineering, and mathematics; and
- a 'cross-disciplinary approach to teaching' with the goals of increasing student interest in STEM and improving students' problem solving and critical thinking skills.

This definition reflects the different views of STEM education in the research literature: that *STEM education* can refer to both individual STEM disciplines and an integrated or interdisciplinary approach.

Integrated STEM Education

There are arguments for and against the integration of the STEM disciplines and debate exists about the contribution of each STEM discipline and the ways they interconnect. For example, the role of mathematics in STEM can be viewed as foundational to science, technology, and engineering; an equal partner; or even an underappreciated silent element (Baldinger et al., 2020; Just & Siller, 2022). This debate flows into the teaching of each discipline in schools, and how, when, or why they might be integrated. Although Maass et al. (2019) argued there is no broad agreement on whether the term 'STEM education' refers to learning individual disciplines or an interdisciplinary approach, arguments for STEM education using an integrated approach have been prevalent internationally for some time (Millar, 2020). There are also strong arguments for the inclusion of arts and humanities in a broadened approach known as STEAM (see Perales & Aróstegui, 2021). While including the arts and humanities has important educational, social, and democratic implications (Perales & Aróstegui, 2021), a focus on STEAM education is beyond the scope of this chapter.

Integrated STEM education engages students in authentic contexts, motivating and interesting them, and provides opportunities to develop an understanding of the nature of the disciplines that comprise STEM, the relationships among them, and the interdisciplinary nature of real-world problems (Just & Siller, 2022). In a rapidly changing world, many students will likely have future jobs not yet in existence—or even imagined. Hence, there is an argument that students need STEM knowledges and skills to participate in workplaces of the future (see Maass et al., 2019; Tytler, 2020). It has also been argued that an interdisciplinary approach closely aligns with the ways in which the STEM disciplines are used in contemporary settings, such as work and research (Tytler et al., 2019). It is impossible to address the challenges of climate change; food, water, and energy security; poverty; pandemic management; and waste reduction without applications of STEM and especially without integrated applications of mathematics and technology, such as mathematical modelling, statistics, and data science (Reynante et al., 2020).

Integrated STEM education does not require that all four disciplines always be integrated. It has been described as the integration of at least two of the disciplines of STEM (Jackson et al., 2021) and as an approach that engages students in the big ideas of at least two of the STEM disciplines to collaboratively address real-world problems (Moore et al., 2020; Roehrig et al., 2021). These views emphasise that STEM differs from its constituent disciplines by addressing contexts and problems that cannot be solved by any one discipline in isolation. At the same time, it is important to recognise that each discipline is distinct, with its own knowledge structures, methods, theories, practices, and skills (Millar, 2020; Reynante et al., 2020) and because of this, it is argued that there is merit in learning each separately. Even so, Steen and Madison (2011) lamented that the focus on traditional mathematics in secondary school "designed to support science (aka STEM) disciplines, does little to help high school graduates become quantitatively literate citizens" (pp. 2–3) and Reynante et al. (2020) observed that the lines between science, pure mathematics, and technology are becoming increasingly blurred.

Not all students will work in STEM-related fields focused on solving STEM-related challenges and yet, all work and workplaces are influenced by factors such as "globalisation, technological progress, and demographic change" (OECD, 2017, p. 2). Regardless of occupation, these factors impact *all* citizens, and it is therefore imperative to equip them with the capacity to act and think critically and logically, to make informed social, ethical, and moral decisions, and to navigate the immense volume of data and claims that are made using and related to STEM. This requires them to have the knowledges, skills, and competences of STEM literacies.

The Literacies of the STEM Disciplines

In this section we present descriptions of the literacies of the individual STEM disciplines before considering the meaning of *STEM literacy*.

Scientific Literacy

The definition of scientific literacy has been debated, redefined, and expanded for decades. The American Association for the Advancement of Science (AAAS, 1989) described scientific literacy as multifaceted and recognised the interdependence of science, mathematics, and technology. In 2003, the OECD (2003) defined scientific literacy as "the capacity to use scientific knowledge, to identify questions and to draw evidence-based conclusions in order to understand and help make decisions about the natural world and the changes made to it through human activity" (p. 133). Unlike the description from the AAAS, this description emphasises scientific knowledge without recognising the interdependence of science on other disciplines, especially mathematics. A central element of scientific literacy absent from these definitions is an understanding of the nature of science [NOS], which not only enhances students' understanding of science but of the ways in which scientists work. NOS focuses on an understanding of scientific inquiry and practices and the methods through which scientific knowledge is developed (National Science Teachers' Association [NSTA], 2020). It assists citizens to make informed decisions about scientifically-based issues and is essential for building public trust in science and in scientists—something that has been challenged during the recent pandemic and the ongoing climate crisis. This stance aligns closely with the argument that all members of society need scientific literacy to be able to participate in democracy and decision making about global issues (Valladares, 2021).

Technology and Engineering Literacy

According to the National Assessment Governing Board (NAGB, 2020), it is easier to describe technology and engineering literacies by first considering what is meant by *technology* and *engineering*. Technology (as opposed to digital technology, which is a subset) is defined as "any modification of the natural world to fulfil human needs or desires" (NAGB, 2020). Technology focuses on product (Hoepfl, 2020) and encompasses diverse outputs from pen and paper to satellites and the internet, including the infrastructure needed to design and manufacture technological outputs. In contrast, engineering focuses on process—engineers may not be directly involved in manufacturing outputs, but they develop and design the objects, processes, and systems that meet human needs and desires.

According to Jackson et al. (2021), there is no globally agreed definition of engineering literacy; however, the NAGB (2020) described *technology and engineering literacy* as the capacity to use, understand, and evaluate technology and to understand technological principles and strategies that are required to develop solutions and achieve goals. This literacy involves the mastery of a set of tools needed to participate intelligently and thoughtfully in society (NAGB, 2020). From this perspective, it has much in common with scientific and mathematical literacy and much to contribute to responsible citizenship and participation in democratic society.

Mathematical Literacy

Mathematical literacy, also known as quantitative literacy or numeracy, involves the use of mathematics in context. Geiger et al. (2015) described numeracy as the "knowledge and capabilities required to accommodate the mathematical demands of private and public life, and to participate in society as informed, reflective, and contributing citizens" (p. 531). This definition identifies the key elements of numeracy as mathematical dispositions, knowledges and skills, and competence in using mathematical tools and interpreting representations within personal, social, work, and societal contexts. It also highlights the critical orientation that underpins the capacity for numerate individuals to evaluate issues and situations and to respond accordingly (Maass et al., 2019). This critical orientation is central to the importance of mathematical literacy in strengthening democracy. As emphasised by Bredberg (2020), without critical thinking, a person is unlikely to be able to make use of facts and will "be more inclined to be influenced by propaganda" (p. 519). Mathematical literacy allows individuals to use, evaluate, and interpret mathematical findings and claims for making decisions in personal and work situations, and about political and global issues (Bredberg, 2020).

STEM Literacy

Before discussing the meaning of *STEM literacy,* it is useful to consider the meaning of the term *literacy* more broadly. The OECD (n.d.) describes literacy as a continuum of learning that enables individuals to achieve goals, develop knowledge and potential, and participate in society. Current global challenges require high levels of literacy to be able to navigate an increasingly complex world. According to Zollman (2012) literacy is deictic—constantly changing; for example, as technologies and languages develop, our ways of being literate adapt and change. This argument applies equally to the literacies of the individual STEM disciplines and also to STEM literacy, especially with the increasing emergence of new technologies and the diversification of applications of STEM.

A common theme among the definitions of the literacies of the STEM disciplines is the need for knowledge, skills, and dispositions to equip individuals to understand information and engage critically with issues in a range of contexts. An understanding of the individual disciplinary literacies contributes to a deeper understanding of the nature of STEM literacy. STEM literacy is more than the sum of the four component silos of STEM (Jackson et al., 2021; Zollman, 2012); however, the descriptions of the literacies of the STEM disciplines reflect their close interrelationships. Zollman (2012) argued that STEM literacy should highlight the literacies of each of the STEM disciplines and focus on their use to address personal, societal, and economic needs. These ideas are incorporated in the definition by Mohr-Schroeder et al. (2020, p. 33):

> STEM literacy is a dynamic process and ability to apply, question, collaborate, appreciate, engage, persist, and understand the utility of STEM concepts and skills to provide solutions for STEM-related personal, societal, and global challenges that cannot be solved using a single discipline'.

To this definition we would add that STEM literacy includes the capacity to understand and use the ideas, knowledge, processes, and skills from relevant disciplines to evaluate and justify claims and to make informed decisions. It allows citizens to better determine when and how to rely on or trust the solutions that are provided by STEM and the arguments for and against them (consider the debates around vaccinations, stem cell research, genetic modification, artificial intelligence).

The Importance of STEM to Democracy

The importance of STEM education is well-established in the research literature (e.g., Tytler, 2020) and in global policies (e.g., OECD, 2017). As emphasised already, we face societal, political, financial, technological, and environmental challenges on local, national, and global levels and clearly, STEM-based solutions will be necessary to address these. While not everyone will play a direct role in solving these problems, the central need for STEM education and STEM literacies, and mathematical literacy in particular, centres on equipping all citizens to engage in discussion and decision-making about the ethical, moral, social, or cultural aspects of such issues and their solutions (Maass et al., 2019). The very concept of democracy relies on an informed and educated population (Kolber & Heggart, 2022) and an emphasis on critical thinking and capacity for informed decision-making equips students to actively engage in democracy as adults (Paterson & Gavrin, 2022). Thus, having strong STEM literacies is a crucial pillar that empowers people in their personal and civic lives and for participation in democracy.

Almost 30 years ago, Bailey (1996) noted that citizens were increasingly drowning in data, predicting an unimaginable increase in the amount of data that would be available as computers became more powerful. Steen (2001) argued that there exist strong links between mathematics, numeracy, and democracy. Reflecting on the need for mathematical literacy education, Steen stated that, "quantitative literacy empowers people by giving them tools to think for themselves, to ask intelligent questions of experts, and to confront authority confidently. These are skills required to thrive in the modern world" (p. 2). Bredberg (2020) also argued that mathematical literacy is central to developing the capacity to ask questions and critique the vast amounts of mathematical information for validity and reasonableness. In 2001, Steen described the world as 'data-drenched'. More than two decades later, we increasingly see all manner of data and mathematical representations being used to influence and manipulate opinions and persuade citizens about societal, economic, and political issues.

Recent events, including the pandemic, revealed issues that can only be addressed by a rethinking of education (Valladares, 2021). The UNESCO International

Commission on the Futures of Education (2020) emphasised the need for deep reflection on curriculum in response to the need to "struggle against the denial of scientific information and actively fight misinformation" (p. 6). UNESCO argued for increased integration of curricula that are based on problems facing the planet and humanity. This has implications for STEM educators. Given that the challenges facing the world are interconnected and inseparable from each other and that STEM will play an important role in addressing them, understanding how they relate to each of us requires some level of STEM literacy. Perhaps more importantly, as citizens in a democracy, we make choices about who to support in elections and how we can act responsibly—individually and within communities and societies—to address these challenges. As argued by Bredberg (2020), mathematical literacy is a pre-requisite for full and active participation in democratic processes.

Democracy and Threats to Democratic Society

In this section, we describe what we mean by *democracy* and describe some of the threats to strong democracies. We also argue that education is imperative as a means of protecting and strengthening democracy. The term democracy derives from Greek, *demos* (the people) and *krati* (power/rule). In short, it means 'rule by the people' (Bredberg, 2020, p. 518). According to the United Nations (2009), "democracy is a universal value, based on the freely expressed will of the people to determine their own political, economic, social, and cultural systems and their full participation in their lives" (p. 1).

This century, there have been some bleak reminders that democracy and freedom are not to be taken for granted (Burridge & Buchanan, 2022) and researchers have identified numerous threats to democracy. Populist leaders manipulating the media and the truth have led the world into more uncertain times in what Burridge and Buchanan described as a "post-truth era" (2022, p. 81). Political science researcher Keane (2020) identified 'The New Despotism' and described the ways in which governments across the world, especially those with life-term despotic rulers, increasingly use technology to control their own populations and use multiple strategies, often involving technology, to breed global anxieties and threaten democracy. The increasing availability of information in mainstream and social media has made it complex for people to engage and make critical and informed decisions—and there have been some high-profile examples of the (near) derailment of democracy in recent years in prominent democratic countries. Kurlantzick (2013) identified a global decline in representative democracy and the *Freedom House Report* (2021) found that global freedom had declined for the 15th consecutive year based on an annual assessment of political rights and civil liberties.

In the 1990s, Mouffe warned that, "democracy is something uncertain and improbable and must never be taken for granted. It is an always fragile conquest that needs to be defended as well as deepened" (1993, p. 6). The need to value and defend democracy and freedom is not always obvious to those privileged to live in democratic countries, especially younger generations, who may not have experienced the major threats to their way of life that older generations experienced last century (e.g., during World War 1 and World War 2). Bredberg (2020) described the phenomenon of young people in well-functioning democracies taking freedom, and their rights to participate politically, economically, socially, and culturally in society, for granted and Foa and Mounk (2016) observed that young people are increasingly apathetic toward both active citizenship and democracy. Sanchez and Middlemass (2022) identified the explosion of misinformation that has been generated deliberately to erode the democratic process as a key driver of the erosion in public confidence in political systems. They called for an increased focus on developing students' media literacies to counter their susceptibility to misinformation. We argue that STEM literacies, particularly mathematical literacy, are also essential.

The Connection Between STEM Education and Democracy

Bybee (2018) argued for STEM education to develop competences such as the ability to use evidence and formulate arguments based on it, and to engage in civil discourse. From this perspective, people who are educated in STEM subjects have skills with which to interpret and evaluate contemporary issues. In no way can STEM education alone hope to give clarity to all issues, but it gives valuable insight and an entry point into understanding them. For example, a STEM educated person may not understand the minutia and complexity of immunology, but they understand the scientific process upon which advancements such as treatments, vaccinations, and diagnostic technologies are made. Similarly, not everyone has an in-depth understanding of the relationships among ocean currents, El Niño and La Niña, and climate change; however, STEM literate citizens are more likely to be able to engage critically with mathematical and scientific information and representations to make informed decisions about claims.

Valladares (2021) argued that science education needs an orientation that promotes social activism and individual and collective agency, an approach that is aligned with the *Bildung* tradition of northern and central Europe (see Sjöström & Eilks, 2021). According to Sjöström and Eilks, *Bildung*-oriented science education aims to develop students' capabilities for self-determination, participation in a democratic society, and to have empathy and solidarity with other members of society. While this work has focused on science education, these arguments also apply to STEM education and STEM literacies, especially to the role of mathematics in society and for addressing inequity.

Strong knowledge means emancipation from "other people's power" (Graumann, 2005, p. 20) and over the years, mathematics has been championed as essential to democracy. Mathematics education has the capacity to be emancipatory and is crucial for social empowerment and independent thought and decision making. Mathematical literacy is important for personal financial stability and security, health outcomes and well-being, and for personal and societal decision making. According to Frankenstein (2001, as cited by Maass et al., 2019), those at the margins of society are most susceptible to being disadvantaged because they lack the numeracy skills to detect bias or think critically about information as it is presented, whether politically or financially. Atweh (2012) envisioned mathematics education as increasing the capacity of such marginalised groups for democratic participation. Skovsmose and Valero (2001) also argued that mathematics education is a means of providing opportunity to all citizens and for addressing inequity in society and by extension its powerful capacity to promote democracy. These authors also cautioned that mathematics can sometimes be taken as certain and unquestionable, which can lead to it playing a more destructive role. We would counter this stance by arguing that a strong STEM education with a focus on mathematics literacy would equip citizens to be able to make more informed and critical judgments of data and claims related to mathematics (or STEM more broadly) and to take a more sceptical view of information as it is presented.

The following example illustrates the need for numeracy to critique the accuracy of the statements made. During the COVID-19 pandemic, a politician claimed there were more vaccinated people in hospital with COVID than unvaccinated, implying that vaccinations were not effective. The politician was correct from an absolute perspective (because the absolute number of hospitalised vaccinated patients was higher) but incorrect from a relative perspective because the *proportion* of vaccinated people in the population who were hospitalised was lower than that for unvaccinated people. In truth, the rate of people hospitalised and unvaccinated was twice that for those who were vaccinated. In this example, an understanding of proportional reasoning is needed to spot the flaw (or deliberate omission) in the politician's argument and undermine the politician's false narrative.

To this point, we have argued that the development of STEM literacies, particularly mathematical literacy are important elements of democratic participation, and we feel strongly that young people should be empowered to engage in decision making that leads to a more sustainable, ethical, and socially just world. In the following section, we describe a resource that we have used in many contexts to emphasise the importance of STEM for helping to address global problems and to build empathy and promote responsible citizenship.

A Resource for Foregrounding Mathematical and Other STEM Literacies

The resource is known as *Food for Thought* (downloadable at https://populationeducation.org/wp-content/uploads/2018/07/Food-for-Thought.pdf). It was developed by the Population Education organisation, which provides K-12 educators with lesson plans, teaching resources, and professional learning focused on human population growth and its impacts on the environment and human health and wellbeing.

According to Population Education (2023), the resource can be used to focus on demographics, economics, public health, public education, energy consumption, land use, urbanisation, wealth, and poverty. The statistics used in the resource are updated frequently from sources such as the United Nations and World Bank.

We have used this resource for many years with practising teachers, school students, and preservice teachers to develop numeracy across the curriculum and build an understanding of the potential for STEM education to address some important global challenges. It is useful for discussing the distribution of international aid and the relative contributions to that aid by the world's wealthiest countries, and the impacts of inequity, immigration, and refugee crises. The activities engage students in using statistics and proportional thinking to understand global demographics, the impact of population growth on the environment, and quality of life indicators such as health, wealth, resource use, and education. The learning experiences allow students to engage collaboratively in applying STEM knowledge, and especially numeracy within an authentic context.

To begin the activity, circles made from string are placed on a classroom floor to represent the relative land areas of Northern America, Latin America (including Mexico), Europe (including Russia), Africa, and Asia (including the Middle East). Table 15.1 summarises the set-up for a class of 25 students.

Students are placed in the circles in numbers relative to the population of each region (as shown in Table 15.1). Depending on the ages of the students, initial questions can be as simple as comparing different populations, for example, 'the ratio of North America's population to that of Asia is 1:15' or 'there are five times as many people in Asia as there are in Europe'. This focus allows further discussion of population density, relative population distribution, and the issues that may arise from overcrowding. The 'citizens' of each region are asked to use another piece of string to represent the proportion of their region that is arable. At this point, the discussion might focus on inequalities and why disparities might exist. It might identify possible STEM applications to address the continued need for arable land to feed increasing

Table 15.1 Information for organising and discussing the food for thought activity

Region	North America	Latin America	Europe	Africa	Asia
Length of string (m)	7.6	7.6	8.2	9.1	9.4
Number of students	1	2	3	4	15
Percentage of world land area	14	15	17	23	24
Percentage of region's arable land	11	8	12	8	15
Energy consumption (one counter = equivalent of 1 barrel of oil)	53	10	27	3	10
Per capita income (one block = US $1000)	55	7	22	2	6

populations, for example, selective breeding of drought resistant crops or those with higher yield or greater nutritional value.

The teacher uses various props (see Table 15.1) to represent the relative per capita energy consumption and gross national income per capita (taking purchasing power parity into account). Students are asked questions about the relative wealth and energy use of different regions, which requires the use of proportional reasoning and engagement with multiple statistics. This can prompt a discussion about diverse topics, again depending on the age of the students and the teacher's chosen focus for the lesson. For example, which region uses resources at the highest rate? Why? Where are the regions without access to natural resources for energy production and how might they benefit from renewable energy—what are the possibilities and implications? How might STEM provide solutions?

Other materials within the resource include statistics about literacy rates, access to education by region and gender, fertility rates, infant mortality, life expectancy, access to improved sanitation, and provision of nurses and midwives. Invariably, the discussion turns to inequity and the unequal distribution of wealth and resources around the world, which can lead to questions about which regions might be able to provide more opportunities for immigration. Teachers can choose to focus on other global challenges such as environmental degradation; urbanisation and burgeoning informal settlements; spread of disease; or challenges associated with food, water, and energy security.

This activity is useful as the basis of an integrated STEM unit in which students learn about the contributions of STEM and its disciplines (both individually and together) in addressing the problems and challenges raised during the activity. It can also be used to engage students in making predictions about the future, for example if the population of each region doubled (the current doubling time for Asia's population is 70 years). A physical simulation where the number of students within the 'Asia circle' is doubled soon illustrates the impact of such a population increase on the other aspects and issues that have been discussed. Teachers might also discuss the uses of and assumptions of mathematical modelling and the ways in which data are gathered and used to make decisions and predictions in demography, the STEM fields, medicine, and so on.

When we have used this activity with practising and preservice teachers, their comments have indicated an increase in awareness of the ways students can be engaged in learning about global challenges from the perspective of different subject areas. They have also expressed an appreciation of the importance of STEM literacy and even when they teach outside the STEM subjects, there is potential to develop students' numeracy and other skills for active citizenship. Some elementary school teachers have redeveloped the resource to focus on school-level contexts (such as play areas, litter) or to focus on the Australian states and territories rather than global regions.

Because the activity prompts discussion about 'fairness' and equity, there are opportunities to broaden the focus to discuss responsible citizenship, not only on a personal basis but for the good of society and the planet. When students have a stronger understanding of the world, they are more likely to be empathetic and to appreciate the situations in which they live, which brings us back to democracy and the need to emphasise the place of education 'as a democratic necessity' (Kolber & Heggart, 2022, p. 79). We believe strongly in the capacity of STEM education for engaging students and in turn building their capacity for active citizenship democratic participation. However, there are also many challenges and implications if this is to be achieved in the complex contexts of schools.

The Challenges of Integrated STEM Education: Reflections as STEM Educators

Schools play an important role in nurturing democracy and developing the skills needed by future citizens in an increasingly challenging world (Heggart, 2021). Other authors in this book have highlighted the importance of mathematics within STEM education and its potential to build twenty-first century skills and responsible citizenship. If successful, this approach would equip students to interpret and critique information about a wide range of issues, distinguish fact from opinion, determine the reliability of information from various sources, and thus have the skills needed to value, defend, and engage in the democratic process.

If teachers are to provide STEM learning experiences so their students can develop strong STEM literacies, as suggested by many scholars (e.g., Anderson & Li, 2020; Bybee, 2018; Maass et al., 2019), there are challenges to be overcome and these have varied implications. Through our own research and our experience as teachers and teacher educators, we have identified that these challenges are threats to effective STEM education and by inference threats to students' opportunities to develop skills for democratic citizenship.

Challenges for Educators

Often what teachers want to do, and believe they should do, is overridden by forces beyond their control. Economic imperatives, government policy, national and international testing, and external assessment all put pressure on schools and teachers to follow a siloed and sometimes narrow curriculum. The complexity of educators' work and the influence of technology, parent and student expectations, administrative demands, and public perceptions all impact the decisions that teachers can and do make.

Burridge and Buchanan (2022) argued that education has become driven by economic imperatives, which has reduced the focus on the human and humanitarian purposes of education. While public expectations, economics, and preparation of a future workforce often drive educational policy, they are not the only important considerations and should not be the only drivers behind the creation and delivery of the curriculum. If students are to be critical thinkers who are equipped to make decisions about their own futures and those of others, to strive for equity and justice, and defend democracy, teachers need to be empowered to be agents of change (Burridge & Buchanan, 2022; Kolber & Heggart, 2022). This has implications for policy makers and school leaders as well as implications about the design, delivery, and assessment of the current curriculum.

Curriculum and Assessment Challenges

There is a need to support the capacity of teachers to engage students in STEM education. Challenges to achieving this relate to the structure of the curriculum, the nature of external assessment in secondary schooling, and high stakes tests, such as the National Assessment Program—Literacy and Numeracy (NAPLAN) in Australia. Teachers have become increasingly restricted in their capacity to diverge from a prescriptive curriculum or to teach it using integrated approaches. In our work over many years in Queensland, we have seen teachers' freedoms to choose how to teach the curriculum eroded by prescriptive and rigid departmental unit plans (known as Curriculum to Classroom in Queensland); walk-throughs by school leaders expecting all teachers of a given year level to be working on the same content on the same day; and by relentless demands to assess students (every five weeks in some primary schools we have worked with) rather than teach students for deep and connected understanding. This issue is further compounded when pressure is applied to schools to show constant improvement in national testing results and even more so when international test results, such as PISA, are published (Lilley, 2022).

While there may be underlying reasons for these imperatives, they do not auger well for the capacity of teachers to teach integrated curriculum or to engage students in relevant and authentic STEM-related contexts. This has implications for curriculum designers and for educational leaders. The Australian Curriculum is arranged in separate learning areas and does not always emphasise opportunities for cross-curricular connections or the achievement of more humanitarian goals. Within the current curriculum there are cross-curricular priorities, one of which is sustainability—an excellent topic through which to promote STEM literacies—and there are general capabilities, which if emphasised and foregrounded across the learning areas, have the potential to contribute to the skills needed for responsible citizenship, and by extension, democratic participation. While curriculum designers have attempted to illustrate the ways in which certain general capabilities might be integrated into the subject matter, a curriculum audit of any subject area reveals that these are not always authentic and, in some cases, opportunities are missed because they have not been made explicit.

School Organisation and Logistics

It is uncommon in secondary schools without a formal middle years of schooling program for students to have the same teacher for more than one subject. If a teacher teaches only one of the STEM subjects, it can be challenging for them to provide students with integrated STEM learning experiences that are authentic and related to real world issues. In addition, out-of-field teaching is an international phenomenon, which is especially prevalent in mathematics (Goos & Guerin, 2021). A recent Australian report showed that fewer than one in four students have a qualified mathematics teacher in each of their lower secondary years (O'Connor & Thomas, 2019). With these challenges, even with the best of intentions, teachers who are restricted to teaching one subject only, especially if they are out-of-field teachers, will find it difficult to engage students in integrated STEM education. This needs to be addressed at multiple levels, with implications for both initial and practising teacher education.

In primary schools, where teachers are responsible for delivering multiple learning areas to the same class of students, there are increasing demands for teachers to focus their time on literacy and numeracy, often to the exclusion of other subject areas. In some instances, schools have engaged STEM coaches to deliver science and digital technology curricula during the regular classroom teacher's non-contact time. In our research, we have observed that these experiences are often less about integrated STEM education and more about developing students' interest in technologies such as robotics and drones. Further, the 'regular' classroom teacher has little chance of making use of these STEM experiences because they are not present when the children are engaging in them. Of course, the situations described here vary from school to school and even classroom to classroom. The point to be made is that the current organisation of schools and the ways in which demands are placed on teachers are counter-productive to developing an authentic and integrated approach to STEM education, let alone one that has the strength and depth to develop much needed skills for responsible citizenship and democracy.

Challenges and Implications for Teacher Education

The fourth area that impacts strongly on teachers' capacity to deliver STEM education is the need to provide practising teachers with ongoing professional development, especially for teachers who do not teach (or are not qualified to teach) more than one STEM subject. This also has clear implications for initial teacher education [ITE]. In ITE programs, it is common for mathematics preservice teachers to have a second teaching area from within STEM; however, this is not always the case. For example, this second area may be psychology, history, economics, English, or a foreign language. The consequence is that for teachers with limited or no background in a second STEM discipline, it is difficult to understand the nature of integrated STEM education. This might be addressed within ITE programs by providing more opportunity for preservice teachers to develop experience or knowledge of how to embed integrated STEM education into their teaching, perhaps through an additional subject focused on STEM curriculum and pedagogy.

The situation might also be addressed by universities revising entrance requirements for mature-age graduate applicants changing careers. Applicants from STEM fields, such as engineers, doctors, or astronomers may not be admitted to the mathematics education program if entrance requirements stipulate certain areas of mathematics in their previous tertiary backgrounds. This fails to recognise the experience of these applicants in using mathematics throughout their careers. Having worked with teachers who have come from these fields, we strongly believe that denying entry to such applicants creates barriers for developing strong STEM education. These are the very teachers with a breadth of experience and a strong understanding of the essential role of mathematics for identifying and solving authentic STEM-related problems—and they have the capacity to bring these knowledges into their teaching and to assist their colleagues to do so.

Conclusion

This chapter has focused on STEM education, the centrality of mathematics within STEM education, and its importance for active citizenship with a democratic society. While today's democracies grapple with many important issues such as identity, inclusion, and equality, all of which help to make a country more democratic, our collective attention must also remain on the maintenance of democratic freedoms. Democracy can be threatened in many ways; for example, by the influence of authoritarian states, inaccurate claims and failure of the media to report objectively, or through citizens' apathy and inaction. Protection of a democracy relies on active citizens engaging critically and making sound electoral judgements.

The link between the maintenance of democracy and STEM education is that STEM education has the potential to develop students' capacity for critical, logical, and factual thought to counter manipulation. All citizens in a democracy require critical thinking skills to understand the enormous variety of data and information about societal and global issues as well the ability to determine the veracity of the sources and representations of information they use to make judgements and decisions about those issues.

If teachers are to be able to engage their students in meaningful STEM curriculum, there must be a recognition of the importance of developing STEM literacies and a focus on ensuring that all STEM teachers have some knowledge of the component STEM discipline, so that they can foreground the relevance of these other subjects within the one they're teaching. While challenges exist, there are also ways in which integrated STEM can be achieved. We have described one example of a resource that is useful for building mathematical literacy and foregrounding STEM literacies as well as skills for responsible citizenship.

STEM education is a key element in the creation of an educated citizenry. If we are to understand and address the issues facing society and the world, then we need to understand the ways in which STEM is 'done'. It is not possible to imagine solving global challenges without the contributions of mathematics, technology, science, or engineering. In essence, the ways in which the issues facing the world will be addressed can't be separated into one STEM discipline or another and none can exist without the others. While in many instances, STEM education is proposed as a solution to understanding and addressing global issues, the role of mathematics and mathematics literacy is often understated. In this chapter, we have argued that mathematics and mathematical literacy are central to the development of students' skills for active citizenship within a democracy with the goal of engaging with and addressing the complex and diverse issues facing our global community.

References

American Association for the Advancement of Science (AAAS). (1989). *Science for all Americans.* AAAS.

Anderson, J., & Li, Y. (2020). Investigating the potential of integrated STEM education from an international perspective. In J. Anderson & Y. Li (Eds.), *Integrated approaches to STEM education* (pp. 1–12). Springer Nature.

Atweh, B. (2012). Mathematics education and democratic participation between the critical and the ethical: A socially response-able approach. In O. Skovsmose & B. Greer (Eds.), *Opening the cage: Critique and politics of mathematics education* (pp. 325–342). Sense Publishers.

Bailey, J. (1996). *After thought: The computer challenge to human intelligence.* Basic Books.

Baldinger, E. E., Staats, S., Covington Clarkson, L. M., Gullickson, E. C., Norman, F., & Akoto, B. (2020). A review of conceptions of secondary mathematics in integrated STEM education: Returning voice to the silent M. In J. Anderson & Y. Li (Eds.), *Integrated approaches to STEM education, advances in STEM education* (pp. 67–90). Springer Nature.

Bredberg, J. (2020). The role of mathematics and thinking for democracy in the digital society. *Policy Futures in Education, 18*(4), 517–530.

Burridge, N., & Buchanan, J. (2022). Teachers as changemakers in an age of uncertainty. In K. Heggart & S. Kolber (Eds.), *Empowering teachers and democratising schooling* (pp. 81–100). Springer Nature.

Bybee, R. W. (2018). *STEM education: Now more than ever.* NSTA Press.

Education Council (2015). *National STEM school education strategy, 2016–2026.* www.educationcouncil.edu.au/site/DefaultSite/filesystem/documents/National STEM School Education Strategy.pdf

European Commission. (2016). *Promoting citizenship through education.* https://ec.europa.eu/newsroom/home/items/29496/default

Foa, R. S., & Mounk, Y. (2016). The danger of deconsolidation: The democratic disconnect. *Journal of Democracy, 27*(3), 5–17.

Frankenstein, M. (2001). *Reading the world with math: Goals for a critical mathematical literacy curriculum* (p. 53). Australian Association of Mathematics Teachers Inc.

Freedom House. (2021). *Freedom in the world 2021: Democracy under siege.* https://freedomhouse.org/report/freedom-world/2021/democracy-under-siege

Geiger, V., Goos, M., & Forgasz, H. (2015). A rich interpretation of numeracy for the 21st century: A survey of the state of the field. *ZDM, 47*(4), 531–548.

Goos, M., & Guerin, A. (2021). Insights into the pedagogical practices of out-of-field, in-field, and upskilled teachers of mathematics. In Y. H. Leong, B. Kaur, B. H. Choy, J. B. W. Yeo., & S. L. Chin (Eds.), *Excellence in mathematics education: Foundations and pathways* (Proceedings of the 43rd annual conference of the Mathematics Education Research Group of Australasia), pp. 203–210. Singapore: MERGA.

Graumann, G. (2005). General knowledge in mathematics education in a democratic society [Allmänbildning i matematikundervisning i ett demokratiskt samhälle]. *Utbildning & Demokrati, 14*(2), 17–26.

Heggart, K. (2021). *Activist citizenship education: A framework for creating justice citizens.* Springer Nature.

Hoepfl, M. (2020). Defining technological and engineering literacy. *Technology and engineering teacher*, 1–6.

Jackson, C., Mohr-Schroeder, M. J., Bush, S. B., Maiorca, C., Roberts, T., Yost, C., & Fowler, A. (2021). Equity-oriented conceptual framework for K-12 STEM literacy. *International Journal of STEM Education, 8*, 38. https://doi.org/10.1186/s40594-021-00294-z

Just, J., & Siller, H.-S. (2022). The role of mathematics in STEM secondary classrooms: A systematic literature review. *Education Sciences, 12*(9), 629. https://doi.org/10.3390/educsci12090629

Keane, J. (2020). *The new despotism.* Harvard University Press.

Kolber, S., & Heggart, K. (2022). Introduction. In *Empowering teachers and democratising schooling: Perspectives from Australia* (pp. 1–16). Springer Nature.

Kurlantzick, J. (2013). *Democracy in retreat: The revolt of the middle class and the worldwide decline of representative democracy.* Yale University Press.

Lilley, G. (2022). The demise of teacher expertise and agency by the "Evidence-Based Discourse." In K. Heggart & S. Kolber (Eds.), *Empowering teachers and democratising schooling* (pp. 43–59). Springer Nature.

Maass, K., Geiger, V., Ariza, M. R., & Goos, M. (2019). The role of mathematics in interdisciplinary STEM education. *ZDM, 51*, 869–884.

Maass, K., Sorge, S., Romero-Ariza, M., Hesse, A., & Straser, O. (2022). Promoting active citizenship in mathematics and science teaching. International Journal of Science and Mathematics Education, 20, 727–746.

McNamara, K. E., Clissold, R., Westoby, R., Stephens, S., Koran, G., Missack, W., & Bartlett, C. Y. (2023). Using a human rights lens to understand and address loss and damage. *Nature Climate Change.* https://doi.org/10.1038/s41558-23-01831-0

Millar, V. (2020). Trends, issues and possibilities for an interdisciplinary STEM curriculum. *Science and Education, 29*, 929–948.

Mohr-Schroeder, M., Bush, S. B., Maiorca, C., & Nickels, M. (2020). Moving toward an equity-based approach for STEM literacy. In C. Johnson, M. J. Mohr- Schroeder, T. Moore., & L. English (Eds.), *Handbook of research on STEM education* (pp. 29–38). Routledge.

Moore, T. J., Johnston, A. C., & Glancy, A. W. (2020). STEM integration: A synthesis of conceptual frameworks and definitions. In C. C. Johnson, M. J. Mohr-Schroeder, T. J. Moore, & L. D. English (Eds.), *Handbook of research on STEM education* (pp. 3–16). Routledge.

Mouffe, C. (1993). *The return of the political.* Verso.

NAGB. (2020). *Technology and engineering literacy.* https://www.nagb.gov/naep-subject-areas/technology-and-engineering-literacy/framework-archive/2014-technology-framework/toc/ch_0/technology_literacy.html

NSTA. (2020). *Nature of science.* https://www.nsta.org/nstas-official-positions/nature-science

O'Connor, M., & Thomas, J. (2019). *AMSI occasional paper 2: Australian secondary mathematics teacher shortfalls: A deepening crisis.* https://amsi.org.au/

OECD. (2017). *Future of work and skills.* Paper presented at the 2nd meeting of the G20 Employment Working Group. Hamburg, Germany.

OECD. (2003). *The PISA assessment framework—Mathematics, Reading, Science and problem solving knowledge and skills.* https://www.oecd.org/education/school/programmeforinternationalstudentassessmentpisa/33707226.pdf

OECD. (n.d.). *Adult literacy.* https://www.oecd.org/education/innovation-education/adultliteracy.htm

Paterson, C., & Gavrin, M. (2022). Teaching for democracy. In *Empowering teachers and democratising schooling: Perspectives from Australia* (pp. 101–110). Springer Nature.

Perales, F. J., & Aróstegui, J. L. (2021). The STEAM approach: Implementation and educational, social, and economic consequences. *Arts Education Policy Review.* https://doi.org/10.1080/10632913.2021.1974997

Population Education. (2023). *Food for thought.* https://populationeducation.org/resource/food-for-thought/

Reynante, B. M., Selbach-Allen, M. E., & Pimental, D. R. (2020). Exploring the promises and perils of integrated STEM through disciplinary practices and epistemologies. *Science and Education, 29,* 785–803.

Riddle, S., Heffernan, A., & Bright, D. (2022). Empowering teachers for education and democracy. In K. Heggart & S. Kolber (Eds.), *Empowering teachers and democratising schooling: Perspectives from Australia* (pp. vii-viii). Springer Nature.

Roehrig, G. H., Dare, E. A., Ring-Whalen, E., & Wieselmann, J. R. (2021). Understanding coherence and integration in integrated STEM curriculum. *International Journal of STEM Education, 8*(2). https://doi.org/10.1186/s40594-020-00259-8.

Sanchez, G. R., & Middlemass, K. (2022). *Misinformation is eroding the public's confidence in democracy.* https://www.brookings.edu/articles/misinformation-is-eroding-the-publics-confidence-in-democracy/

Sjöström, J., & Eilks, I. (2018). Reconsidering different visions of scientific literacy and science education based on the concept of Bildung. In Y. Dori (Ed.), *Cognition, metacognition, and culture in STEM education* (pp. 65–88). Springer.

Sjöström, J., & Eilks, I., et al. (2021). Reconsidering different visions of scientific literacy and science education based on the concept of *Bildung*. In Y. J. Dori (Ed.), *Cognition, Metacognition, and Culture in STEM Education* (pp. 65–88). Springer.

Skovsmose, O., & Valero, P. (2001). Breaking political neutrality: The critical engagement of mathematics education with democracy. In B. Atweh, H. Forgasz, & B. Nebres (Eds.), *Sociocultural research on mathematics education* (pp. 37–55). Lawrence Erlbaum Associates.

Steen, L. (2001). The case for quantitative literacy. In L. Steen (Ed.), *Mathematics and democracy: The case for quantitative literacy* (pp. 1–22). National Council on Education and the Disciplines.

Steen, L., & Madison, B. L. (2011). Reflections on the tenth anniversary of Mathematics and Democracy. *Numeracy, 4*(1), 1–8.

Tytler, R. (2020). STEM education for the twenty-first century. In J. Anderson & Y. Li (Eds.), *Integrated approaches to STEM education, Advances in STEM Education* (pp. 21–43). Springer Nature.

Tytler, R., Williams, G., Hobbs, L., & Anderson, J. (2019). Challenges and opportunities for a STEM interdisciplinary agenda. In B. Doig, J. Williams, D. Swanson, R. B. Ferri & P. Drake (Eds.), *Interdisciplinary mathematics education: The state of the art and beyond. ICME-13 Monographs* (pp. 51–81). Springer Open.

United Nations. (2009). *Resolution adopted by the General Assembly on 18 December 2009.* https://documents-dds-ny.un.org/doc/UNDOC/GEN/N09/471/13/PDF/N0947113.pdf?OpenElement

International Commission on the Futures of Education. (2020). *Education in a post-COVID world: Nine ideas for public action.* UNESCO.

Valladares, L. (2021). Scientific literacy and social transformation: Critical perspectives about science participation and emancipation. *Science & Education, 30,* 557–587.

Zollman, A. (2012). Learning for STEM literacy: STEM literacy for learning. *School Science and Mathematics, 112*(1), 12–19.

Chapter 16
Contributions of the Pedagogical Action of Ethnomodelling to STEM Education

Milton Rosa and Daniel Clark Orey

Abstract An ethnomodelling-based pedagogical action helps students to demonstrate the development of their mathematical processes as they reason, solve problems, communicate ideas, and choose appropriate representations through the development of daily local mathematical practices. It also recognizes that STEM education involves a teaching and learning process based on four specific disciplines: science, technology, engineering, and mathematics. This involves an innovative interdisciplinary approach that uses mathematical tools for the development of local, global, and dialogic pedagogical actions. An ethnomodelling perspective involves pedagogical actions that connect with STEM education, where students develop diverse ways of thinking, reasoning, inferring, modelling, and developing mathematical knowledge in sociocultural contexts. This kind of pedagogical action proposes a transformative pedagogy aimed at developing students into critical and reflective citizens. At the same time, it enables them to develop a glocalized perspective, which incorporates knowledge systems from local cultural practices and links them to knowledge systems from multiple worldviews. Teachers need to be supported in designing and implementing ethnomodelling pedagogical actions for *STEM* education to prepare students to tackle problems and situations they face in their daily lives.

Keywords Ethnomathematics · Ethnomodelling · Pedagogical action · Citizenship education

Initial Considerations

STEM Education actively engages learners in science, mathematics, engineering, and technology throughout the development of their schooling by providing students with opportunities to learn science and engineering through real-world applications (Chiu et al., 2015). STEM pedagogical actions are related to the development of activities that help students to foster inquiry, creativity, critical and reflective thinking,

M. Rosa (✉) · D. C. Orey
Universidade Federal de Ouro Preto, Ouro Preto, Brazil
e-mail: milrosa@hotmail.com

J. Anderson and K. Makar (eds.), *The Contribution of Mathematics to School STEM Education*, https://doi.org/10.1007/978-981-97-2728-5_16

as well as opportunities to improve cross-cultural collaboration and communication. According to the U.S. National Research Council [NRC] (2014), the development of and interaction with STEM disciplines helps students to interact and engage with members from diverse communities as well to develop skills in innovation, collaboration, cooperation, modelling, and inquiry.

Similarly, ethnomodelling enables students to develop analytical competencies, abilities, and skills by addressing diverse ways to solve problems that encourages them to experience and understand phenomena that occur in their own sociocultural contexts. Yet, to reach this goal, it is necessary that students are taught using pedagogical actions that help them to understand and apply STEM disciplines at both local and global levels (Rosa & Orey, 2018). Thus, the ethnomodelling approach enables students to become aware citizens who apply mathematical knowledge to address issues that contribute to the betterment of their communities (Stephenson, 2012). As well, STEM Education aims to understand the application of diverse ways of knowing that expands knowledge (Lantz, 2009) by helping students to develop skills, attitudes, and habits of mind that helps them to become productive and responsible citizens (Schlechty, 2011).

Thus, the main objective of this chapter is to offer research informed advice supporting the use of local and global mathematical practices through ethnomodelling to address the research question:

> How can ethnomodelling be integrated into STEM education so that teachers can develop pedagogical actions to meet students' needs from diverse cultural groups?

This approach is best developed by elaborating pedagogical actions that connect the school curriculum to sociocultural contexts and link STEM education, ethnomathematics, and ethnomodelling to help students become responsible citizens.

The Contribution of STEM Education to the Development of Responsible Citizenship

Glocalized societies empower members of distinct cultures with a set of core knowledge and abilities to pose and solve problems they face daily. They also evaluate evidence to make sense of information accumulated from their diverse sociocultural contexts to develop their decision-making processes (USA, 2016). In this regard, glocalized societies enable members of distinct cultural groups to develop abilities to adapt knowledge and behaviors that naturally fit into and enrich their own cultures. In addition, they are empowered to describe mathematical procedures, techniques, and practices in local and global communities (D'Ambrosio, 2020).

In addition, developing a glocalized society helps us deal with dialogical processes, that require an ongoing negotiation between local and global mathematical, scientific, technological, and engineering knowledge. For example, Rosa and Orey (2021) state that these processes are conducted through the development of cultural dynamism, which is related to interactions that produce innovative ways of thinking and reasoning and help members of distinct cultures solve societal issues concerning politics, economics, health, and the environment. During this process, global mathematical knowledge is reinvented and adapted to the local reality. Similarly, effective localization requires global mathematical knowledge, just as localization also helps to promote globalization. Hence, cultural dynamics are related to the development of STEM education, ethnomathematics, and ethnomodelling.

According to this perspective, STEM education enables students to develop knowledge, attitudes, and abilities, which encourage them to identify questions taken from real-life situations to draw evidence-based conclusions about daily problems and phenomena. It also involves integrating the STEM subjects by breaking down compartmentalized disciplines and constructing connections to students' daily lives (NRC, 2014). However, as D'Ambrosio (2006) affirms, traditional curricula remain in place, thus impeding the ability of teachers to identify connections between knowledge traditions and students' cultural backgrounds.

Currently, STEM education is recognized as a key for the development of sociocultural opportunities that are fundamental components of a comprehensive and holistic education because it is a meta-disciplinary approach defined as "the creation of a discipline based on the integration of other disciplinary knowledge into a new 'whole'" (Lantz, 2009, p. 1). It promotes the advancement of school curricula with an interdisciplinary approach. Thus, rather than teaching science, technology, engineering, and mathematics as separate and discrete subjects, STEM education integrates them into a cohesive and holistic educational paradigm (Elaine, 2014).

Connecting knowledge when posing and solving real-world problems also supports the development of students' inquiry, creativity, critical thinking, and reflective reasoning abilities (Stephenson, 2012). Glocalized societies also require its members to develop content knowledge, abilities to both analyze and synthesize information, to use technology and communication effectively, and to take ownership of their own learning process, as well to collaborate with members of other cultures.

All communities need citizens who can lead global and local efforts to reduce distrust and fear among members of distinct cultural groups (Zhao, 2010). Currently, glocalized societies expect their members to solve daily problems through processes of investigation, model building, data analysis, presentation of evidence-based reasoning, and communication of findings. In this context, STEM contributes to the development of these competencies by enabling members of distinct cultures to learn how to take team-based approaches towards ongoing challenges through the process of exploration and discovery of local and global mathematical knowledge in an interdisciplinary way. Similarly, Alsina (2020) highlights the importance of connections between disciplines so that mathematical content can be developed alongside science, technology, and engineering.

Therefore, schools must support teachers to develop innovative teaching and learning approaches through active learning coupled with rigorous mathematical knowledge so that students are prepared to live in distinct sociocultural contexts (Gates & Mirkin, 2012). Through the pedagogical actions of teachers, students can become problem-solvers, innovators, inventors, and self-reliant logical thinkers, as well as technologically literate (Lantz, 2009). These competencies can be achieved through the development of STEM education, ethnomathematics, and ethnomodelling.

Ethnomodelling, STEM Literacy, and Citizenship Education

Ethnomodelling is the study of mathematical ideas, techniques, and procedures elaborated by members of distinct cultural groups involving locally developed mathematical practices. This enables them to study mathematical content as a system bounded to their own contextual reality and reveals equal effort is required to understand of all components of these systems as well the interrelationship among them. For example, Ruggiero et al. (2022) stated ethnomodelling adds the cultural aspects of mathematics to the development of the mathematical modelling process by helping students to translate diverse mathematical procedures, techniques, and practices found in their own sociocultural contexts.

Results of the study conducted by Grewall (2023) showed that making mathematical connections is central to STEM education activities, which includes mathematical knowledge developed by members of distinct cultures. The connections that branch out from mathematics to science connect to history, culture, and indigenous ways of knowing in local and global perspectives through dialogue. Thus, "within the context of an ethnomathematics/STEM framework, ethnomodelling was used to develop lessons and study mathematical phenomena within cultures" (Grewall, 2023, p. 50). Furthermore, D'Ambrosio (2015) states that these "systems of knowledge reveal not only their convenience to explain reality, facts, and phenomena, but also, they are important strategies to cope with daily situations and problems not only for the individual but for society and the people in general" (p. 42).

Others have promoted the development of STEM literacy. For example Hallström and Schönborn (2019) reported STEM education requires the development of teachers' pedagogical actions using modelling to promote STEM literacy and the transfer of knowledge and skills between contexts that enable students to communicate and collaborate with members of distinct cultural groups. The U.S.A National Governors Association [NGA] (2011) described STEM literacy as the ability to:

- adapt to and accept changes driven by innovative technologies,
- work with others across borders,
- anticipate the multilevel impact of their actions,
- communicate complex ideas to a variety of audiences, and
- find measured creative solutions to problems faced by society.

If STEM disciplines were created and developed to close the gap between theory and practice (Martín Páez et al., 2019), then STEM literacy helps students to solve real-world problems (Maas et al., 2019) faced in their own sociocultural contexts through the development of contextualized pedagogical actions (Rosa & Orey, 2021). According to D'Ambrosio (2020), a more transformative education favors a pedagogical action that helps to develop student-centered methodologies to solve real-world problems related to contextualized daily phenomena.

Thus, appropriate pedagogical actions can promote connections between day-to-day mathematical knowledge and the one systematized by the school through the promotion of ethnomodelling. Similarly, Schlechty (2011) affirms that "engaging students in work that results in their need to learn material that is essential to their education as citizens in a democracy and to their right to claim to be well educated human beings is the primary business of schools" (p. 8). Appropriate teacher pedagogical actions that connect STEM subject knowledge encourages a deeper understanding of the STEM disciplines (Bybee, 2010).

Consequently, it is important that students have practical STEM experiences so that they will participate in their own communities as empowered and transformative citizens. For example, Bybee (2010) states that teachers must be aware of how STEM disciplines are shaped by the material, intellectual, social, technological, and cultural environments. Similarly, in their literature review, Nicol et al. (2023) identified and examined potentialities as well as problematic issues of mathematics in STEM education for local and global economies, human needs, and the importance of place and land, and stated that "critical, social justice, and ethnomathematical research offer possible contexts for mathematics education within STEM" (p. 1234).

As in many other countries, appropriate teaching and learning processes related to STEM disciplines are not widespread in the school curricula in Brazil (Rosa & Orey, 2021). Marginson et al. (2013) affirm that there is a growing demand for STEM education-based activities to meet the diverse needs of glocalized societies. A strategy that supports the development of citizenship education through a STEM education perspective involves mathematical modelling.

There is a need for researchers and teachers to identify appropriate teaching and learning practices and develop contextualized activities based on ethnomathematics and modelling. An important goal of STEM education-based activities is to support the development of students' competencies and to develop their critical thinking and reflective reasoning in a diversity of contexts. According to Volmert et al. (2013), this view of education challenges the perpetuation of stereotypic ideas of cultures that fail to recognize the benefits of higher level critical thinking abilities and lifelong civic engagement, that help students bridge their own cultures to the school context.

The elaboration of STEM education-based activities aim to build curricular content and understanding through the development of interdisciplinary applications (Lantz, 2009). The use of ethnomathematics and modelling approaches in the school curricula promotes the study of mathematical procedures, techniques, and practices developed by members of distinct cultures by connecting everyday local knowledge with mathematical content studied in the classrooms (Rosa & Orey, 2021).

Thus, ethnomodelling invites students to develop their creativity, discovery, experimentation, and simulation in the pedagogical actions developed in the classrooms. For example, Furuto (2016) states that an important influence of these learning environments is that STEM cultural practices are recovered and integrated in curricular activities that help students to develop a global basis for the development of mathematical knowledge through the pedagogical action of ethnomathematics. Similarly, D'Ambrosio (2016) states that ethnomathematics encourages the investigation and adaptation of these concepts both within and outside of classrooms because it is a program in the history and philosophy of mathematics with pedagogical implications.

Teachers must be supported to incorporate teaching and learning experiences emanating both from in and outside of schools (Rosa & Orey, 2018). This is because the more real, familiar, and humanized STEM disciplines become to students, the more likely they are to picture themselves in these interdisciplinary fields. Culturally, it is necessary that STEM disciplines are connected to the sociocultural context of the students to enable them to perceive the holistic nature of learning with experiences that occur outside of the classrooms. This perspective encourages the investigation and adaptation of STEM disciplines within formal and informal environments through the development of sociocultural capacities by developing pedagogical actions contextualized in daily tasks.

Additionally, members of many indigenous cultures developed engeneering concepts by using modelling techniques to make canoes in response to their need to travel and to transport goods on water (D'Ambrosio, 2020). According to Furuto (2019), STEM disciplines are related to the construction and navigation of the canoes because it incorporates scientific, technological, mathematical, and engineering knowledge developed locally. Concepts of STEM disciplines are also found in the construction of tipis (Orey, 2000) and hemispherical lodges (Shockey & Mitchell, 2017).

Another example is related to the determination of the volume of a truncated cone motivated by the application of techniques learned by the ancestors of Mr. Joaquim, a wine producer in Ijuí, in the state of Rio Grande do Sul, Brazil, which is an important example of the comprehension of an ethnomathematical practice and its natural encounter with mathematical modelling through ethnomodelling and STEM Education (Bassanezi, 2002). It is also important to mention three Brazilian investigations that focused on how members of distinct cultures apply their mathematical techniques to solve daily problems, which helped researchers to develop pedagogical actions related to students own sociocultural contexts. These studies were associated with coffee production, peripheral communities, and ethnomathematics trails (Ruggiero et al., 2022).

Similarly, the results of the study conducted by Grewall (2023) indicated that teachers perceived that cultural contexts related to an ethnomodelling/STEM framework supported students to develop mathematical connections between their communities and the school context. Thus, to help students develop their understandings of fractions, teachers elaborated contextualized curricular activities connected to history, science, technology, and engeneering in an interdisciplinary way. Accordingly, Grewall (2023):

> The ethnomathematics/STEM framework is a non-Eurocentric interdisciplinary approach that uses ethnomathematics in place of western mathematics in Science, Technology, Engineering, and Mathematics (STEM) education. Ethnomathematics promotes a humanistic mathematics that includes ancestral and indigenous perspectives and practices within its broader definition, which entails mathematics and culture (p. 56).

The importance of ethnomathematics and modelling regarding STEM education is related to the expansion of its practical implications in schools because it proposes the development of curricular innovation in an effort to counter the arrogance, inequity, and bigotry in society since its main objective is to transform students into critical and reflective citizens through the development of enriched contextualized activities that seek to improve connections between mathematical content and societal critical issues (D'Ambrosio, 2016).

For example, results of the study conducted by Rosa et al. (2023) showed that students who are residents in a peripheral community in Brazil understood and proposed solutions to the lack of basic sanitation in their own sociocultural environment through ethnomodelling. The participants developed competencies that encouraged them to analyze, decide, and judge solutions through the elaboration of ethnomodels. These results also showed that ethnomodels are cultural artifacts used to enable students to understand and comprehend systems that compose the reality of members of distinct cultures. This pedagogical action promoted the development of students' citizenship as well as the acknowledgment that the denial and exclusion process of peripheral communities' members still prevails in modern society.

The results of a study conducted by Rodríguez-Nieto and Alsina (2022) shows that contributions of a network developed between ethnomathematics and STEM education are related to:

(a) intradisciplinary connections, which enables the introduction of mathematics as an integrated whole, rather than as a set of isolated form of content knowledge,
(b) interdisciplinary connections between knowledge from different disciplines (STEM) that promotes the connections between each other, and
(c) connections between mathematical knowledge practiced by members of distinct cultures with institutionalized mathematics.

Ethnomathematics has a strong role in promoting STEM education because scientific and technological views lead to the development of innovative ideas. In addition, locally developed educational materials can build a strong connection to a functional mathematics curriculum (Adenegan et al., 2016). Based on the investigations conducted by Rosa and Orey (2018) and Orey et al. (2020), and Nicol et al. (2023), research is demonstrating how a curriculum based on STEM and ethnomathematics education can help to build students' personal, cultural, and community knowledge.

An Ethnomodelling Pedagogical Action for STEM Education

Ethnomodelling helps students to improve competencies, abilities, and skills related to STEM education because it enables members of distinct cultures to engage in the process of studying mathematical ideas developed in their own sociocultural context. In this regard, the pedagogical action of ethnomodelling translates STEM ideas and procedures that are embedded in local cultural practices to their contemporary equivalents (Eglash et al., 2006).

Hence, STEM activities based on ethnomodelling have the potential to restore students' cultural identity, and dignity, and develop their citizenship. Experiences with such activities allow students to reason, solve problems, communicate ideas, and choose appropriate mathematical representations based on their own sociocultural contexts. The absence of this opportunity in mathematics classrooms means students may only perform calculations that are unrelated to their daily experiences (Rosa et al., 2023). Ethnomodelling has resulted in students changing their positions in relation to their community and increasing their concern for the social/community problems they face daily (Rosa et al., 2023).

Ethnomodelling experiences can support the acquisition of a STEM-based criticality by making use of both emic (local) and etic (global) school knowledge. Emic knowledge is essential for an intuitive and empathic understanding of mathematical ideas of a culture. It is essential for conducting effective ethnographic fieldwork. It is concerned with the meaning that members of distinct cultures ascribe to events, phenomena, relations, behaviors, experiences, and practices they develop to deal with their own sociocultural contexts (Ruggiero et al., 2022). Emic knowledge aims to understand the cultural meaning and significance of a particular behavior or practice, as it is understood by members who engage in it (Mostowlansky & Rota, 2020).

Etic (global) knowledge is an outsider's perspective, who looks at a culture from the perspective of the external observers, such as researchers and teachers. This approach tends to focus on the observable events, phenomena, relations, behaviors, experiences, and practices developed by members of distinct cultures (Ruggiero et al., 2022). It is essential for cross-cultural comparisons that demand standardized units, measures, or frameworks, and categories to compare different cultures, and it may involve the use of concepts and theories from other disciplines to facilitate communication and collaboration (Mostowlansky & Rota, 2020). A dialogical (glocal) approach uses both emic (local) and etic (global) knowledge through processes of interaction.

An Example of Ethnomodelling—An Ivory Hair Pin from Africa

One example of ethnomodelling can be found in an ivory hair pin developed by members of Mangbetu culture, from the Uele River area in the northeastern region of the Democratic Republic of Congo, Africa. By examining the geometric pattern involved in the production of this cultural artifact, students can explore the mathematization procedures involved in this process. According to Ruggiero et al. (2022) mathematization is a process in which members of distinct cultural groups develop specific local mathematical tools to solve problems present in their sociocultural context. Figure 16.1 shows the elaboration of a dialogical ethnomodel of the Mangbetu ivory sculpture, which is the decorative end of an ivory hair pin composed of four scaled similar heads that shows a scaling design.

Eglash (1999) affirms that this particular ivory hairpin is a cultural artifact and "gives explicit instructions for generating a particular set of spatial patterns" (p. 61). The creation of a Mangbetu design reflects the artisans' desire to "make it beautiful and show the intelligence of the creator" (Schildkrout & Keim, 1990, p. 100) by adhering to angles that are multiples of 45 degrees. Thus, Eglash (1999) explains that this is only one part of an elaborated geometric aesthetic based on multiples of 45-degree angles, that are used in many Mangbetu designs. This mathematization process also shows a geometric relation in the sculpture iterative square structure representing the construction of an indefinite algorithm.

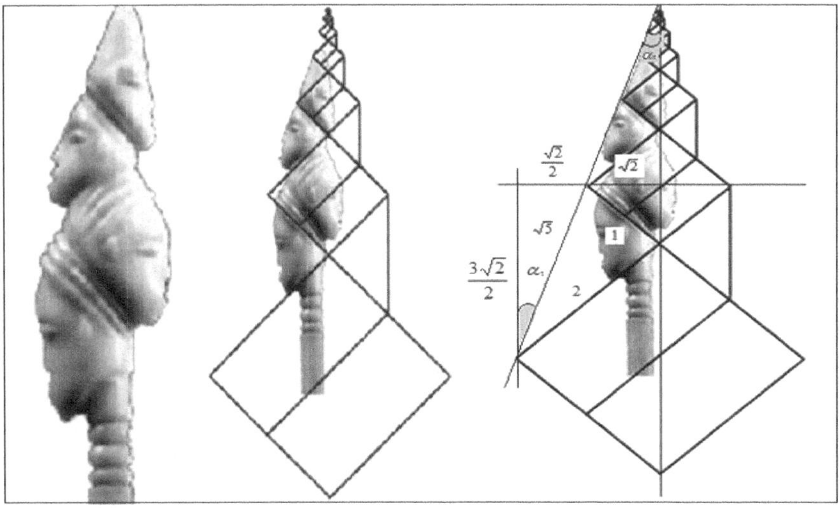

Fig. 16.1 Mathematization process of the elaboration of a dialogical ethnomodel of the Mangbetu ivory hatpin. *Source* Eglash (1999, p. 61)

By analyzing this process through an ethnomodelling and STEM lens, an emic (local) observation seeks to understand mathematical practices used to carve this sculpture from the perspective of internal dynamics and relationships of the members of the Mangbetu culture. An etic (global) perspective provides a cross-cultural contrast and comparative perspective. By using academic mathematics this phenomenon can be understood by others from different cultural backgrounds, or the outsiders, that is grounded in the mathematical knowledge of the insiders.

Both emic and etic perspectives are essential to understanding local mathematical techniques applied in this ancestral cultural artifact, which is made with sophisticated technologies and designs (Rosa & Orey, 2023). These technologies and engineering are highly evolved because they are the product of thousands of years of expertise, oral traditions, change, and continuity. There is nothing crude or rudimentary about these cultural artifacts because they are made with advanced technologies that require intelligence, local knowledge, practice, skills, and patience (Prindle, 1999).

This ethnomathematical approach aims to reduce the effects of colonization, prejudice, inequity, and harm due to the disconnections between knowledge as practiced globally in the academy (etic) and its local and practical use in everyday life (emic) (Rosa & Orey, 2018). Thus, it is necessary that approaches to STEM education consider diverse forms of pedagogical actions that engage students in making sense of mathematical and non-mathematical practices developed in diverse contexts. Nicol et al. (2023) state that it is important to "address inequities and increase access to STEM fields for historically underrepresented students, ethnomathematics teachers are looking to integrate STEM education as an approach to develop mathematical competencies in students' own communities" (p. 3). Similarly, Furuto (2016) highlights the importance of using ancestral engineering to develop innovative technological resources relevant to local community.

An Example of Ethnomodelling—Roller Carts from Brazil

The second example of ethnomodelling involves the Brazilian cultural practice of designing roller carts, whose construction is simple and diverse. Orey et al. (2020) affirm that the history of these roller carts is not well known, but the first examples of these carts were built in cities such as São Paulo, Rio de Janeiro, and Belo Horizonte, between the end of 1960s and the beginning of 1970s. This tradicional street race activity takes over asphalted streets of the towns and cities in Brazil. Since roller carts have become a relevant street activity, this race is part of the Brazilian cultural heritage. Races with roller carts, which amuse children and make adults nostalgic, is one of the oldest and most amusing street activities Brazilian people know and play (see Fig. 16.2).

According to Orey et al. (2020), the main material for the construction of roller carts is the bearings that are very popular in all Brazilian states, which are obtained in automobile maintenance workshops, and they vary in size and shape, as well in the use of diverse materials. Because roller carts are handcrafted, they require a deep

Fig. 16.2 Brazilian roller cart built with wood and steel bearings. *Source* Authors' personal file

dedication from runners and, in addition to creativity, it is necessary to be innovative in their designs and materials, which makes them even faster. The construction process of these carts by the students enables them to learn how to use simple tools along with recycled materials especially related to the wood and bearings.

Basically, the carts are constituted of a wooden body with a movable axis in front, called a guide, and that three or four bearings that can be used. It is important to state that "one of the most common types of Brazilian roller carts are built with a wooden frame and discarded steel bearings from automotive repair shops" (p. 200). This local (emic) knowledge regarding the construction of the carts was diffused from generation to generation. Considering that roller carts use a push/propulsion only at the start given by riders, and the mathematical rigor (etic knowledge) helps students to perform calculations that influence the development and the acceleration achieved by the carts with the elaboration of ethnomodels. The mathematization process of the elaboration of ethnomodels deals with calculations related to mass of the roller cart, circumference diameter of the blades, number of blades, the geometric shapes of the fixed and movable parts of the structure of this cart, the width of the axes between the blades, the length of the main axis and the symmetry of the cart and its aerodynamics (Orey et al., 2020).

Mathematical techniques and procedures used for the construction of the carts are handed down to the students from family members, which are translated for the elaboration of ethnomodels. This environment enables students to explain, understand, manage, analyze, and critically reflect on this cultural practice (Orey et al.,

2020). As a result, the roller cart activity helps students to: (a) design, build, and test the equipment (roller carts), (b) develop training tools, (c) regulate and standardize measurements, (d) ensure the safety of participants, (e) analyze and interpret data to understand the performance of participants and their interaction with the equipment. Figure 16.3 shows the plan of one the roller carts developed by students.

This context enabled Orey et al. (2020) to argue about the importance of developing data based on informed reflections on reality, which dealt with the construction of roller carts. This pedagogical action became a transformative activity that enabled students to explain their reasoning, share their understanding, and develop their abilities to organize, manage, and find solutions to this mathematical practice. In this regard, Ayunani, et al. (2020) argued that students need to make STEM connections to expand the use of their procedures and techniques by associating cultural practices to school mathematical concepts. In addition, Fritz and et al., (2011, p. 3) suggest a:

> (…) key element is to establish engineering as a cultural activity, and to recognize the differences in practice between cultures. A second goal is to determine whether alternative engineering design methods can be effectively applied to the solution of current societal challenges, especially with respect to the sustainable use of resources.

The challenge is for teachers to develop activities that arise from the interest of students because this action is related to the transformation of school mathematics (etic) into a living subject that is connected to real-world situations (emic) (Orey et al., 2020). Such opportunities allow learners to critically analyze and reflect upon mathematical phenomena that occurs in their own communities and to experience the connections between science, mathematics, engineering, and technology to their cultural heritage. Consequently, teachers need to be able to translate the school curriculum through ethnomodelling experiences for students for them to understand the connection between both traditional and non-traditional applications.

Final Considerations

Educational systems around the world have persistent inequities related to student participation in STEM activity. These disparities threaten the government's ability to close gaps in education or maintain the preeminence of mathematical and scientific research and technological innovation. As well, many schools worldwide are challenged to promote the widespread adoption of emerging best practices related to a growing diversity of STEM Education/ethnomodelling pedagogical actions that encourage active students' participation.

Contexts provided in this chapter help to describe how ethnomodelling can be integrated into STEM education-based activities to develop pedagogical actions for all students. These kinds of pedagogical action enables teachers to help a diversity of students to build "knowledge of the nature and history of science, mathematics, and technology, an understanding of common themes that cut across the disciplines, and

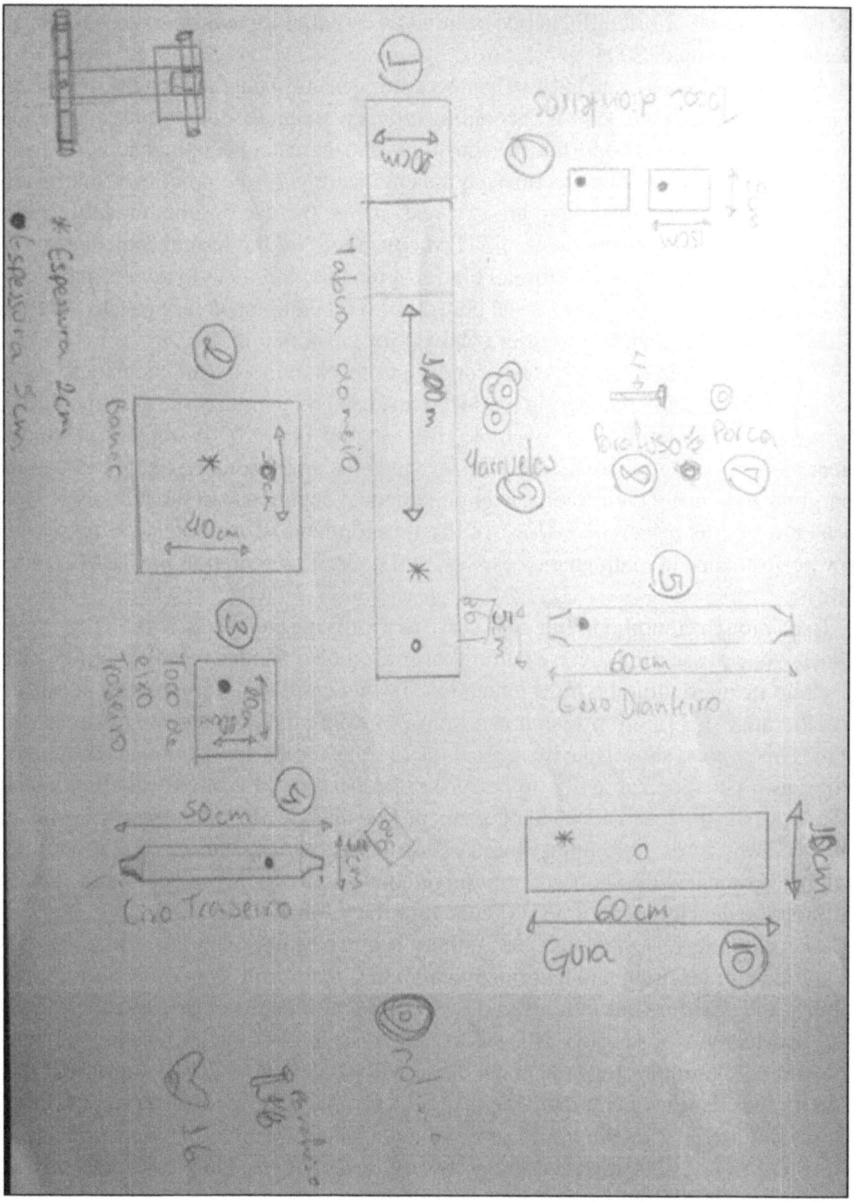

Fig. 16.3 Draft of roller carts. *Source* Authors' personal file

the development of scientific habits of mind as essential aspects of science literacy" (Kesidou & Koppal, 2004, p. 5).

Although relatively rare in Brazil, integrating aspects of engineering into pedagogical actions in the classrooms is "becoming more prevalent in schools and can provide great problem-solving opportunities for students to learn about mathematics, science, and technology while working through the engineering design process" (Stohlmann et al., 2012, p. 30). Thus, there are two reasons for the use of ethnomodelling and ethnomodels in the development of STEM education: (a) the formal context ensures students an unrestricted educational platform by allowing them to access and share school content, and (b) the informal context plays a complementary role to practice what they are learning, and to offer contexts for greater autonomy.

Thus, STEM education creates conditions to examine the mathematics, technology, science, and engineering produced and used by members of distinct cultures through ethnomodelling. For example, Buxton and Lee (2010) suggest an implementation of pedagogical alternatives for students who are more likely to become marginalized during their traditional mathematical experience in the schools, which is manifested as a lack of interest in the development of mathematical activities, low performance in mathematics classes, and underrepresentation in STEM-related careers.

Even though ethnomodelling may serve as a unifying element across STEM disciplines while possessing diverse cultural origins, it also underscores the significance of ethno or intercultural STEM education. In this case, it is important to delineate specific areas for future research and policy considerations related to this research field. This context shows that there are three reasons for the application of ethnomodelling as a pedagogical action to be further investigated: (a) ethnomodelling as an effective path to reach local mathematical knowledge, (b) ethnomodelling as an effective way to develop intercultural classroom activities, and (c) ethnomodelling as a pedagogical action to transform the relations between mathematics and society through the development of STEM education-based activities.

According to D'Ambrosio (2016), there is a need to develop new directions for the process of teaching and learning mathematics that result from procedures, techniques, and practices that emerge in distinct sociocultural contexts. These ideas indicate that the new generation of teachers, scientists, engineers, and mathematicians need broader attitudes towards mathematics. Rosa and Orey (2018) affirm that this calls for the creation of new mathematical professionals who mediate between mathematics, engineering, technologies, and applied science. The links between ethnomodelling and STEM Education certainly show the possibility for innovative pedagogical actions for all students that help them to develop critical thinking skills applied to their school and everyday lives.

References

Adenegan, K. E., Ipinlaye, A. B., & Adewoye, R. A. (2016). Ethnomathematics and indigenous mathematics: Implications for science technology engineering and mathematics education. *Science & Technology, 2*(6), 251–256.

Alsina, A. (2020). Conexiones matemáticas a través de actividades STEAM en Educación Infantil. *Unión: Revista Iberoamericana de Educación Matemática, 16*(58), 168–190.

Ayunani, D., Mardiyana, M., & Indriati, D. (2020). Analyzing mathematical connection skill in solving a contextual problem. *Journal of Physics Conference Series, 1511*(1), 1–10.

Bassanezi, R. C. (2002). *Ensino-aprendizagem com modelagem matemática [Teaching and learning with mathematical modelling].* São Paulo, SP, Brazil: Editora Contexto.

Buxton, C., & Lee, O. (2010). *Diversity and equity in science education*: New York, NY: Teachers College Press.

Bybee, R. W. (2010). Advancing STEM education: A 2020 vision. *Technology and Engineering Teacher, 70*, 30–35.

Chiu, A., Price, A. C., & Ovrahim, E. (2015). *Supporting elementary and middle school STEM education.* Museum of Science and Industry.

D'Ambrosio, U. (2006). The program ethnomathematics and the challenges of globalization. Cir*cumscribere: International Journal for the History of Science, 1*(1), 74–82.

D'Ambrosio, U. (2015). Mathematical modelling as a strategy for building-up systems of knowledge in different cultural environments. In G. A. Stillman, W. Blum & M. S. Biembengut (Eds.), *Mathematical modelling in education research and practice*: cultural, social and cognitive influences (pp. 35–44). Cham, Switzerland: Springer.

D'Ambrosio, U. (2016). Change in space, urban culture and ethnomathematics. In Z. Babaci-Wilhite (Ed.), *Human rights in language and STEM Education* (pp. 207–219). Sense Publishers.

D'Ambrosio, U. (2020). Sobre las propuestas curriculares STEM y STEAM y el Programa de Etnomatemática [About the STEM and STEAM curricular proposals and the Ethnomathematics Program]. *Revista Paradigma, 41*, (Edición Cuadragésimo Aniversario: 1980–2020), 151–167.

Eglash, R. (1999). *African fractals: Modern computing and indigenous design.* Piscataway, NJ: Rutgers University Press.

Eglash, R., Bennett, A., O'Donnell, C., Jennings, S., & Cintorino, M. (2006). Culturally situated designed tools: Ethnocomputing from field site to classroom. *American Anthropologist, 108*(2), 347–362.

Elaine, J. H. (2014). *What is STEM Education?* New York, NY: Beta Live Science.

Fritz, D. W., Strong, D. S., & Bryant, J. T. (2011). *Ancestral engineering*: Towards the integration of archaeology and engineering. Proceedings of the Canadian Engineering Education Association (pp. 1–6). CEEA/ACEG.

Furuto, L. H. L. (2016). Lessons learned: Strengths-based approaches to mathematics education in the Pacific. *Journal of Mathematics and Culture, 10*(2), 55–72.

Furuto, L. H. L. (2019). We never voyage alone. In H. Aikau & V. Gonzalez (Eds.). *Detours: A decolonial guide to Hawai'i.* Durham, NC: Duke University Press.

Gates, S. J., & Mirkin, C. (2012). *Encouraging STEM students is in the national interest.* The Chronicle of Higher Education.

Grewall, T. K. (2023). *Students' mathematical understandings of fraction division with an ethnomathematics/STEM framework.* Dissertation Doctor of Philosophy. The Graduate School—Education. Reno, NV: University of Nevada.

Hallström, J., & Schönborn, K. J. (2019). Models and modelling for authentic STEM education: Reinforcing the argument. *International Journal of STEM Education, 6*(1), 1–10.

Kesidou, S., & Koppal, M. (2004). Supporting goals-based learning with STEM outreach. *Journal of STEM Education: Innovations and Research, 5*(3/4), 5–16.

Lantz, H. B. (2009). *Science, technology, engineering, and mathematics (STEM) education: What form? What function?* Charlotte, NC: SEEN Southeast Network.

Maas, K., Geiger, V., Ariza, M. R., & Goos, M. (2019). The role of mathematics in interdisciplinary STEM education. *ZDM—Mathematics Education, 51*(6), 869–884.

Marginson, S., Tytler, R., Freeman, B., & Roberts, K. (2013). *STEM: country comparisons*. Report for the Australian Council of Learned Academies. Melbourne, Australia: Australian Council of Learned Academies.

Martin-Páez, T., Aguilera, D., Perales-Palacios, F. J., & Vílchez-González, J. M. (2019). What are we talking about when we talk about STEM education? A review of literature. *Science Education, 103*(4), 799–822.

Mostowlansky, T., & Rota, A. (2020). Emic and etic. In: Stein, F. (Ed.). *The open encyclopedia of anthropology*. Cambridge, England: The Cambridge Encyclopedia of Anthropology.

NGA. (2011). *Building a science, technology, engineering and math agenda*: an update of state actions. New York, NY: National Governors Association.

Nicol, C., Thom, J. S., Doolittle, E., Glanfield, S., & Ghostkeeper, E. (2023). Mathematics education for STEM as place. *ZDM, 55*(4), 1231–1242.

NRC. (2014). *STEM integration in K-12 education*: *status, prospects, and an agenda for research*. National Research Council. Washington, DC: The National Academy Press.

Orey, D. C. (2000). The ethnomathematics of the Sioux tipi and cone. In H. Selin (Ed.), *Mathematics across culture: The history of non-western mathematics* (pp. 239–252). Kluwer Academic Publishers.

Orey, D. C., Rosa, M., & Soares, R. B. (2020). Mathematical modelling as a learning environment to transform a street activity into a sport practice. In G. A. Stillman, G. Kaiser, & C. E. Lampen (Eds.), *Mathematical modelling education and sense-making* (pp. 199–207). Springer.

Prindle, T. (1999). *What is Ethnotechnology?* Vernon, CT: NativeTech: Native American Technology and Art.

Rodríguez-Nieto, C. A., & Alsina, A. (2022). Networking between ethnomathematics, STEAM education, and the globalized approach to analyze mathematical connections in daily practices. *EURASIA Journal of Mathematics, Science, and Technology Education, 18*(3), 1–22.

Rosa, M., & Orey, D. C. (2010). Ethnomodelling: A pedagogical action for uncovering ethnomathematical practices. *Journal of Mathematical Modelling and Application, 1*(3), 58–67.

Rosa, M., & Orey, C. C. (2018). STEM education in the Brazilian context: An ethnomathematical perspective. In R. Jorgensen & K. Larkin (Eds.), *STEM education in the junior secondary: The state of play* (pp. 221–248). Springer.

Rosa, M., & Orey, D. C. (2021). An ethnomathematical perspective of STEM education in a glocalized world. *BOLEMA, 35*(70), 840–876.

Rosa, M., & Orey, D. C. (2023) Interactions between sociocultural perspectives of mathematical modelling and ethnomathematics in an ethnomodelling approach. *Vidya, 43*(2), 113–132.

Rosa, M., Orey, D. C., & Mesquita, A. P. S. S. (2023). An ethnomodelling perspective for the development of a citizenship education. *ZDM, 55*(4), 1–13.

Ruggiero, E. D. D., Rodrigues, J., Mesquita, A. P. S. S., & Rosa, M. (2022). Ethnomodelling as a pedagogical action in diverse contexts by using a dialogical knowledge. In M. Rosa, R. Cordero, D. C. Orey, & P. Carranza (Eds.), *Mathematical modelling program in Latin America: A collaborative context for social construction of knowledge for educational change* (pp. 93–116). Springer.

Schildkrout, E., & Keim, C. A. (1990). *African reflections: Art from Zaire with contributions by Didier, Demolin, John Mack, Thomas Ross Miller, and Jan Vansina*. University of Washington Press.

Schlechty, P. C. (2011). *Engaging students: The next level of working on the work*. Jossey-Bass.

Shockey, T., & Mitchell, J. B. (2017). An ethnomodel of a Penobscot lodge. In M. Rosa, L. Shirley, M. Gavarrete, & W. Alangui (Eds.), *Ethnomathematics and its diverse approaches for mathematics education* (pp. 257–281). Springer.

Stephenson, N. (2012). *Introduction to inquiry based learning: Teach Inquiry*. Round Rock, TX: Education Solution.

Stohlmann, M., Moore, T. J., & Roehrig, G. H. (2012). Considerations for teaching integrated STEM education. *Journal of Pre-College Engineering Education Research, 2*(1), 28–34.

USA. (2016). *STEM 2026: A vision for innovation in STEM education.* U.S. Department of Education.

Volmert, A., Baran, M., Kendall-Taylor, N., & O'Neil, M. (2013). *You have to have the basics down really well: Mapping the gaps between expert and public understanding of STEM learning.* Frame Works Institute.

Zhao, Y. (2010). *Catching up or leading the way.* Alexandria, VA: ASCD

Chapter 17
Developing Responsible Citizenship Through Integrated STEM Education Activities

Lisa Steffensen

Abstract In this chapter, I focus on how pre-service teachers used important societal issues to engage school students in developing responsible citizenship. Theoretical perspectives from critical mathematics education are used to investigate the experience of a group of pre-service teachers during their practicum with seventh-grade students. These teachers used the refugee crisis to raise the students' awareness and enable them to be(come) critical, responsible citizens. Three situations from their practicum are used to discuss how responsible citizenship developed within an integrated STEM curriculum approach. In the first situation, involving boat refugees, students estimated the number of refugees in a boat. In the second situation, involving refugees in Europe, students used technology to identify and process relevant information. In the third situation, involving refugees worldwide, students interpreted information in statistically empathetic ways to develop an informed understanding of the refugee crisis as responsible citizens. The results from this study indicate integrated STEM curriculum approaches should include challenging societal issues emphasising their ethical, social, and cultural aspects to foster responsible citizenship.

Keywords Responsible citizenship · Critical mathematics education · Addressing societal issues · Interdisciplinary · Democracy

Introduction

The refugee crisis is a societal challenge affecting individuals and society worldwide and can create controversy among citizens, politicians, and others. Decision-making on these issues is not straightforward: various concerns must be weighed against each other, and citizens must understand the issues and need to be actively engaged. Competencies based on science, technology, engineering, and mathematics (STEM)

L. Steffensen (✉)
Western Norway University of Applied Sciences, Bergen, Norway
e-mail: Lisa.steffensen@hvl.no

J. Anderson and K. Makar (eds.), *The Contribution of Mathematics to School STEM Education*, https://doi.org/10.1007/978-981-97-2728-5_17

are essential for dealing with data and information about such complex challenges. STEM-field professionals can contribute to solutions with engineering and design that address refugees' needs, such as for low-cost, sustainable housing and systems for clean water, energy, and sanitation in refugee camps. They can provide digital access to education through quality learning platforms and mobile apps for language learning. Additionally, STEM professionals can work alongside decision-makers to provide access to data on patterns and trends, migration flows, demographic profiles, and social impact.

The STEM field is sometimes referred to as hard science, defined as "science that is based on measuring and observing physical facts or events and is not influenced by personal feeling or opinions" (Oxford Learner's Dictionaries). Such a clinical definition disconnects it from human values and can exacerbate the lack of consideration for ethical and moral aspects of STEM in STEM teaching and learning. However, it is crucial to make ethically sound decisions when dealing with the refugee crisis. Given the important role of STEM professionals in addressing societal issues, there are arguments for including ethical, moral, cultural, and social aspects of STEM in teaching and learning in the STEM school curriculum, and perspectives from critical education may be relevant (Maass et al., 2019, 2023). To accommodate these arguments, changes have been made to the Norwegian curriculum.

In the revised Norwegian curriculum (Ministry of Education & Research, 2017), democracy and citizenship are introduced as an interdisciplinary topic, with students being expected to engage with current societal challenges. They are intended to gain insight into challenges and dilemmas, understand how solutions can be found through knowledge and collaboration, and learn about various actions and their consequences. In the mathematics curriculum (Ministry of Education & Research, 2019a, p. 4), the democracy and citizenship topic emphasises that students should develop the competence to "explore and analyse findings from real datasets" and "evaluate the validity of such findings" to enable participation in public debate using informed arguments. Furthermore, they should develop competency in being "conscious of the underlying conditions and premises for the mathematical models which form the basis for decisions in their own lives and society" (p. 4). In the natural science curriculum, the democracy and citizenship topic emphasises that students should understand and be "critical about argumentation in public debates" and be "active citizens and contribute to technological development and sustainability" (Ministry of Education & Research, 2019b, p. 4). Hence, the subject-specific curricula emphasise interdisciplinary teaching of democracy and citizenship.

Thus, given the importance of STEM competencies for dealing with societal challenges and the need for responsible citizenship, it is worth examining how teachers engage students in such issues. This chapter focuses on *how pre-service teachers (PTs) can engage students in responsible citizenship through an interdisciplinary modelling activity.* To investigate this question, I use theoretical perspectives from critical mathematics education and empirical data from a mathematics course in teacher education, in which a group of PTs introduced a refugee modelling activity for their seventh-grade students during practicum.

Using STEM Modelling Tasks to Develop Responsible Citizenship

Subject-based school practices and structures can hinder integrated approaches to interdisciplinary teaching and make them challenging for subject teachers (Anderson, 2020). Within teacher education, subject-based teaching organisation means the practice of integrating STEM approaches focusing on responsible citizenship can be a challenge for teachers and PTs. However, focusing on real-world problems early on has the potential to support effective STEM curriculum approaches (Kurup & Li, 2022). Teacher education is critical for supporting practising teachers and PTs in the design of STEM tasks and activities to develop responsible citizens.

STEM tasks may involve individual or group problem-solving activities for students. Well-designed STEM asks are interdisciplinary, have real-world relevance, and foster competencies such as critical thinking and creativity. Mathematical modelling refers to processes of representing real-world phenomena by way of mathematical representations and techniques. In a STEM modelling task, students engage in modelling to understand problems based on real-world phenomena. One promising path for developing citizenship competencies is the inclusion of socio-scientific issues that contain ethical elements, inquiry-based learning, and student-centred learning approaches (Maass et al., 2022). The interdisciplinary nature of modelling activities can make this a fruitful approach to developing responsible citizenship among students while investigating real-world challenges (English, 2023; Maass et al., 2019, 2023; Skovsmose, 2023).

Teacher education programmes are essential for advancing ideas about integrated STEM approaches to responsible citizenship. Maass et al. (2022) investigated how teachers can be supported in connecting mathematics and science with citizenship education in professional development courses for teachers by using contexts relevant to society. The authors argued that, given global challenges such as increasing migration, our society needs to find adequate solutions whilst respecting fundamental values such as freedom, democracy, and human rights. In addition, they argued ethical, moral, social, and cultural dimensions are seldom made explicit in STEM modelling activities.

In order to make ethical decisions and judgements in a changing world, Maass et al. (2019) emphasise the need for social, civic, and intercultural understanding and competency. Although issues of responsible citizenship need to be addressed in teacher training in STEM education, some subject teachers appear reluctant to engage with these issues (Maass et al., 2019). It is essential to support teachers' inclusion of controversial issues that can help students to develop responsible citizenship. Such support can involve evidence-based teaching materials, assessment strategies, and reflection on the degree of scaffolding. Theoretical perspectives from critical mathematics education, may be suitable for supporting teachers, PTs, and students when they engage with societal issues important for responsible citizenship.

The Role of Critical Mathematics Education

Critical mathematics education (CME) emphasises that students and teachers should engage with social issues as critical citizens. According to Skovsmose (2023), CME should foster critical thinking, social awareness, and understanding of mathematics as a social practice that can promote equity and social justice. In CME, mathematics is not considered neutral and value-free but as embedded in social context, and awareness of mathematics' apolitical nature can promote its political, cultural, and ideological dimensions. CME seeks to empower students to be(come) responsible citizens, in terms not only of their computational competencies but also their understanding of the role of mathematics in shaping—or formatting—society. Skovsmose (2023) describes how mathematical representations (e.g. figures, statistics, and graphs about refugees) not only bring awareness of important aspects of societal issues but can also format our perceptions of, and actions towards, inclusive citizenship. Ernest (2015) emphasises that mathematics education should empower students socially and politically so that they become numerate, critical citizens who can use their knowledge in social and political realms for their own and society's improvement. Similarly, Mukhopadhyay and Greer (2007) argue that mathematics education should empower individuals with critical competencies and the dispositions to use them as a tool for social change. While such critical perspectives have played a part in forming research within mathematics education, do these ideas have something to offer integrated STEM curriculum approaches?

Practices in the STEM field can reinforce society's social inequalities and power structures. Critical and reflective thinking is essential for critiquing the role of science, technology, engineering, and mathematics in society. As responsible citizens, it is important to question the assumptions, values, and biases underlying mathematical, technical, and scientific concepts, methods, and applications. Previous research connects theoretical perspectives from CME to integrated STEM education. Wolfmeyer et al. (2018) and Wolfmeyer et al. (2015) consider how the interdisciplinary STEM space can take inspiration from ideas in CME to provide fruitful settings for exploring societal issues. Rather than framing STEM education in terms of workforce readiness and greater student access to STEM careers, they position STEM education within a framework that critically examines the role of STEM and proposes critical work within it. Similarly, Nicol et al. (2019) describe how mathematics educators across educational levels and national borders have collaborated on task development and their understanding of teaching STEM for social justice. They designed tasks involving water usage, electricity consumption, air quality, affordable housing, homelessness, and so on. However, they found it less challenging to include 21st-century skill competencies such as critical thinking, communication, and collaboration than STEM disciplinary content since mathematical enquiry did not necessarily occur in the other STEM disciplines or in the social issue under consideration.

Mathematics can play an important role in STEM education (Anderson, 2020; English, 2023) although the mathematics is not always visible. It is therefore vital that mathematics teachers and researchers working with problems requiring competencies from various fields make mathematics explicit (Shaughnessy, 2013). Maass et al. (2019) suggest three interdisciplinary approaches that can advance the role of mathematics: twenty-first-century skills, mathematical modelling, and education for responsible citizenship.

Modelling activities emphasising a critical understanding of the role of mathematics in society are sometimes called socio-critical modelling (Barbosa, 2009). The context involves societal issues, particularly about advancing responsible citizenship. Several studies have focused on socio-critical modelling and responsible citizenship in teacher education. Almeida and Carreira (2019) use the context of world wealth distribution to critically reflect on inequality and use mathematics to unveil social and political phenomena and promote informed critical positioning among PTs. Similarly, Julie and Mudaly (2007) had students engage in addressing salary-related equity issues in South Africa with in-service teachers. Environmental issues provide the context of several modelling research studies in teacher training. For instance, Zapata-Cardona and Martínez-Castro (2021) engaged PTs in Colombia in global warming issues, and Villarreal et al. (2015) had PTs investigate issues of their own choosing such as trash and recycling, lack of water, and water and energy consumption in Argentina. Although these examples of socio-critical modelling include interdisciplinary topics and, to various degrees, aspects of responsible citizenship, they do not focus on integrated STEM approaches.

Within CME research, some studies include the refugee context. Skovsmose et al. (2023) argue that it is vital that students learn to take the perspective of others by recognising the challenges experienced by refugees. Mukhopadhyay and Greer (2007) discuss how mathematics-based descriptions (e.g. graphs) of refugees crossing the US-Mexico border can be accompanied by images of coffins to illustrate the gravity of the situation and avoid a stance of neutrality. This can encourage what they call "statistical empathy": the ability to relate statistical data to the reality they represent (p. 119). They argue that it can be problematic if decision-makers lose their ability to relate to what the data represent. Renert (2011) expresses similar concerns, although he refers to another crisis—environmental sustainability—and our inability to feed an increasing population. Bousalis and Furner (2020) make use of the US history of immigration in their investigation of middle school students' reflections while challenging anti-immigrant discourse. Vorhölter et al. (2019) describe a modelling activity in which students suggest the distribution of 1.5 million refugees in Germany in 2015. In these research examples, different approaches are taken to critically investigate the refugee crisis from an educational perspective. The study by Mukhopadhyay and Greer (2007) focuses on connecting numbers to actual crises—the number of refugee deaths. The harsh reality can be hard for students to take in, and teachers must know their students well and take precautions. The study by Bousalis and Furner (2020) examines these issues through a historical lens to reflect on present issues. This approach may be less disturbing to students, but they may not transfer historical situations to current situations. Additionally, students probably observe the severity of present-day situations in the media. In a classroom environment with peers, they have a space where they can process and reflect on these challenging issues without losing their ability to relate to what the numbers represent.

The Norwegian Context and Background for This Study

Critical perspectives have a tradition in Norwegian teacher education, inspired for instance by the work of Mellin-Olsen (1987) and Skovsmose (2023). To support students as citizens, classrooms must offer an environment that allows them to democratically engage with societal issues. Norwegian students generally do well on international surveys of civics and citizenship education (Huang et al., 2017). For instance, students aged 14 to15 scored well on competencies related to, and on understanding democracy, social systems, and democratic principles. They trust political institutions and have positive attitudes about fulfilling their role as responsible citizens. In schools, students have an environment open to discussion and where they can debate and disagree. Nine out of ten school students participate in school elections (e.g. for class representatives), and six out of ten contribute to decision-making on how the school is run (e.g. participating in planning, implementation, and assessment of education).

Given the various approaches internationally to organising STEM education, a brief overview of Norwegian compulsory education (grades 1–10) is provided. Science is taught as an integrated subject (natural science) consisting of biology, chemistry, geology, and physics. Mathematics is taught explicitly and includes statistics. Technology and engineering are not subjects on their own but are integrated into other subjects. Technology is explicitly integrated into natural science, mathematics, and social studies. It occupies a prominent position in natural science and is one of the five core elements; it is explicitly stated that students should understand, develop, and use technology (Ministry of Education & Research, 2019b). In mathematics, students are required to understand technology so they are able "to make responsible life choices" (Ministry of Education & Research, 2019a, p. 4). Regarding social studies, students are required to discuss the implications of technology for individuals, society, and nature (Ministry of Education & Research, 2019c). Engineering is not explicitly described in the curriculum before grades 11 to 13, and in the senior years, it can be a component of an elective subject.

In addition, all compulsory education subjects focus on five basic skills: oral, writing, reading, numeracy, and digital. For instance, numeracy refers to "the ability to collect, adapt and present relevant statistics" and "the ability to compare, assess and argue whether calculations, results and presentations are valid" in natural science (Ministry of Education & Research, 2019b, p. 5), and in social studies, to "the ability to compile, work with, analyse and assess statistics and big data in light of the context, and to take a position on societal issues" (Ministry of Education & Research, 2019c, p. 6). These understandings of numeracy focus on reflective and participating citizens rather than on more instrumental competencies.

The study reported in this chapter investigated how PTs in their second year of study of a compulsory mathematics course for first-to-seventh-grade engaged students in responsible citizenship through an interdisciplinary modelling activity. I used anonymised data from a larger study *Learning about Teaching Argumentations for Critical Mathematics Education in Multilingual Classrooms* (https://prosjekt.hvl.no/latacme/en/), a Western Norway University of Applied Sciences (HVL) project. The teacher training is at the master's level, and all PTs take similar courses during their first two years, regardless of what they later specialise in. PTs were required to design and implement a mathematical modelling activity during their seventh-grade practicum and write a report reflecting on their experience and findings as a component of their assessment. In Norway, PTs typically work in groups during practicum. Data were collected from six classes of around 30 students in line with National Research Ethics Committees' (2023) guidelines.

This research focuses on one group, consisting of Vilde, Saga, Anna, and Mai (pseudonyms), who were chosen on the basis of their topic "modelling and refugees". At the time the PTs had not yet decided on their subject specialisation for their masters. The PTs had their practicum at a school where there was a focus on the refugee crisis. They took an integrated approach to mathematics, social studies, Norwegian, and music with 24 students. The school invited external plenary speakers to talk about their experiences as refugees, and parents were invited to participate. The PTs asked questions such as:

What do you think of when you hear 'Forced to flee' [På flukt]?

Have you heard anything on the news lately about people forced to flee?

Where are people fleeing from?

The students discussed these, and later the PTs made plenary mind maps based on the responses. They showed the movie *Refugees: The Shared Story of Harry and Ahmed*,[1] which relates two different stories from a child's perspective. They asked the students to discuss it in pairs and imagine what it would be like to be forced to flee their country. They also asked the students to process their thoughts by creating a poem or a story titled "Forced to flee".

The empirical data included the teacher educator's teaching material, the PTs' lesson plans, the written report, students' and PTs' notes, and a transcribed follow-up interview. The data were thematically analysed in NVivo by way of combined inductive and deductive coding (based on theoretical perspectives from CME in particular) and concept mapping. The data yielded various themes, such as "posing questions", "feeling the numbers", "taking actions", "modelling for lived democracy", and "approaches to solving the question" (see also Steffensen & Kasari, 2023). The themes were discussed with colleagues from the larger project.

The following analysis and discussion focuses on three situations in the modelling activity: boat refugees, refugees in Europe, and refugees worldwide. These examples

[1] *Refugees: The Shared Story of Harry and Ahmed*: https://www.youtube.com/watch?v=APWyeCuvk5Q.

were chosen because they demonstrate how responsible citizenship can occur within an integrated STEM approach.

The STEM Modelling Activity

The PTs were inspired by a pedagogical approach called "mathematics in three acts" (Meyer, 2011; Zhou & Hansen, 2022). In the first act, teachers introduce the central conflict visually to inspire students to wonder and pose questions using as few words as possible. In the second act, students come up with approaches to investigate the situation and consider what tools to use. In the last act, the students present their findings. The PTs began by showing four images to the school students. One such image shows asylum seekers travelling by boat off the coast of Africa in the Mediterranean (https://www.worldpressphoto.org/collection/photo-contest/2015/massimo-sestini/1). The other images include refugee camps in Bulgaria, Ethiopia, Venezuela, and Myanmar.

After seeing the images, the school students decided which questions to ask in groups. To support them, the PTs asked questions such as.

What does this make you wonder about?

Why does this problem exist?

What is the cause?

Who benefits from the situation?

Who is affected?

How does this influence us and our personal histories?

The PTs helped the students to pose questions by broadening the context and providing them with emotional support (Steffensen & Kasari, 2023).

After the group discussions, the PTs summarised the questions on the whiteboard in plenary before continuing with group work. Two groups focused on boat refugees, two on refugees in Europe, and one on refugees worldwide. During the students' investigations, the four PTs and the practicum teachers supported them by asking questions and providing technical assistance. The students presented their findings twice, first in an oral plenary discussion and then with a written poster.

Enquiring About Refugees in Overcrowded Boats

The image of the overcrowded boat made an impact on the school students. They live close to the sea and most of them are aware of the threat to life of too many people in a boat at sea. On seeing these images, the boat design, number of people on board, buoyancy, length of journey, conditions at sea, access to life jackets, fresh water, food, warm clothing, fuel, and so on were all relevant issues for them to investigate and discuss. Two groups decided to estimate the number of boat-refugees shown in the picture. One group counted everyone in the image one by one, which may work when numbers are small. However, counting large numbers of people is challenging and time-consuming. It is also easy to lose track of who has already been counted. The students used the image on the whiteboard, which may have made it difficult to count. If the PTs had provided a printed image, it might have facilitated the counting process. Students could have used different colours to group the refugees, making counting easier. This approach could provide an estimate of 354 individuals (17 groups of 20 plus 14), and it does not require advanced mathematical competency and can be used in the early school years. Although this approach has clear limitations regarding larger crowds, counting people in this way can be used as an initial step in coming up with an estimate.

Another group counted the number of people in two directions (length and width), then multiplied these numbers to find the answer. A width count of 15 people and length count of 40 gives an estimate of around 600. The school students simplified and mathematised the situation by using suitable mathematical formulas that they were familiar with. Although they recognised that the boat was not rectangular, an essential feature of the students' problem-solving strategy was that they thought of using a formula in a new situation.

The PTs indicated that after a while, the school students became confused using the counting technique. In addition, these two groups had different estimates when they presented their results in the oral plenary session. Whereas one had counted around 600, the other group had a lower count. The students reacted to this dissimilarity. According to Mai, the discrepancy provided an opportunity to dwell on the possible causes. They discussed how the fact that the people were sitting in a boat might make it challenging to estimate their number. They reflected on the challenges in counting, their different approaches, and the uncertainty of their findings. Both groups used the same number (around 600) when presenting their poster.

Estimating the number of refugees is important for several reasons, such as in the allocation of various forms of humanitarian aid (e.g. medical and technological aid) or in political or military interventions. However, estimates can also serve political interests. Greer (2009) discusses how the low estimates of Iraqi deaths in 2003 were used to downplay the effects of the invasion. If citizens were under informed about the extent of casualties, their political opposition to the occupation might be different than if they were well-informed, and Greer emphasises that this poses a challenge to democratic functions. Estimates of gatherings and protests have become politically loaded, as in the controversy around the Trump inauguration in 2017 or the earlier demonstrations in Hong Kong (Watson & Yip, 2011). In the 2015 European migrant crisis, estimates were used to draw a picture of out-of-control immigration, which led to the strengthening of border controls in many countries. If students learn various approaches to estimating for politically tense situations, as in the refugee crisis, this can provide them with competencies essential for responsible citizenship. Consideration of the ethical, social, cultural, and economic aspects of such estimations is in line with what Maass et al. (2023) describe as a potential way of promoting responsible citizenship among students on a serious societal topic.

The discrepancies among the school students' estimates and how they dealt with them are like the findings of Elicer et al. (2021) and Triantafillou et al. (2018). Triantafillou et al. (2018) described how ninth-grade students estimated the number of people at an antiracism demonstration in Greece. Although the students focused on various estimates, they insisted on asking what the correct answer was. This apparent search for one solution suggested that students may found it difficult not to provide "one correct answer". It also suggested that they were moving towards a mutual understanding, which could be beneficial in a real-life situation. Elicer et al. (2021) connected an estimation task to students' protests in Chile by encouraging twelfth-grade students to participate in discussions about how decision-making drew upon mathematical arguments. By acknowledging that all the students had different estimates, the students made sense of statistical variation. However, reflection on the issue and the use of mathematics were challenging for the students and teachers involved, since they needed to experience embedded uncertainty as an essential feature of estimating.

By using this image, the PTs drew their school students' attention to a severe aspect of the refugees' situation; thus, teachers can play an essential role in bringing critical societal issues to the fore. The PTs deliberately focused on this issue, and invited the school students to design a task, simplify and mathematise information, estimate, calculate, validate, and present their findings, which were all essential to developing modelling competencies in line with the ideas of CME.

Using images to estimate the number of boat refugees is difficult for several reasons. One of these is the serious question of why the refugees were risking their lives in small boats. The school students were wondering how many of them fell overboard. There are important stories behind such images. Therefore, throughout the lessons, the school students processed their thoughts holistically in various ways together with peers, and their regular teacher, who knew them well. Another estimating challenge was the limited reference points in the image that could be used to support the calculation of the area of the boat, and the school students had few tools to help them. This contrasts with images of street demonstrations, as in the studies by Elicer et al. (2021) and Triantafillou et al. (2018), where Google Maps could be used for measuring. However, the use of technology to estimate crowd size presents ethical challenges (connected with for example the use of surveillance cameras, facial recognition programs, or mobile phone signals to map individuals) provides an opportunity for reflection in STEM education.

Responsible Citizenship, STEM, and the Refugee Situation in Europe

Refugees who come to Europe often start their journey in refugee camps (e.g. in Greece, Italy, or Bulgaria). Various countries have different policies regarding refugees, and where they later end up can vary. In integrated STEM approaches, it may be relevant to investigate the conditions in refugee camps, for instance by enquiring about issues such as access to clean water, the sanitary conditions, housing, beds, education, medical help, and so on. Students can also be supported to design solutions to challenges. The school students in this research wanted to find information about refugees in Europe. One group searched randomly online to find relevant articles and websites. Another group used a recommended interactive statistics website showing global refugee figures from the Norwegian Refugee Council (https://www.flyktninghjelpen.no/global/statistikk/statistikk/ and https://www.flyktninghjelpen.no/global/flyktningregnskapet-landprofiler/). The website offers abundant information, and the students navigate it and select relevant figures from interactive maps.

The ability to access, critically understand, and interact with various media are highlighted as a key competency of citizenship (Geiger et al., 2023). According to Geiger et al. (2023), terms such as "data literacy" and "information literacy" draw not only on mathematical and statistical components but also on ideas and concepts beyond these disciplines. When socio-political issues are included in STEM education and political parties might deliberately try to exert their influence, students need the competency to judge whether the information they find online is reliable, which is an aspect of responsible citizenship. The Internet is not a straightforward source, and students need a search strategy and source awareness competencies. For instance, fake news is often accompanied by scientific "facts" or numbers that look authentic and are used to influence (Hauge et al., 2019). Source awareness and critique are particularly important when the topic is controversial and causes public debate, as in the refugee crisis.

Both groups of school students identified and processed relevant data from websites recommended by the PTs. Online searches could be time-consuming and fruitless, and difficult for young students to judge the quality of their findings. Anna reflected on whether they made it too easy for the students when recommending websites: "So, one can discuss whether this approach fosters students' critical thinking." The group that randomly searched online continually asked the PTs whether sites were appropriate. Their use of the PTs to confirm the authenticity and validity of websites shows that they knew not everything on the Internet is reliable, which can be viewed as a first step towards source criticism. For teachers, it can be expedient to recommend a suitable website, as the PTs did. However, the PTs could have involved the students in judging how trustworthy the websites were to develop the students' source criticism skills. They could have asked students questions about who the author was, what the author's goals might have been, what was being said, and what the purpose was. Depending on the students' age and experience, teachers need to find a balance between suggesting suitable Internet websites and supporting students with questions that advance their critical thinking skills to encourage independent and reflective judgement.

Teachers play an essential role in helping students to master various technologies. It may be appropriate to simplify procedures during modelling activities, for instance, or when working with large data sets or computationally intensive activities. The school students decided to create a spreadsheet of the number of refugees from various countries and summarise the numbers. While making the spreadsheet, Mai noticed that the school students had manually summarised the numbers instead of using the spreadsheet tools. She indicated that she was "really surprised", because the boy, in her opinion, was "very digitally smart". Mai showed the students how to use formulas in the spreadsheet and how they can use various tools such as graphs. Although students may appear very digitally smart, this does not necessarily mean that they have mastered specific programs (e.g. spreadsheets or apps) or Internet search skills (e.g. the use various engines, specific keywords, quotation marks, avoiding commercials, etc.).

Digital technology can assist students who are investigating, experimenting, simulating, organising, calculating, and creating visualisations to communicate results (Greefrath et al., 2018; Siller & Greefrath, 2010). The school students in this research used technology mainly to retrieve, organise, and process information (e.g. in spreadsheets and shared online Word documents) and to communicate their findings. To make the activity more cognitively demanding, the PTs could have supported the school students in the use of spreadsheets to estimate the potential number of refugees arriving over time, the impact on local resources, and so on. Foreseeing students' approaches to modelling activities where they formulate their own problems is not straightforward, nor is integrating perspectives from the four STEM subjects (or other subjects) (Nicol et al., 2019). However, it is important for teachers consider these aspects in advance so that they are better prepared to support students by asking questions and leading them to more challenging activities with the potential to include integrated approaches. According to Geiger et al. (2023), the increased level of media and digital resources requires a focus on digitally enabled data literacy to empower citizens to access and select data sources and to analyse and apply quality assurance principles to the data.

Anna and Vilde indicated that the school students became more engaged when they were enquiring about the facts as opposed to when it was "us just telling how many refugees it is". They reflected on how this process influences school students' knowledge of refugees. Vilde said, "I think many gained much greater insight into everything that has to do with refugees." Anna immediately followed up by saying, "So, I feel they are more involved in the task because it is something they may care about or have begun to care about because of this project." These statements indicate that they think the refugee context made a difference, either as something real they already cared about or as something they became engaged with during their project. It is essential to care about other human beings when making decisions for society. Getting students engaged through integrated STEM approaches to enquire about societal issues can help to develop students' competency in investigating and being informed by relevant information and also allows them to care for other citizens in crisis, which may be crucial when gaining a standpoint as a responsible citizen.

Responsible Citizenship and the Refugee Situation Worldwide

The PTs deliberately included images of refugees from various parts of the world, such as Venezuela, Ethiopia, and Myanmar. As a result, several student posters included statements such as "In Venezuela, 4,878,561 million people are fleeing" and "In Mali, there are 326,000 refugees in the country, and the military has arrested the president, prime minister and defence minister." Teachers have an important role to play in broadening the context to include perspectives beyond a Norwegian or European perspective, because most refugees do not reach Norway or Europe. Instead, they flee to neighbouring countries in Africa, Asia, or South America. As responsible citizens, we should be concerned not only with what happens in our neighbourhood but also globally, as are these school students. Such concerns align with what Geiger et al. (2023) say about global citizenship, namely that educational programmes should focus on domestic and international issues. Also, they problematise how programmes related to global citizenship tend to be in social studies subjects, which overlooks the potential for valuable contributions from other curriculum areas.

One group investigated the refugee situation in several countries worldwide. It began by identifying the substantial number of refugees worldwide, currently over 100 million.[2] According to Mai, they calculated the percentage so that they would understand better, "because the numbers of refugees are so large, and they are too large to understand". During their work, the school students found a video[3] about education and refugees on one of the recommended websites (UN Refugee Agency, 2022).

The video presents percentages of children attending school. For instance, information about that although 91% of all children worldwide attend school, only 50% of refugee children do. In their poster, the group wrote "50% of refugee children attend school. However, that means that 50% do not attend school and do not have a good life." Thus, the school students emphasised the meaning by inverting the first sentence to show that 50% of refugee children do not attend school. They investigated how the refugee crisis affects refugee children. They were interpreting what the numbers meant. Perhaps they related to the refugee children and felt sympathetic towards them because they could not attend school. One might say they were statistically empathetic (as described by Mukhopadhyay & Greer, 2007). They did not take a stance of neutrality. Instead, they stressed the seriousness of the refugee children not having a good life. Large numbers can be challenging to understand (Renert, 2011). How one chooses to represent refugee numbers can make a difference. Making representations more understandable (e.g. as percentages) can help students better understand what the numbers mean.

By including this important message from the video in their poster, the students took a stand on issues relevant to their lives. One could say they were acting as responsible citizens. They used mathematics (amongst other skills) to communicate

[2] This figure is from June 2022; at the time the students investigated it, it was 82.4 million.

[3] *Refugee education in crisis: three standout statistics* https://www.youtube.com/watch?v=OQcpG8OCG2U&t=12s.

a message that was an expression of responsible citizenship. Scientific facts and mathematical representations like these not only bring about awareness, they can format our actions in the refugee crisis (Skovsmose, 2021, 2023). I will conclude this discussion with a statement from one of the school students as an example of taking a stand as a responsible citizen:

> I think it is very stupid that they [countries/governments] do not accept anyone [refugees] since so many people are fleeing. All countries should take in refugees since there are so many of them. Everyone must contribute to the world, or the world will not be a better place to live in for those who come after us.

Concluding Comments and Practical Implications

To understand how PTs can engage students in responsible citizenship, this chapter investigated three situations from a modelling activity in a PT practicum with seventh grade students, where the context was the refugee crisis. The first situation involved students' estimating the number of refugees in a boat. The second situation involved their using digital technology to investigate refugees in Europe. In the third situation, the students identified relevant information about refugees worldwide, interpreted it, and took a stand on what the figures represented. Although 'Democracy and Citizenship'was recently introduced as an interdisciplinary topic in the revised Norwegian curriculum and is new to PTs and teacher educators, this chapter shows some relevant aspects of how integrated STEM education activities can contribute to the development of responsible citizenship.

Practical implications of this chapter for teacher educators, researchers, and teachers in the STEM field include the need to include social-political topics in the teaching and learning of integrated STEM. The PTs had the school students engage with the refugee crisis, which provided opportunities for them to advance their ethical, political, and social reasoning alongside perhaps more traditional STEM-based argumentation (Maass et al., 2022, 2023). It also provided opportunities for school students to take a stand on the refugee crisis and communicate this to peers, which is an essential part of responsible citizenship. Practical implications also include the design of the activity and the role of PTs. The PTs inspired the activity by showing the images and inviting school students to investigate by posing questions of their own, resulting in various questions for classroom exploration. They facilitated the school students' use of digital technology, which helped to develop their information literacy as part of citizenship (Geiger et al., 2023). The PTs broadened the context to include global concerns over the refugee situation, which aligns with ideas about global citizenship (Geiger et al., 2023).

This research adds to and supports the research suggesting that mathematics teacher education should provide opportunities for integrated STEM approaches (Anderson, 2020; Anderson et al., 2020; Kurup & Li, 2022; Maass et al., 2019). This could occur in a school-based practicum, as in this research, where PTs had the opportunity to teach integrated STEM approaches during their teacher training (see also

Kurup & Li, 2022). Mathematical modelling is an effective way of demonstrating the relevance of mathematics in these issues. It can include (politically loaded) estimations, the use of technology, and the interpretation of scientific facts and figures and related implications. However, a focus on mathematics should not preclude focusing on other STEM subjects, as emphasised by Venville et al. (2002) or problematised by Nicol et al. (2019). The PTs in this study did not explicitly include science as one of the integrated subjects. Future research should investigate mathematical modelling involving responsible citizenship and focus explicitly on connecting it to all four STEM disciplines.

References

Almeida, L., & Carreira, S. (2019). The configuration of mathematical modelling activities: A reflection on perspective alignment. In U. T. Jankvist, M. v. d. Heuvel-Panhuizen & M. Veldhuis (Eds.), *Eleventh congress of the European society for research in mathematics education* (pp. 1112–1119). ERME. https://hal.science/hal-02408663/document

Anderson, J. (2020). The STEM education phenomenon and its impact on school curriculum. *Curriculum Perspectives, 40*(2), 217–223.

Anderson, J., English, L., Fitzallen, N., & Symons, D. (2020). The contribution of mathematics education researchers to the current STEM education agenda. In J. Way, C. Attard, J. Anderson, J. Bobis, H. McMaster, & K. Cartwright (Eds.), *Research in mathematics education in Australasia 2016–2019* (pp. 27–57). Springer.

Barbosa, J. C. (2009). Mathematical modelling, the socio-critical perspective and the reflexive discussions. In M. Blomhøj & S. P. Carreira (Eds.), *Proceedings from topic study group 21 at the 11th international congress on mathematical education in monterrey, Mexico* (pp. 133–143). Roskilde Universitet.

Bousalis, R., & Furner, J. M. (2020). Linking middle school mathematics and social studies through immigration issues. *Middle School Journal, 51*(1), 19–28.

Elicer, R., et al. (2021). A crowd size estimation task in the context of protests in Chile. In G. A. Nortvedt, N. Buchholtz, J. Fauskanger, F. Hreinsdóttir, M. Hähkiöniemi, & B. E. Jessen (Eds.), *Bringing Nordic mathematics education into the future* (pp. 33–40). Svensk Förening för Matematikdidaktisk Forskning.

English, L. D. (2023). Ways of thinking in STEM-based problem solving. *ZDM—Mathematics Education*. 55(7), 1219–1230. https://doi.org/10.1007/s11858-023-01474-7

Ernest, P. (2015). The social outcomes of school mathematics: Standard, unintended or visionary? *International Journal of Education in Mathematics, Science and Technology, 3*(3), 187–192.

Geiger, V., Gal, I., & Graven, M. (2023). The connections between citizenship education and mathematics education. *ZDM—Mathematics Education, 55*(5), 923–940. https://doi.org/10.1007/s11858-023-01521-3

Greefrath, G., Hertleif, C., & Siller, H.-S. (2018). Mathematical modelling with digital tools—A quantitative study on mathematising with dynamic geometry software. *ZDM Mathematics Education, 50*(1–2), 233–244.

Greer, B. (2009). Estimating Iraqi deaths: A case study with implications for mathematics education. *ZDM Mathematics Education, 41*(1–2), 105–116.

Hard Science. (2023, June). In Oxford Learner's Dictionaries. Retrieved June 19, 2023 from https://www.oxfordlearnersdictionaries.com/definition/english/hard-science

Hauge, K. H., Kacerja, S., & Lilland, I. E. (2019). Xenophobia and numbers in the media –discussing mathematics education in the post-truth era. *Philosophy of Mathematics Education Journal,* (35).

Huang, L., Ødegård, G., Hegna, K., Svargård, V., Helland, T., & Seland, I. (2017). *Unge medborgere: Demokratiforståelse, kunnskap og engasjement blant 9.-klassinger i Norge. The International Civic and Citizenship Education Study (ICCS 2016)*. IEA, NOVA.

Julie, C., & Mudaly, V. (2007). Mathematical modelling of social issues in school mathematics in South Africa. In W. Blum, P. L. Galbraith, H. Henn, & M. Niss (Eds.), *Modelling and applications in mathematics education* (pp. 503–510). Springer.

Kurup, P., & Li, X. (2022). Building preservice teachers' confidence for teaching and learning STEM, incorporating innovative, interdisciplinary, and integrated (3I) initiatives. In M. A. Peters (Ed.), *Encyclopedia of teacher education*. Springer.

Maass, K., Geiger, V., Ariza, M. R., & Goos, M. (2019). The role of mathematics in interdisciplinary STEM education. *ZDM—Mathematics Education, 51*(6), 869–884.

Maass, K., Sorge, S., Romero-Ariza, M., Hesse, A., & Straser, O. (2022). Promoting active citizenship in mathematics and science teaching. *International Journal of Science and Mathematics Education, 20*(4), 727–746.

Maass, K., Zehetmeier, S., Weihberger, A., & Flößer, K. (2023). Analysing mathematical modelling tasks in light of citizenship education using the COVID-19 pandemic as a case study. *ZDM—Mathematics Education, 55*(1), 133–145.

Mellin-Olsen, S. (1987). *The politics of mathematics education*. Reidel.

Meyer, D. (11 February 2011). The three acts of a mathematical story. https://blog.mrmeyer.com/2011/the-three-acts-of-a-mathematical-story/

Ministry of Education and Research. (2017). *Core curriculum—values and principles for primary and secondary education*. https://www.regjeringen.no/contentassets/53d21ea2bc3a4202b86b83cfe82da93e/core-curriculum.pdf

Ministry of Education and Research. (2019a). *Curriculum for mathematics year 1–10*. https://www.udir.no/lk20/MAT01-05?lang=eng

Ministry of Education and Research. (2019b). *Curriculum in natural science*. https://www.udir.no/lk20/nat01-04?lang=eng

Ministry of Education and Research. (2019c). *Curriculum for social studies*. https://www.udir.no/lk20/SAF01-04?lang=eng

Mukhopadhyay, S., & Greer, B. (2007). How many deaths? Education for statistical empathy. In B. Sriraman (Ed.), *International perspectives on social justice in mathematics education* (pp. 119–136). IAP.

National Research Ethics Committees. (11 October 2023). Guidelines for research ethics in the social sciences and the humanities. https://www.forskningsetikk.no/en/guidelines/social-sciences-humanities-law-and-theology/guidelines-for-research-ethics-in-the-social-sciences-humanities-law-and-theology/

Nicol, C., Bragg, L. A., Radzimski, V., Yaro, K., Chen, A., & Amoah, E. (2019). Learning to teach the M in/for STEM for social justice. *ZDM, 51*(6), 1005–1016.

Renert, M. (2011). Mathematics for life: Sustainable mathematics education. *For the Learning of Mathematics, 31*(1), 20–26.

Shaughnessy, J. M. (2013). Mathematics in a STEM context. *Mathematics Teaching in the Middle School, 18*(6), 324–324.

Siller, H.-S., & Greefrath, G. (2010). Mathematical modelling in class regarding to technology. In V. Durand-Guerrier, S. Soury-Lavergne & F. Arzarello (Eds.), *Proceedings of the sixth congress of the European society for research in mathematics education* (pp. 2136–2145).

Skovsmose, O. (2021). Mathematics and crises. *Educational Studies in Mathematics, 108*(1), 369–383.

Skovsmose, O. (2023). *Critical mathematics education*. Springer.

Skovsmose, O., Moura, A. Q., & Carrijo, M. (2023). Inclusive citizenship through mathematics education: A conceptual investigation. *ZDM—Mathematics Education, 55*, 941–951.

Steffensen, L., & Kasari, G. (2023). "Forced to flee"—mathematical modelling in primary school. *Journal of Mathematics and Culture, 17*(4), 264–284.

Triantafillou, C., Psycharis, G., Bakogianni, D., & Potari, D. (2018). Enactment of inquiry-based mathematics teaching and learning: The case of statistical estimation. In E. Bergqvist, M. Österholm, C. Granberg & L. Sumpter (Eds.), *Proceedings of the 42nd Conference of the International Group for the Psychology of Mathematics Education* (Vol. 4, pp. 291–298).

UN Refugee Agency (UNHCR). (2022). *Figures at a Glance*. Retrieved February 18, 2023 from https://www.unhcr.org/figures-at-a-glance.html

Venville, G. J., Wallace, J., Rennie, L. J., & Malone, J. A. (2002). Curriculum integration: Eroding the high ground of science as a school subject? *Studies in Science Education, 37*(1), 48–84. https://doi.org/10.1080/03057260208560177

Villarreal, M. E., Esteley, C. B., & Smith, S. (2015). Pre-service mathematics teachers' experiences in modelling projects from a socio-critical modelling perspective. In G. Stillman, W. Blum, & M. Salett Biembengut (Eds.), *Mathematical modelling in education research and practice* (pp. 567–578). Springer.

Vorhölter, K., Greefrath, G., Borromeo Ferri, R., Leiß, D., & Schukajlow, S. (2019). Mathematical modelling. In H. N. Jahnke & L. Hefendehl-Hebeker (Eds.), *Traditions in German-speaking mathematics education research* (pp. 91–114). Springer.

Watson, R., & Yip, P. (2011). How many were there when it mattered? *Significance, 8*(3), 104–107. https://doi.org/10.1111/j.1740-9713.2011.00502.x

Wolfmeyer, M., Chesky, N., & Lupinacci, J. (2015). STEM: Friend or foe to critical mathematics education? In S. Mukhopadhyay & B. Greer (Eds.), *Proceedings of the Eighth International Mathematics Education and Society Conference* (pp. 139–144). Ooligan Press, Portland State University, Oregon, US.

Wolfmeyer, M., Lupinacci, J., & Chesky, N. (2018). Inserting critical mathematics into STEM education. In R. Jorgensen & K. Larkin (Eds.), *STEM education in the junior secondary: The state of play* (pp. 273–294). Springer Singapore.

Zapata-Cardona, L., & Martínez-Castro, C. A. (2021). Statistical modeling in teacher education. *Mathematical Thinking and Learning, 25*(3), 64–78.

Zhou, S., & Hansen, R. (2022). When mathematics in three acts meets mathematical modelling. In J. Hodgen, E. Geraniou, G. Bolondi & F. Ferretti (Eds.), *Proceedings of the Twelfth Congress of the European Society for Research in Mathematics Education (CERME12)* (pp. 1–8). Free University of Bozen-Bolzano and ERME.

Chapter 18
Developing Teacher's Capacity to Implement Interdisciplinary STEM Tasks Specifically Targeted to Responsible Citizenship Education

Katja Maass and Oliver Straser

Abstract The world is facing severe global challenges such as climate change, food security, rising migration, social justice, or Pandemic. In these times, citizenship education seems more important than ever. How can this citizenship education relate to mathematics and science learning? In this paper we present an interdisciplinary international design research study MaSDIV in which we developed teaching and professional development materials connecting mathematics and science education to citizenship education. We will present interdisciplinary classroom materials from STEM education that serve the purpose of citizenship education as well as the principal design features of a related PD. The evaluation results of the accompanying evaluation of this PD course, which was implemented by partners in six countries to support teachers in connecting mathematics and science education with citizenship education were promising. For example, our results show that in general, most participating teachers reported a high overall satisfaction with the course. They also indicate that teachers' self-efficacy beliefs about using realistic contexts relevant to society, their learning-related beliefs about the benefits of using such contexts as well as their own teaching practice changed significantly after participating in the MaSDiV PD course.

Keyword Responsible citizenship · Teacher professional development · Classroom materials · Realistic contexts · Teachers' self-efficacy beliefs

K. Maass (✉) · O. Straser
International Centre of STEM Education, University of Education, Freiburg, Germany
e-mail: maass@ph-freiburg.de

J. Anderson and K. Makar (eds.), *The Contribution of Mathematics to School STEM Education*, https://doi.org/10.1007/978-981-97-2728-5_18

The Need for Integrated STEM Education for Citizenship Education

The world is facing severe global challenges such as climate change, food security, rising migration, social justice issues and pandemics. Such challenges affect fundamental values of freedom, democracy as well as human rights, and therefore we need a society capable of finding adequate solutions for global challenges whilst respecting fundamental values. In this context, it is vital that citizens across the world learn how to use the mathematical and scientific competencies to evaluate what is being said about the challenges mentioned above, the data pointing to the challenges and the solutions to them. They must learn to differentiate between scientific facts, results based on science but influenced by other aspects like economics, ethics etc. and narratives which are not based on facts. Mathematics and science education can have a say in these demands and many of the solutions to the challenges mentioned above are based in mathematics and science.

Unfortunately we face—amongst others—one major challenge in this respect: science (including mathematics) education has focused on the learning *of* science (Hazelkorn et al., 2015), on pure science detached from societal implications. This focus can be contrasted with learning *of* and *about* science (Osborne & Dillon, 2008). Learning *about* science includes the social, cultural and ethical dimensions of science (e.g., decision-making in genetic engineering), which are often neglected in classrooms (Maass et al., 2019a, 2019b). Learning *of* and *about* science also fosters young people's understanding of nature, applications and implications of science (Hazelkorn et al., 2015). Consequently, by learning *of* and *about* science, students learn principles vital in democratic and pluralistic European societies. In this sense, science and mathematics education can also become part of citizenship education.

However, implementing innovative teaching approaches or even such a change of perspective on STEM teaching is not an easy task. Research and experience reveal that innovative teaching approaches promoted by educational research differ significantly from the day-to-day practices of teachers in many countries (Van den Akker, et al., 2006; Burkhardt, 2018; Burkhardt & Schoenfeld, 2003; Cobb & Jackson, 2015, Krainer, 2011).

To overcome this gap in relation to citizenship education in STEM, we developed materials for a long-term professional development (PD) course designed within the project *MaSDiV* (Supporting mathematics and science teachers in addressing diversity and promoting fundamental values, 2017–2020, Erasmus + KA3). This project involved six countries. From each country a higher education institution and an educational authority participated in the project, to design, pilot and optimize the course. The course was designed to last at least 14 h. The design of the PD course was based on a comprehensive theoretical basis.

Theoretical Basis for the Course Design

Connecting Science and Mathematics Education with Citizenship Education

The idea of connecting STEM education with citizenship education is not new. In mathematics education, mathematical modelling has been discussed for decades. Mathematical modelling is an approach to bringing extra-mathematical contexts into lessons (for an overview see e.g. Burkhardt, 2018). In science education, research proposes the engagement of socio-scientific issues (SSI) as a path to developing citizenship competences. SSI are controversial, socially relevant, real-world problems that are informed by science and often include an ethical component (Sadler et al., 2007, Zeidler & Nicols, 2009). When implementing both modelling and SSI in day-to-day lessons, we can use inquiry-based learning (IBL), a student-centered learning approach, in which students are actively involved in inquiry-related processes. We now examine the three approaches—mathematical modelling, socio-scientific issues, and inquiry-based learning.

Mathematical Modelling—There are many different definitions of mathematical modelling (Kaiser & Sriraman, 2006). In this paper, we regard mathematical modelling as solving an extra-mathematical problem from the real world by carrying out a modelling process (Niss et al., 2007) and take the modelling process as outlined in Fig. 18.1 as a basis.

Mathematical modelling can motivate students to engage more deeply in mathematics and develop a realistic perspective on mathematics (Kaiser, 1995), foster mathematical and scientific literacy (Steen, 2001), have a positive impact on students'

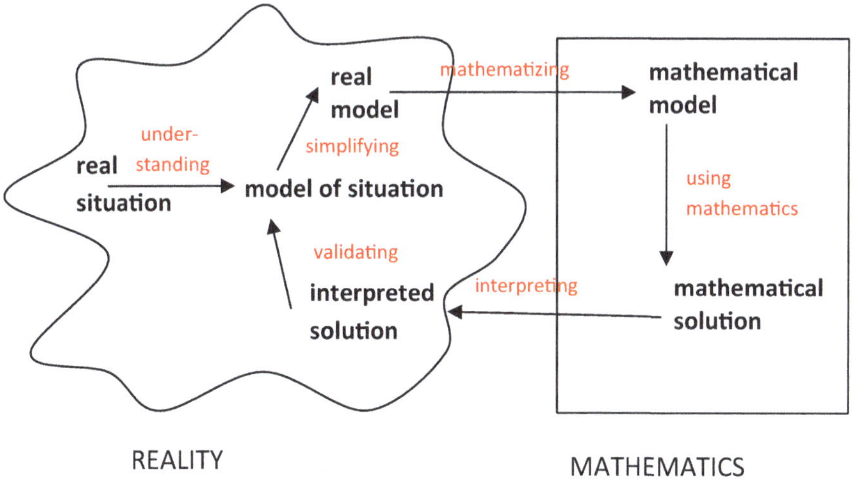

Fig. 18.1 An idealized scheme of the modelling process (adapted from Blum & Leiß, 2007)

competence in applying mathematics to complex situations (see e.g. Maass, 2007), mathematical competencies (e.g. English & Watson, 2018) and on students' attitudes towards mathematics (e.g., Mischo & Maass, 2013).

Citizenship education as described earlier requires the selection of appropriate situations and problems. Such problems should include decision making and ethical, moral, social or cultural aspects. Unfortunately, these aspects are often not explicitly mentioned in discussions about mathematical modelling (though not excluded). Discussions about decision-making and controversy are explicitly allocated to socio-scientific issues (Sadler, 2011).

Socio-Scientific Issues—Socio-scientific issues (SSIs) engage students in dialogues, discussions, and debates based on science. They have a controversial nature and students need to form an opinion and decide on this issue. This decision might include taking moral, ethical or social aspects into account (Zeidler & Nichols, 2009). When deciding on such matters, your decisions do not only rely on unambiguous natural sciences but also on personal values and attitudes. Naturally, such issues serve primarily the purpose of educating for scientific citizenship (Owen et al., 2012). A typical example of an SSI is the question whether to shutdown to overcome COVID-19 was reasonable.

Research has shown that SSIs can be used as contexts for learning scientific content (Applebaum et al., 2006; Walker, 2003; Zohar & Nemet, 2002), for understanding the nature of science (Osborne & Dillon, 2008), and for citizenship education (Herman et al., 2018; Radakovic, 2015; Sadler et al., 2007).

Theoretical lenses such as modelling and SSI in mathematics or science education can provide pathways for choosing relevant problems but they do not refer to specific pedagogies that should be implemented in science and mathematics classrooms. One approach that has shown to be appropriate for dealing with relevant problems is inquiry-based learning (IBL) (Knippels & van Dam, 2017). Consequently, combining modelling and SSI with IBL approaches may have the potential to promote active citizenship in mathematics and science education.

Inquiry-Based Learning (IBL)—IBL is a student-centered learning approach, in which students are involved in inquiry-related processes such as observing phenomena and creating their own questions, selecting mathematical or scientific approaches, creating representations to clarify relationships, seeking explanations, interpreting and evaluating solutions as well as communicating those solutions (for details see Maass et al., 2021).

This active stance that students take, as promoted in IBL, serves the purpose of citizenship education. It supports the development of critical thinking and decision making skills, learning to consider ethical, social and cultural aspects, and learning to deal with controversy (Geiger et al., 2015; Zeidler & Nichols, 2009).

Modelling, SSI and IBL—Based on similar reflections, Maass et al., (2019a, 2019b) proposed extending the mathematical modelling cycle to include a focus on decision-making, especially in dealing with controversial real-world problems that require the consideration of moral, ethical or social issues. Both climate change and the COVID-19 pandemic are good examples of modelling applications that require careful analysis of these issues. When following this modelling process, students

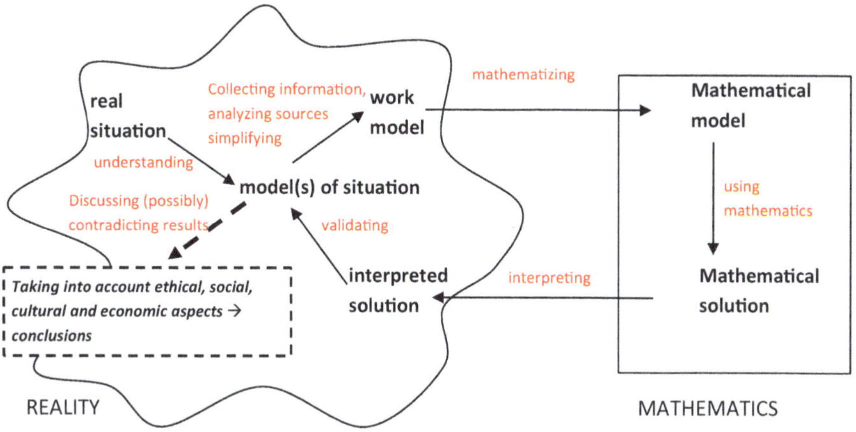

Fig. 18.2 Extended modelling process for controversial real-world problems (Maass et al., 2019a, 2019b)

learn that decisions on SSI are not only based on scientific evidence but also depend on personal values and attitudes, which is an important learning result for students (Fig. 18.2).

The reflections above show that citizenship education can be connected to science education and mathematics education by using the concepts of mathematical modelling, socio-scientific issues and inquiry-based learning. However, when we aim at designing a PD course, we also need to consider the characteristics of effective PD.

Professional Development of Teachers

When talking about teachers' PD, we relate it to growth in teachers' professional knowledge (Shulman, 1986). Additionally, PD should also attend to teachers' classroom practice (Clarke & Hollingsworth, 2002), as well as their motivation, self-regulation, and beliefs (Baumert & Kunter, 2013). Teachers' beliefs comprise implicit and explicit notions about teaching and learning (Pajares, 1992). Teachers' motivation and self-regulation are strongly connected to teachers' self-efficacy beliefs (Bandura, 1997). Self-efficacy beliefs describe the beliefs about one's own capability to execute a given task despite possible difficulties (Tschannen-Moran & Woolfolk Hoy, 2001). Both teachers' learning-related beliefs and self-efficacy beliefs are key aspects of their professional competencies that serve as amplifiers and filters in bringing the professional knowledge of teachers into action (e.g. Gess-Newsome, 2015).

In recent decades, considerable research has been carried out on effective PD. PD courses should take into account teachers' needs (Guskey, 2002) and the challenges that teachers face (Maass, 2011). They should also combine phases of learning in seminars with phases of learning at school (Lipowsky & Rzejak, 2015) and initiate cooperation between teachers (McLaughlin & Talbert, 2006). Naturally, they need to be relevant to teaching practice (Clarke, 1994) and should foster teachers' reflection on their beliefs about mathematics and science teaching as well as on their teaching experiences (Tirosh & Graeber, 2003). These criteria for PD courses have proven effective in practice (see e.g. Maass & Engeln, 2018, 2019) and provide guidelines for the design of PD courses on a general level. To reach out to teachers at scale with this PD concept, we need to consider various approaches to instructional improvement at scale.

General Approaches to Instructional Improvement at Scale

Innovative teaching approaches cannot be implemented by addressing teachers' knowledge and understanding alone. To overcome the theory–practice-gap, Maass and et al., (2019a, 2019b) suggest a framework for instructional improvement at scale that can involve more stakeholders from across an education system. This framework lists the following six general approaches:

1. **Research**: filling gaps on pressing issues, practitioners must deal with in their teaching.
2. **Resources**: designed for different contexts (for classroom teaching, professional development) and tailored to the respective target groups.
3. **Professional development:** starting from teachers' needs, developed in cooperation with schools, in different formats (e.g. workshops, coaching, online), considering scaling-up.
4. **Contexts**: analyzing the context and considering the different contexts (e.g. size of a country, curriculum, teacher education and diversity of students).
5. **Cooperation**: between different stakeholders, supported by sustainable cooperation structures.
6. **Communication**: via target group-oriented channels and language to spread scientific results and transfer knowledge to an audience which can make use of it and to promote research results and raise awareness to a wide public audience.

Approaching implementation from this perspective aims to overcome prior predominant research structures, such as the linear research—development—dissemination model with different groups of experts taking responsibility for each of these three stages (Begg et al., 2003). In the following, we outline the design of the PD course based on our theoretical philosophy, and give insight into the course we used for piloting. We also outline how we designed the PD course for instructional improvement at scale and provide implications for practice.

Design of the Professional Development Course Within the Project *MaSDiV*

Design Process of the PD Materials

In order to develop tangible materials of high quality that are usable across the project consortium of six countries, we decided to design the PD materials within an iterative, cyclic process of design, consultation, re-design and trial (Gravemeijer & Cobb, 2006). As we had to (1) design materials on an international level and then (2) adapt them for use in the individual national contexts, we split the design into two phases.

During design phase 1, three cycles of design, review and optimization of the international MaSDiV course took place. In the first cycle, the Dutch design team formulated the initial aims of the PD course and then designed the first version of the course. This version was screened by project consortium members. In the second round, an international expert team apart from the partnership reviewed the designed materials and provided feedback. Finally, in the third cycle, the PD course was evaluated by project partners with their national teams (consisting of teacher educators, teachers, researchers, policy makers). Thus, in the three cycles the materials were reviewed from different perspectives, namely from a geographical perspective (diverse European countries), from a disciplinary perspective (experts in science, mathematics and general education), and from a stakeholder perspective (designers and researchers, teacher educators as well as policy makers and representatives from education ministries). In each cycle experts were asked whether the design team should keep, change, or delete individual activities of the PD course and to provide arguments for their choice. Our main criteria for evaluation in research phase I were; the theoretical basis, consistency (the intervention is logically designed and supports the intended learning goals) and practicality.

In design phase 2, the international course was adapted to each national context and piloted to investigate usability and implementation of the underlying theoretical ideas in practice (van den Akker et al., 2006). The national project teams discussed necessary changes with their teacher educators, representatives of high-level school authorities, colleagues from universities and teachers and adapted the materials before piloting them.

This means, we designed the professional development course in cooperation with different stakeholders and we considered both the context and teachers' needs. We developed resources for classroom teaching and PD and implemented the course. So, our design process considered four of six general approaches (resources, professional development, context and cooperation) to instructional improvement at scale (Maass et al., 2019b).

The Design Results: The MaSDiV Professional Development Concept and an Example Used in the Intervention

The course was based on the theoretical basis as outlined earlier. To this end, our materials included controversial questions such as

Do we improve the environmental situation of our planet by buying food in zero-waste shops?
Should judges solely rely on DNA-tests in criminal cases in court?
Do we fight social exploitation with buying fair-trade products?

The aims of this module were to:

(1) understand the value of using a modelling, SSI and IBL approach in science and mathematics and apply this in classroom teaching;
(2) understand how modelling, SSI and IBL in science and mathematics can support active citizenship; and
(3) understand the nature, applications and implications of science and mathematics, including moral and ethical issues, for societies.

After introducing IBL, modelling and different reality-based contexts to participants, they were asked to discuss reasons for using contexts in STEM and compared different types of contexts. Afterwards, we addressed the question of "Can the earth feed us—is their enough food?" by given participants the following text:

Meat or vegetables

Most people in the world are vegetarians. The main crops are wheat, rice, corn and potatoes. Eating meat is one of the rich Western diets, and this is questionable from an environmental perspective. Cattle herds in the world use larger and larger land-areas, more water and energy.

An increasing proportion of the world's cereal production is used to feed animals. But the animals have a low efficiency. When the grain is passing through animals, 96% of the calories are 'lost' and only 10% of the protein from the grain remains in the animal. Conclusion: the more animal products mankind eats, the less people can be satisfied. (Svanfeldt & Svensson, 2000, p. 77)

Participants were asked e.g.: What subject-specific content and concepts (and questions and activities) can be related to this context? What role can the context have and what are the benefits of using this context? What opportunities and challenges does this context present for IBL, promoting scientific literacy and for being relevant for society? At a later stage, participants were asked to provide a STEM-based solution to this question and then to design a lesson suitable for their students.

Participants were also asked to select further contexts, suitable for citizenship education, discussing them in similar ways and to design their own tasks and related lessons. Finally, they discussed challenges and issues related to using such tasks for citizenship education and approaches to addressing them.

Considering the time needed for effective change to day-to-day teaching and considering the importance of both learning in the seminar and at school, we offered long-term PD courses (Lipowsky & Rzejak, 2015). Several PD days were spread over a longer period. We initiated cooperation between teachers (McLaughlin & Talbert, 2006) as follows. First, *during the seminar*, teachers worked on contexts as explained above that made them familiar with the challenges of connecting STEM education with citizenship education. Methods used in the module allowed teachers to work in groups, to investigate problems, to reflect on issues, and to discuss open questions. *In the implementation phase at school*, teachers were encouraged to adapt and use the activities in their own classrooms. We encouraged them to visit and observe each other's lessons to give further opportunities for feedback and reflection. Finally, *back in the PD course*, teachers shared their classroom experiences and discussed the pedagogical implications. This cyclic process was repeated at each PD event, as new pedagogical issues were addressed.

Methodology

Field Trials

The MaSDiV PD courses were implemented in all six partner countries. National teams adapted the international model where necessary. Subsequently, the national implementations varied to some extent (mainly regarding the organizational framework), whilst supporting the major topics based on the same materials and major pedagogical principles. The PD course included at least 14 in-person hours for the participating teachers. In Spain, teachers were also offered online-follow-up support for several months focusing on transferring the pedagogical approach to teachers' practice. In 2018 and 2019, 453 voluntary teachers took part in the PD courses.

Design of the Evaluation Questionnaire

To assess the development of participating teachers during the MaSDiV PD, a quantitative pre-post-design was used. Teachers filled out a questionnaire at the beginning of the first PD course session and at the end of the last session. The questionnaires included descriptive background variables, items on the perception of the PD program and on teachers' self-efficacy beliefs, teachers' learning-related beliefs, and teaching practices (Baumert & Kunter, 2013; Pajares, 1992). The background variables comprised demographic information such as age, gender, subject, and previous formal learning experience of IBL. Items were assessed on a four-point Likert-scale whenever appropriate.

Perception of the MaSDiV PD. In the post-questionnaire, a total of 15 items were designed for this purpose. Following the theoretical discussion on effective PD, the items on the perception of PD were carefully chosen to cover key quality criteria such as the satisfaction with the experience, the utility and application of the content

Table 18.1 Overview of the sample and subject distribution (abbreviations are country-codes)

Country	CY	GE	MT	NL	SP	TR
N	35	35	81	57	53	50
Mathematics	38%	100%	51%	31%	29%	56%
Science	62%	0%	49%	69%	71%	44%

(adapted based on Grohmann & Kauffeld, 2013), collaboration during professional development (based on Cordingley et al., 2005), and the role of materials as a resource (Donna & Hick, 2017; Grossman & Thompson, 2008).

Teachers' self-efficacy beliefs, learning-related beliefs, and practices. These aspects are key outcome measures of effective PD and drivers of sustainable change and were consequently assessed with a paper–pencil test in the pre- and post-test (for details see Sorge, 2020). See the appendix.

The construction of the four items related to self-efficacy beliefs (SEB) focused on assessing the confidence in one's own capability to implement contexts in classrooms successfully (e.g. Bandura, 1997; Pajares, 1996). The scale for teachers' learning-related beliefs (LRB)was also newly developed and focused on teachers' explicit notions about effective teaching and learning (e.g. Kleickmann et al., 2016) with a specific focus on the use of contexts relevant to society. Teaching practices (TP) regarding the use of contexts relevant to society were assessed with a newly developed self-report instrument of three items. All scales show good reliabilities ($\alpha_{LRB} = 0.80$, $\alpha_{SEB} = 0.79$, $\alpha_{TP} = 0.74$).

Sample

From the total of 453 teachers who participated in the PD, matching pre-post data sets were available for N = 311 mathematics and science teachers from the six European countries Malta, Turkey, the Netherlands, Spain, Cyprus, and Germany. Two thirds of the teachers were female with a mean age of 39 years (SD = 11 years). The average teaching experience was 17 years (SD = 12 years). The overview of the sample is shown in Table 18.1. Overall, 49% of the participants were mathematics teachers and 51% taught science at school.

Analysis Strategies

To analyze the degree teachers' self-efficacy beliefs, learning-related beliefs, and practices changed across the MaSDiV PD and which factors contributed to that change, we used paired t-tests as well as multiple regression analyses. Three different multiple regression analyses were specified with teachers' self-efficacy beliefs, learning-related beliefs, and practices as dependent variables, and characteristics from the teachers and the PD as independent variables (for details see Maass et al., 2021).

Results and Discussion

The MaSDiV PD program was designed to support mathematics and science teachers in developing their capacity to implement interdisciplinary STEM tasks specifically targeted to responsible citizenship education. We investigated i) how teachers experienced the program, ii) how the program impacted core aspects of teachers' professional competence and practice, and iii) what factors contributed to the impact of the program.

The $N = 341$ teachers that answered the post-test questionnaire showed a high overall satisfaction with their experiences during the professional development program, with a mean value of 3.11 ($SD = 0.52$). Furthermore, as indicated in Fig. 18.3a, 90% of the teachers rated their experiences above a value of 2.5. Teachers also rated the quality and number of applicable materials with a mean value of 2.75 ($SD = 0.60$). Figure. 18.3b shows that 69% of the participating teachers agreed or strongly agreed with the quality and number of applicable materials. Accordingly, 31% of the teachers rate the quality and number of applicable materials below 2.5 (i.e., as not sufficient enough).

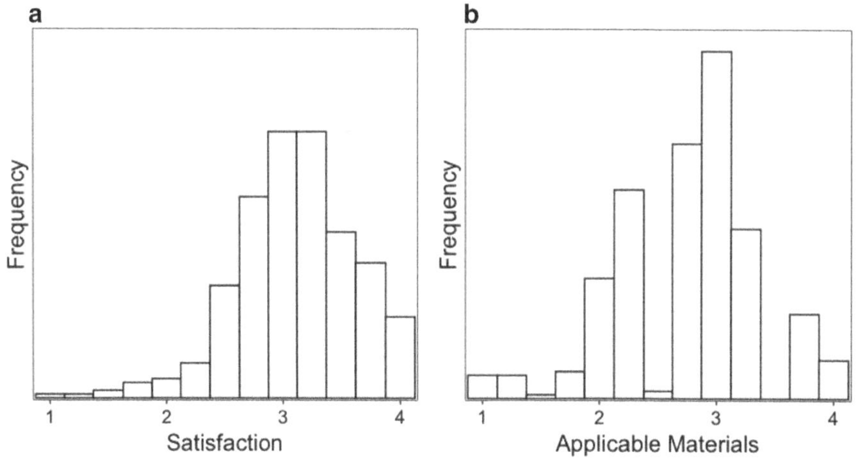

Fig. 18.3 The perception of the PD course from participating teachers addressing **a** general satisfaction and **b** the provision of applicable materials (Maass et al., 2021)

Summing up, 90% of all teachers rated that they were satisfied or even highly satisfied with the PD course and their own growth. This shows that the design process according to design research as outlined above actually lead to a PD course targeted to teachers' needs and wishes. The feedback on the quality and availability of applicable materials as one quality indicator of the PD was more mixed compared to the feedback on general satisfaction. While a majority of 69% of the teachers also agreed or strongly agreed that they received high-quality materials, a substantial number of teachers were less satisfied with the provided materials.

To investigate changes in teachers' self-efficacy beliefs, learning-related beliefs, and practices after the completion of the professional development program, we used data from $N = 311$ teachers from the pre- and post-test. Table 18.2 summarizes the findings from the comparison between pre- and post-test.

Our results show that teachers' self-efficacy beliefs about using relevant contexts, their learning-related beliefs about the benefits of using relevant contexts and their own teaching practice changed significantly after participating in the PD course. The strongest impact of the course was found for teachers' self-efficacy beliefs. The change in teachers' learning-related beliefs was significant too, although with a small effect size. This is not surprising as teachers' learning-related beliefs are considered to be stable over time (Törner, 2002) and the participating teachers already had rather strong beliefs about the benefits of using contexts for student learning, and thus, less room for improvement. The latter is not surprising as teachers applied to participate in the PD course and therefore were interested in the topic.

Importantly, our pre-post-test comparison showed that participating teachers indeed reported that they use socially relevant contexts more frequently compared to the beginning of the PD course. Consequently, a well-planned PD can support teachers in applying complex teaching approaches such as citizenship education in mathematics and science.

In a third step, we investigated which influencing factors regarding the teachers, their school, and the professional development, contributed to the outcomes of the PD program. Table 18.3 shows how characteristics of teachers (prior beliefs, prior IBL learning and their perception of cultural diversity at their school) and characteristics of the professional development program impact the influence of teachers' self-efficacy, their learning-related beliefs and their practice.

Table 18.2 Teachers' self-efficacy beliefs, learning-related beliefs, and teaching practice from the pre- and post-test including t-value from paired t-test and Cohen's d

Scale	Pre	Post	*t*	*d*
Self-efficacy beliefs	2.80 (0.48)	2.99 (0.43)	7.19***	0.41
Learning-related beliefs	3.25 (0.42)	3.32 (0.45)	3.13**	0.18
Teaching practice	2.18 (0.63)	2.38 (0.66)	5.63***	0.32

Note *** $p < 0.001$, ** $p < 0.01$

Table 18.3 Multiple regression model investigating the impact of teacher and PD characteristics on teachers' self-efficacy beliefs, learning-related beliefs, and practices

	Self-efficacy beliefs	Learning-related beliefs	Practices
Prior beliefs/practices	0.45***	0.38***	0.48***
Prior IBL learning	0.07	− 0.11*	0.01
Cultural diversity at school	− 0.11*	− 0.14**	-0.01
Spanish concept with online support	0.03	0.11*	0.12*
Satisfaction with PD	0.14*	0.05	0.06
Applicable materials	0.06	0.11	0.17**
R^2	0.25	0.24	0.32

Note *** $p < 0.001$, ** $p < 0.01$, * $p < 0.05$

Teachers' satisfaction of the course correlated with the degree teachers' self-efficacy increased during the PD. Teachers who felt that they learnt something during the PD and had valuable opportunities to discuss their ideas with other teachers also reported significant higher self-efficacy beliefs after the PD course. Consequently, teachers' satisfaction also impacts teachers' self-efficacy, which is another reason, why a careful design process is important.

Through the combination of teacher, school, and PD characteristics, we were able to explain a substantial amount of 32% of the variance for teachers' post-test answers on their practices. For teachers' self-efficacy beliefs and learning-related beliefs, the regression model explains 25% and 24% of the variance respectively.

Our results show that the MaSDiV PD impacted teachers' self-efficacy beliefs, learning-related beliefs, and practices differently based on their personal characteristics, their school characteristics, and the specifics of the PD course itself. The change in teachers' learning-related beliefs differed based upon teachers' prior learning-related beliefs and their opportunities in relation to IBL. The MaSDiV PD course had a substantial impact for teachers with limited prior knowledge on the topics at hand. Here, specialized courses for advanced teachers or ongoing online support for questions in relation to implementation into practice (see Spain) might be supportive.

Another factor contributing to the impact of the PD course was the perceived cultural diversity at the teachers' own school. Teachers who reported a high cultural diversity in their own school had significant weaker self-efficacy beliefs and learning-related beliefs compared to teachers with less cultural diversity. One reason might be that teachers in schools with high perceived diversity focused less on the use of contexts for citizenship education, but rather on basics in mathematics and science education, since contexts often require complex speech comprehension, which is known to be one of the main error sources (Radatz, 1979).

Returning to the applicable materials, we saw that whereas a majority of 69% of the teachers also agreed or strongly agreed that they received high-quality materials, a substantial number of teachers were less satisfied with the provided materials. Our regression analysis (Table 18.3) also shows that teachers who felt that they received high-quality applicable teaching materials, also used contexts in their own classroom more frequently. On the one hand, this feedback may suggest that there is still a need to increase the number of ready-to-use materials in the course. This is in line with other research that highlights that teaching and curricular materials affect instructional practice and student learning (Maass et al., 2019a, 2019b; Ross et al., 2003; Schoen et al., 2003). An increase of applicable materials for the PD course could therefore have an impact on teachers' actual practice in their own classrooms. On the other hand, externally designed classroom materials do not always meet the requirements of the respective student groups and their achievement levels. Also, for reasons of capacity building, it seems to be worthwhile to enable teachers to design their own tasks (Davis et al., 2011; McFadden & Roehrig, 2017; Voogt et al., 2016). In conclusion, it seems to be worthwhile balancing between given exemplary materials and also encouraging teachers to design their own materials.

Our results also indicate that the course in Spain had an additional impact on teachers' learning-related beliefs and practices. A possible explanation could be that Spain provided substantially longer online support of the PD-program, with a special emphasis on transfer to teaching practices as well as promoting the exchange of ideas and experiences among teachers in a professional learning community. For the next PD course, such an online support in all countries might be impactful.

Our results underline the need for long-term PD courses. Teachers who attended the course in Spain with the long online support had significantly stronger learning-related beliefs about the benefits of using contexts and reported a more frequent use of contexts in their own classrooms. Seemingly, this extended support and trial phase enabled the teachers to change their teaching practice consistently and also reflect on those experiences, which caused an additional change in teachers' learning-related beliefs (Guskey, 2002). Therefore, one measure to enlarge the impact of a PD course on teachers might be to provide teachers with extended support and opportunities for 'trying-out' in order to change their teaching practice towards citizenship education.

Implications for Practice

As mentioned before, teachers participated voluntarily in the PD courses in all countries. Thus, our results must be interpreted with caution since voluntarily participating teachers are expected to have a strong motivation to work on their own teaching practice. Consequently, our results cannot be generalized to all science and mathematics teachers.

Second, we used a paper–pencil test that relies heavily on the perception of teachers' own practice as well as their reflection on their own beliefs. While such a paper–pencil test certainly provides valuable insights and is an ecological assessment of teachers' beliefs and practices, additional research is necessary. This could involve using students' assessments of teaching practices as well as qualitative data to gather a more complex picture of teachers' belief systems (Luft & Roehrig, 2007).

Despite those limitations, our study provides implications regarding how science and mathematics instruction can be used to support citizenship education.

- To impact teaching practice in relation to citizenship education in mathematics and science, it is important to design related materials for classroom teaching and for professional development based on design research.
- To develop materials of high quality it is important to design materials in iterative, cyclic process of design, consultation, re-design and trialling.
- Within this design process, we designed the *PD course* in *cooperation* with different stakeholders and we took into account the *context* as well as teachers' needs.
- The results show how important it is to design materials tailored to the target groups' needs as this has an impact on the self-efficacy, learning-related beliefs and the teaching of teachers.
- Within the course there is a need to balance between providing best practice exemplary teaching materials and having teachers design their own materials.
- It seems important to develop *resources* for classroom teaching and *professional development* when implementing the programme.
- An ongoing online support for teaching in their teaching practice seems to be helpful as the Spanish example shows. This support is not as time-consuming as ongoing face-to-face courses and therefore addresses teachers' time limitations.
- More specialized courses for advanced teachers might be helpful to further support them.

Last but not least, we know from research (e.g. Burkhardt & Schoenfeld, 2003; Maass et al., 2019a, 2019b; Maass & Artigue, 2013) that the development of one PD course and its piloting in six countries with 400 + teachers will not change teaching practice at scale nor will it lead to more PD courses about this topic if not actively followed up.

As a result, to provide long term support and connections in relation to innovation in STEM education, we established the permanent structure of the *International Centre of STEM Education* (ICSE; icse.eu) in 2017. ICSE is an internationally connected research centre with the aim to help improving STEM education across Europe through practice-related research and its transfer into practice. As ICSE considers international collaboration as key to continuous improvements in European STEM education they have initiated the foundation of the ICSE consortium. The ICSE Consortium to date comprises 16 higher education and research institutes from across Europe.

Following on from the need for *communication* as a general approach to instructional improvement at scale, we use ICSE's communication channels (social media channels, newsletters) and ICSE's website to promote the idea of citizenship education in STEM long term. The idea of *cooperation* with different stakeholders of STEM education (ministries, community institutions, school, out-of-school learning) to improve STEM education which was used in MaSDiV for the design of the PD course and was followed up in ICSE and now supports spreading the idea of citizenship education among the different key actors.

To provide online-availability of the materials on a larger platform, ongoing PD activities on this topic and ongoing PD of the PD providers, ICSE and the ICSE consortium have set up a project called ICSE Academy (2022–2025, Erasmus+, Teacher Academy, https://icse.eu/icse-academy/). The ICSE Academy intends to innovate teacher education based on existing best practices and transnational exchange strategies. The ICSE Academy partners are higher education researchers, STEM initial teacher education (ITE) and PD providers, educational policy makers, and schools from 13 countries learning with, from and about each other through specific innovative collaborative structures. The ICSE Academy will offer and evaluate PD in different formats which are not only supposed to support teachers' professional learning but also the mutual learning of the PD providers (such as job-shadowing—ITE/CPD providers attend each other's courses to learn from each other—or an interdisciplinary European workshop series (for teachers across Europe, run jointly by all ITE/CPD providers). Topics of the PD will vary but will also include activities on citizenship education in STEM. The ICSE Academy has also set up an international material platform including materials for classroom teaching, teacher education and professional development for several relevant topics and particularly for citizenship education.

By this measure, the ICSE Academy as a new project of ICSE (Consortium), will continue to work on the connection of citizenship education and STEM. Our work as outlined in this paper is an example of how innovative instruction can impact schools.

Acknowledgements The project MaSDiV received funding from the European Union Erasmus+ Programme under grant agreement n° 582943-EPP-1-2016-2-DE-EPPKA3-PI-POLICY. This paper reflects only the authors' views and the European Union is not liable for any use that may be made of the information contained herein.

Appendix

Learning-related beliefs

To what extend do you agree with the following statements?

	Strongly disagree	Disagree	Agree	Strongly agree
Science and math instruction needs to address issues relevant to society to foster students' values	☐	☐	☐	☐
Using contexts relevant to society allows students to develop deeper understanding	☐	☐	☐	☐
Instruction needs to provide students with opportunities to discuss issues relevant to society	☐	☐	☐	☐

(continued)

(continued)

Situating activities in contexts relevant to society will help improve students' ability to make justified decisions	☐	☐	☐	☐

Self-efficacy beliefs

To what extend do you agree with the following statements?

I feel confident that …	Strongly disagree	Disagree	Agree	Strongly agree
…I can incorporate issues relevant to society in my teaching	☐	☐	☐	☐
…I can develop teaching activities addressing issues relevant to society	☐	☐	☐	☐
…I can use inquiry-based activities to engage students in dealing with issues relevant to society	☐	☐	☐	☐
…I can facilitate classroom discussions about issues relevant to society	☐	☐	☐	☐

Inquiry-based teaching practice

How often do the following activities take place in your lessons?

In my teaching …	Never or hardly ever	Some lessons	Most lessons	Almost every lesson
…my students discuss controversial issues relevant to society	☐	☐	☐	☐
…I use contexts to promote fundamental values (e.g. democratic values)	☐	☐	☐	☐
…I use contexts relevant to society	☐	☐	☐	☐

References

Applebaum, S., Barker, B., & Pinzino, D. (2006). Socioscientific issues as context for conceptual understanding of content. *Paper presented at the National Association for Research in Science Teaching.*

Bandura, A. (1997). *Self-efficacy: The exercise of control*: Freeman.

Baumert, J., & Kunter, M. (2013). The COACTIV model of teachers' professional competence. In M. Kunter, J. Baumert, W. Blum, U. Klusmann, S. Krauss, & M. Neubrand (Eds.), *Cognitive Activation in the mathematics classroom and professional competence of teachers* (pp. 25–48). Springer.

Begg, A., Davis, B., & Bramald, R. (2003). Obstacles to the dissemination of mathematics education research. In A. J. Bishop, M. A. Clements, C. Keitel, J. Kilpatrick, & F. K. S. Leung (Eds.), *Second International Handbook of Mathematics Education* (pp. 593–634). Kluwer Academic Publishers.

Burkhardt, H., & Schoenfeld, A. (2003). Improving educational research: Towards a more useful influential and better-funded enterprise. *Educational Researcher, 32*(9), 3–14.

Burkhardt, H. (2018). Ways to teach modelling—a 50 year study. *ZDM, 50*, 61–75. https://doi.org/10.1007/s11858-017-0899-8

Blum, W., & Leiß, D. (2007). How do students and teachers deal with modelling problems? In: C. Haines, W. Blum, P. Galbraith & S. Khan (Eds.), *Mathematical modelling (ICTMA 12): Education, engineering and economics* (pp. 222–231). Chichester, UK: Horwood.

Clarke, D. (1994). Ten Key Principles from research for the professional development of mathematics teachers. In D. B. Aichele & A. F. Coxford (Eds.), *Yearbook / National Council of Teachers of Mathematics: Vol. 1994. Professional development for teachers of mathematics* (pp. 37–47). National Council of Teachers of Mathematics.

Clarke, D., & Hollingsworth, H. (2002). Elaborating a model of teacher professional growth. *Teaching and Teacher Education, 18*(8), 947–967. https://doi.org/10.1016/S0742-051X(02)00053-7

Cobb, P., & Jackson, K. (2015). Supporting teachers' use of research-based instructional sequences. *ZDM, 47*, 1027–1038.

Cordingley, P., Bell, M., Thomason, S., & Firth, A. (2005). *The impact of collaborative continuing professional development (CPD) on classroom teaching and learning: Review: How do collaborative and sustained CPD and sustained but not collaborative CPD affect teaching and learning?* (Research Evidence in Education Library). Retrieved from EPPI-Centre, Social Science Research Unit, Institute of Education, University of London website: http://eppi.ioe.ac.uk/cms/Portals/0/PDF%20reviews%20and%20summaries/cpd_rv2.pdf?ver=2006-03-02-124802-077

Davis, E. A., Beyer, C., Forbes, C. T., & Stevens, S. (2011). Understanding pedagogical design capacity through teachers' narratives. *Teaching and Teacher Education, 27*(4), 797–810.

Donna, J. D., & Hick, S. R. (2017). Developing elementary preservice teacher subject matter knowledge through the use of educative science curriculum materials. *Journal of Science Teacher Education, 28*(1), 92–110. https://doi.org/10.1080/1046560X.2017.1279510

English, L. D., & Watson, J. (2018). Modelling with authentic data in sixth grade. *ZDM, 50*(1–2), 103–115. https://doi.org/10.1007/s11858-017-0896-y

European Commission/EACEA/Eurydice. (2016). *Promoting citizenship and the common values of freedom, tolerance and non-discrimination through education: Overview of education policy developments in Europe following the Paris Declaration of 17 March 2015*: Publications Office.

Geiger, V., Goos, M., & Forgasz, H. (2015). A rich interpretation of numeracy for the 21st century: A survey of the state of the field. *ZDM, 47*(4), 531–548. https://doi.org/10.1007/s11858-015-0708-1

Gess-Newsome, J. (2015). A model of teacher professional knowledge and skill including PCK: Results of the thinking from the PCK Summit. In A. Berry, P. J. Friedrichsen, & J. Loughran (Eds.), *Teaching and learning in science series. Re-examining pedagogical content knowledge in science education: [October 2012, PCK Summit Working conference in Colorado Springs]* (1st ed., pp. 28–42) Routledge.

Grohmann, A., & Kauffeld, S. (2013). Evaluating training programs: Development and correlates of the questionnaire for professional training evaluation. *International Journal of Training and Development, 17*(2), 135–155. https://doi.org/10.1111/ijtd.12005

Grossman, P., & Thompson, C. (2008). Learning from curriculum materials: Scaffolds for new teachers? *Teaching and Teacher Education, 24*(8), 2014–2026. https://doi.org/10.1016/j.tate.2008.05.002

Guskey, T. R. (2002). Professional development and teacher change. *Teachers and Teaching, 8*(3), 381–391. https://doi.org/10.1080/135406002100000512

Gravemeijer, K., & Cobb, P. (2006). Design research from a learners' perspective. In J. v. d. Akker, K. Gravemeijer, S. Mc Keeney & N. Nieveen (Eds.), *Educational design research* (pp. 17–51). Oxford: Routledge Chapman & Hall.

Hazelkorn, E., Ryan, C., Beernaert, Y., Constantinou, C. P., Deca, L., Grangeat, M., Karikorpi, M., Lazoudis, A., Casulleras, R. P., & Welzel-Breuer, M. (2015). *Science education for responsible citizenship*: Report to the European Commission of the expert group on science education: Publications Office. Retrieved from http://ec.europa.eu/research/swafs/pdf/pub_science_education/KI-NA-26-893-EN-N.pdf

Herman, B. C., Sadler, T. D., Zeidler, D. L., & Newton, M. H. (2018). A socioscientific issues approach to environmental education. In G. Reis & J. Scott (Eds.), *International Perspectives on the Theory and Practice of Environmental Education: A Reader* (pp. 145–161) Springer. https://doi.org/10.1007/978-3-319-67732-3_11

Kaiser, G. (1995). Realitätsbezüge im Mathematikunterricht: Ein Überblick über die aktuelle und historische Diskussion. *Materialien für einen realitätsbezogenen Mathematikunterricht*, 2, 66–84. Retrieved from https://publishup.uni-potsdam.de/opus4-ubp/frontdoor/index/index/docId/26937

Kaiser, G., & Sriraman, B. (2006). A global survey of international perspectives on modelling in mathematics education. *ZDM, 38*(3), 302–310. https://doi.org/10.1007/BF02652813

Kleickmann, T., Tröbst, S., Jonen, A., Vehmeyer, J., & Möller, K. (2016). The effects of expert scaffolding in elementary science professional development on teachers' beliefs and motivations, instructional practices, and student achievement. *Journal of Educational Psychology, 108*(1), 21–42. https://doi.org/10.1037/edu0000041

Knippels, M. -C., & van Dam, F. (2017). PARRISE, promoting attainment of responsible research and innovation in science education, FP7. *Impact, 2017*(5), 52–54. https://doi.org/10.21820/23987073.2017.5.52

Krainer, K. (2011). Teachers as stakeholders in mathematics education research. In B. Ubuz (Ed.), *Proceedings of the 35th conference of the international group for the psychology of mathematics education* (Vol. 1 pp. 47–62). Ankara, Turkey: Middle East Technical University.

Lipowsky, F., & Rzejak, D. (2015). Key features of effective professional development programmes for teachers. *Ricercazione, 7*(2), 27–51.

Luft, J. A., & Roehrig, G. H. (2007). Capturing science teachers' epistemological beliefs: The development of the teacher beliefs interview. *Journal of Science Education, 11*(2), 38–63.

Maass, K. (2007). Modelling in class: What do we want students to learn. In C. Haines, P. Galbraith, W. Blum, & S. Khan (Eds.), *Mathematical modelling (ICTMA 12): education, engineering and Economics. Proceedings from the twelfth international conference on the teaching of mathematical modelling and applications* (pp. 63–78) Elsevier.

Maass, K. (2011). How can teachers' beliefs affect their professional development? *ZDM, 43*, 573–586. https://doi.org/10.1007/s11858-011-0319-4

Maaß, K., & Artigue, M. (2013). Implementation of inquiry-based learning in day-to-day teaching: A synthesis. *ZDM—The International Journal on Mathematics Education, 45*(6), 779–795.

Maass, K., & Engeln, K. (2018). Impact of professional development involving modelling on teachers and their teaching. *ZDM, 50*(1–2), 273–285. https://doi.org/10.1007/s11858-018-0911-y

Maass, K., & Engeln, K. (2019). Professional development on connections to the world of work in mathematics and science education. *ZDM, 51*(6), 967–978. https://doi.org/10.1007/s11858-019-01047-7

Maass, K., Doorman, M., Jonker, V., & Wijers, M. (2019a). Promoting active citizenship in mathematics teaching. *ZDM, 51*(6), 991–1003. https://doi.org/10.1007/s11858-019-01048-6

Maass, K., Sorge, S., Romero Ariza, M., Hesse, A. & Straser, O. (2021). Promoting active citizenship in mathematics and science teaching. *International Journal of Science and Mathematics Education*.https://doi.org/10.1007/s10763-021-10182-

Maass, K., Cobb, P., Krainer, K., et al. (2019b). Different ways to implement innovative teaching approaches at scale. *Educational Studies in Mathematics, 102*(3), 303–318. https://doi.org/10.1007/s10649-019-09920-8

Maaß, K., Artigue, M., Burkhardt, H., Doorman, M., English, L. D., Geiger, V., Krainer, K., Potari, D., & Schoenfeld, A. (2022). Mathematical modelling—A key to citizenship education. In N. Buchholtz, B. Schwarz, & K. Vorhölter (Eds.), *Initiationen mathematikdidaktischer Forschung* (pp. 31–50). Springer.

McFadden, J. R., & Roehrig, G. H. (2017). Exploring teacher design team endeavours while creating elementary-focused, STEM-integrated curriculum. *International Journal of STEM Education, 4*, 1–22.

McLaughlin, M. W., & Talbert, J. E. (2006). Building school-based teacher learning communities: Professional strategies to improve student achievement. *The series on school reform: Teachers College Press.* Retrieved from http://search.ebscohost.com/login.aspx?direct=true&scope=site&db=nlebk&db=nlabk&AN=158409

Mischo, C., & Maaß, K. (2013). The effect of teacher beliefs on student competence in mathematical modelling—An intervention study. *Journal of Education and Training Studies, 1*(1), 19–38.

Niss, M., Blum, W., & Galbraith, P. (2007). Introduction. In W. Blum, P. L. Galbraith, H.-W. Henn, & M. Niss (Eds.), *New ICMI Study Series: Vol. 10. Modelling and Applications in Mathematics Education: The 14th ICMI Study* (pp. 3–32) Springer.

Osborne, J., & Dillon, J. (2008). *Science Education in Europe: Critical Reflections.* A Report to the Nuffield Foundation.

Owen, R., Macnaghten, P., & Stilgoe, J. (2012). Responsible research and innovation: From science in society to science for society, with society. *Science and Public Policy, 39*(6), 751–760. https://doi.org/10.1093/scipol/scs093

Pajares, F. (1996). Self-efficacy beliefs in academic settings. *Review of Educational Research, 66*(4), 543. https://doi.org/10.2307/1170653

Pajares, M. F. (1992). Teachers' beliefs and educational research: Cleaning up a messy construct. *Review of Educational Research, 62*(3), 307–332. https://doi.org/10.3102/00346543062003307

Radakovic, N. (2015). "People can go against the government": Risk-based decision making and high school students' concepts of society. *Canadian Journal of Science, Mathematics and Technology Education, 15*(3), 276–288. https://doi.org/10.1080/14926156.2015.1062938

Radatz, H. (1979). Error analysis in mathematics education. *Journal for Research in Mathematics Education, 10*(3), 163. https://doi.org/10.2307/748804

Ross, J. A., McDougall, D., Hogaboam-Gray, A., & LeSage, A. (2003). A survey measuring elementary teachers' implementation of standards-based mathematics teaching. *Journal for Research in Mathematics Education, 34*(4), 344. https://doi.org/10.2307/30034787

Sadler, T. D. (2011). *Socio-scientific Issues in the Classroom: Teaching, learning and research* (Vol. 39): Springer Netherlands.

Sadler, T. D., Barab, S. A., & Scott, B. (2007). What do students gain by engaging in socio-scientific inquiry? *Research in Science Education, 37*(4), 371–391. https://doi.org/10.1007/s11165-006-9030-9

Schoen, H. L., Cebulla, K. J., Finn, K. F., & Fi, C. (2003). Teacher variables that relate to student achievement when using a standards-based curriculum. *Journal for Research in Mathematics Education, 34*(3), 228. https://doi.org/10.2307/30034779

Shulman, L. S. (1986). Paradigms and research programs in the study of teaching: A contemporary perspective. In M. C. Wittrock (Ed.), *Handbook of research on teaching: A project of the American Educational Research Association* (3rd ed.) Macmillan:

Sorge, S. (2020). *Report on the evaluation of the project supporting mathematics and science teachers in addressing diversity and promoting fundamental values (MaSDiV).* Results of policy experimentation. Retrieved from International Centre for STEM Education website: https://icse.eu/wp-content/uploads/2018/SonstigePDFs/MaSDiV_D5.1_WP5_Evaluation-Report.pdf

Steen, L. A. (Ed.) (2001). *Mathematics and democracy: The case for quantitative literacy:* NCED.

Svanfeldt & Svensson. (2000). *Medan jorden snurrar, Naturkunskap kurs A*, Natur & Kultur

Tirosh, D., & Graeber, A. O. (2003). Challenging and changing mathematics teaching classroom practices. In A. J. Bishop, M. A. Clements, C. Keitel, J. Kilpatrick, & F. K. S. Leung (Eds.), *Second International Handbook of Mathematics Education* (pp. 643–687) Springer Netherlands. https://doi.org/10.1007/978-94-010-0273-8_22

Törner, G. (2002). Mathematical beliefs—A search for a common ground: some theoretical considerations on structuring beliefs, some research questions, and some phenomenological observations. In G. C. Leder, E. Pehkonen, & G. Törner (Eds.), *Beliefs: A Hidden Variable in Mathematics Education?* (pp. 73–94) Springer Netherlands. https://doi.org/10.1007/0-306-47958-3_5

Tschannen-Moran, M., & Woolfolk Hoy, A. (2001). Teacher efficacy: Capturing an elusive construct. *Teaching and Teacher Education, 17*, 783–805.

Van den Akker, J., Gravemeijer, K. P. E., McKeeney, S., & Nieveen, N. (2006). Introducing educational design. In J. v. d. Akker, K. Gravemeijer, S. Mc Keeney, & N. Nieveen (Eds.), *Educational design research* (pp. 3–7). Routledge Chapman & Hall.

Voogt, J. M., Pieters, J. M., & Handelzalts, A. (2016). Teacher collaboration in curriculum design teams: Effects, mechanisms, and conditions. *Educational Research and Evaluation, 22*(3–4), 121–140.

Walker, K. A. (2003). *Students' Understanding of the Nature of Science and Their Reasoning on Socioscientific Issues: A Web-based Learning Inquiry (Unpublished dissertation).* University of Florida, Tampa, FL.

Zeidler, D. L., & Nichols, B. H. (2009). Socioscientific issues: Theory and practice. *Journal of Elementary Science Education, 21*(2), 49–58. https://doi.org/10.1007/BF03173684

Zohar, A., & Nemet, F. (2002). Fostering students' knowledge and argumentation skills through dilemmas in human genetics. *Journal of Research in Science Teaching, 39*(1), 35–62. https://doi.org/10.1002/tea.10008

Part VI
A Reflection

Chapter 19
Reflecting on the Contribution of Mathematics to STEM Education

Merrilyn Goos

Abstract It is commonly claimed that mathematics is the discipline that underpins all of STEM. However, mathematics often seems to occupy a "servant" role in integrated STEM education, acting as a tool to support learning in the other STEM disciplines. In this chapter, I first consider three other established approaches to interdisciplinary mathematics education—curriculum integration, mathematical literacy, and mathematical modelling—to identify theoretical and practical resources for integrated STEM education that pushes mathematics to the fore. I then organise my commentary on the contributions to this volume around five key research themes: interdisciplinary curriculum models and approaches; student experiences and outcomes; teacher preparation and professional development; classroom implementation and task design; and policy, structures, and leadership. This analysis suggests three perspectives on the role of mathematics in integrated STEM education that need to work together to enrich students' learning: mathematics as the *goal*, mathematics as a *tool*, and mathematics as an *integrative bridge* between the STEM disciplines.

Keywords Integrated STEM education · Mathematical literacy · Mathematical modelling · Mathematics education

Introduction

STEM education research is a relatively young field that has seen a dramatic increase in scholarly publications since about 2010 (Li et al., 2020). It is likely that most researchers who are publishing in this field have their primary background in one of the constituent S-T-E-M education disciplines—including mathematics education. In fact, it is commonly claimed that mathematics is the discipline that underpins all of STEM. Yet, analysis of the literature on integrated STEM education has found few examples of studies in which the role of mathematics was foregrounded: instead,

---

M. Goos (✉)
University of the Sunshine Coast, Sippy Downs, Australia
e-mail: mgoos@usc.edu.au

J. Anderson and K. Makar (eds.), *The Contribution of Mathematics to School STEM Education*, https://doi.org/10.1007/978-981-97-2728-5_19

in most of these studies, students used mathematics they already knew to support learning or problem solving in the other STEM disciplines (Baldinger et al., 2020; Just & Siller, 2022). It seems that we still have much to learn about how new mathematical concepts and practices can be developed through integrated STEM experiences and how mathematics can enrich understanding of the other STEM disciplines (English, 2016; Fitzallen, 2015)—hence, the existence of this book.

Why should we be concerned that mathematics often seems to be incidental, invisible, or even devalued in integrated STEM education? I believe that mathematics matters in integrated STEM education because we are preparing young people for a world that is inherently interdisciplinary and in which mathematics plays a central role in helping us solve complex social, ethical, and scientific problems (Geiger et al., 2023). In addition, technological advances due to digitalisation and automation of previously manual or routine tasks have increased, rather than diminished, the mathematical skills requirements in the workforce of the future in many areas such as health care, finance, engineering, and retail (Sakurai & Goos, 2023). Making sense of the world in which we live demands an interpretive lens that integrates, rather than keeps separate, the forms of disciplinary knowledge we have at our disposal—including mathematical knowledge.

In reflecting on our current understanding of the contribution of mathematics to STEM education, it is not my intention to review each of the chapters in this book. Instead, I begin by pointing to various approaches to mathematics education that connect mathematics with other subjects in the school curriculum or with real-world problems as a way of identifying theoretical and practical resources for integrated STEM education. I then introduce a framework for organising the literature on the role of mathematics in integrated STEM education and use it to show how the work presented in this book highlights key issues for the field, as well as gaps that remain to be investigated.

Interdisciplinary Mathematics Education

Integrating mathematics with other disciplines is not a new idea in school education. In the USA, the National Council of Teachers of Mathematics (2000) explained that students' understanding of mathematics is enhanced by recognising mathematical connections to other subjects. More recently, the *Australian Curriculum: Mathematics* stated that students should apply mathematics to model situations in various fields and disciplines (Australian Curriculum, Assessment and Reporting Authority [ACARA], 2023). One of the barriers to implementing integrated STEM education, however, lies in the common organisation of school curricula around separate subject areas. Nevertheless, we can gain insight into how STEM integration might be achieved by looking at the history of interdisciplinary approaches in mathematics education. For example, the work of the ICME Topic Study Group on interdisciplinary mathematics education, first introduced at ICME-13 in 2016, deals with integration amongst both STEM and non-STEM disciplines (Doig et al., 2019; Williams et al., 2016). Going back further in time, curriculum integration encompassing various degrees of interdisciplinarity has long been proposed as a way of enhancing student engagement in mathematics, developing richly connected mathematical knowledge, and revealing how this knowledge is used in real-world contexts—despite the many practical challenges involved in implementing such an approach (Berlin & Lee, 2005; Venville et al., 1998).

A second context for integrating mathematics with other disciplines – including the STEM disciplines—arises from the notion of mathematical literacy, or numeracy, as a twenty-first century competency (Maass et al., 2019). The definition of mathematical literacy used in the Programme for International Student Assessment (OECD, 2018) emphasises mathematical reasoning and problem-solving in real-world contexts, but the approaches that teachers are expected to use for developing students' mathematical literacy vary from one country to another. In Singapore, for example, support for mathematical literacy is the responsibility of mathematics teachers and there is no expectation that mathematics should be integrated with other subjects in the curriculum. In contrast, in some other countries all teachers are considered responsible for developing students' mathematical literacy (or numeracy) in the subjects they teach (e.g., ACARA, n.d.; Ministry of Education, New Zealand, 2023; Norwegian Directorate for Education & Training, 2012). The latter approach requires teachers to recognise the inherent numeracy demands of subjects across the curriculum, including subjects in the STEM areas.

Mathematical modelling provides yet another way to engage students in learning mathematics through an integrated approach that connects mathematics to real world situations and often to other school subjects as well. While various perspectives on mathematical modelling have been identified in the literature (Kaiser & Sririman, 2006), the goal of educational modelling is to improve students' understanding of the world while they learn mathematical concepts and methods. The contexts for educational modelling can come from disciplines other than mathematics, such as science, geography, art, sport, or economics. In the STEM education literature, there is already a strong emphasis on mathematical modelling as a tool for problem solving and a process for integrating the STEM disciplines (e.g., Leung, 2020; Stohlmann, 2020).

As I read the chapters in this book, I could hear echoes of each of the approaches described above: curriculum integration, mathematical literacy/numeracy across the curriculum, and mathematical modelling. STEM educators and researchers are already drawing on some of these resources to conceptualise integration, and there is surely value in exploring the "family connections" between these approaches to illuminate future work on meaningful integration of mathematics into STEM education.

A Framework for Reflecting on the Contribution of Mathematics to STEM Education

At around the same time as author teams were writing their chapters to be published in this volume, I was co-editing a special issue of *ZDM Mathematics Education* (with Susana Carreira and Immaculate Namukasa) on mathematics and interdisciplinary STEM education. In the survey paper that introduced the journal issue, we modified the STEM integration framework of Honey et al. (2014) to enable us to classify literature sources according to the issues that were most relevant to the role of *mathematics* in integrated STEM education (Goos et al., 2023). The thematic clusters are:

1. Interdisciplinary curriculum models and approaches;
2. Student experiences and outcomes;
3. Teacher preparation and professional development;
4. Classroom implementation and task design;
5. Policy, structures, and leadership.

This framework provides a starting point for me to reflect on the contributions this volume makes to research on mathematics in STEM education.

Interdisciplinary Curriculum Models and Approaches

Proponents of interdisciplinary mathematics education warn that progress in the field is hampered by confusion over key concepts and terminology used to describe the nature and scope of integration (Doig & Williams, 2019). Similarly, in integrated STEM education, there are different ways of classifying and labelling subject connections and disciplinary emphases. The classification devised by Vasquez et al. (2013) is probably the best known: it sets out a continuum of subject connections that increase in strength from multidisciplinary to interdisciplinary and transdisciplinary integration. How a particular example of STEM integration is classified depends on whether the concepts and skills of the connected disciplines are learned separately or together and the extent to which they are merged into a seamless whole. But questions remain about the methods and rationales used to connect *mathematics* with the other STEM disciplines. With the role of mathematics in mind, we might ask:

- *What* is to be integrated in integrated STEM education?
- By *what mechanism* is this integration process to be achieved?

One common way to answer the *What?* question involves connecting disciplinary knowledge across two or more STEM subjects taught at school, for example, by matching the content and learning outcomes of the respective curricula of these subjects. An example of disciplinary knowledge integration is shown in Table 19.1, taken from a study in which a mathematics teacher collaborated with a science teacher to design a sequence of integrated mathematics and science units for lower secondary school students (Goos & Askin, 2005). In this study, however, as in many others, mathematics became a tool for representing concepts and solving problems in science, and the mathematics teacher felt that science learning goals dictated the integrated units' content and structure.

A different approach to STEM integration involves blending discipline-specific forms of *epistemic knowledge*, that is, understanding of how knowledge is built through the distinctive inquiry and problem-solving practices of each of the STEM disciplines. Leung (2020) proposed that the signature forms of epistemic knowledge in S, T, E, and M are respectively scientific inquiry, computational thinking, engineering design thinking, and mathematical modelling. He then developed an integrated curriculum model that mapped these forms of knowledge onto a common epistemic frame based on Polya's problem-solving cycle. Figure 19.1 shows what the mapping would look like for integration of mathematics and science. Similar to the example of integrating forms of disciplinary knowledge presented above, this epistemic integration approach relies on collaboration between teachers who possess deep knowledge within their subject specialisms for effective integration to occur. Yet this may be an unrealistic expectation in contexts where teachers are poorly prepared or where out-of-field teaching is prevalent (see Hobbs & Porsch, 2022).

In my view, none of the integration models discussed and illustrated in this section adequately represents relationships between the constituent STEM disciplines or the mechanisms by which integration can be achieved, especially in relation to the role of mathematics. Perhaps we could think of these as being "first-generation" integration models. To move the field forward, we need to look beyond questions about which, and how many, disciplines to integrate and create meaningful STEM education experiences that make the contribution of mathematics both necessary and visible. Across the chapters in this volume (in particular, the chapters by Ärlebäck and Albarracin, Fitzpatrick et al., and Larsen et al.), I can see three perspectives on what this contribution might look like: mathematics as the *goal*, mathematics as a

Table 19.1 Integrating mathematics and science disciplinary knowledge in the lower secondary school curriculum

Mathematics content and outcomes	Science content and outcomes
Topics: • Variables, patterns and algebraic equations, measurement and graphing	Topics: • Gas Laws Boyle's Law $P_1V_1 = P_2V_2$ Charles's Law $\frac{V_1}{T_1} = \frac{V_2}{T_2}$ Ideal Gas Equation $PV = nRT$
Outcomes: • Interpret and model trends in data and solve problems using graphs, formulae and equations • Analyse data displayed in graphs and assess any unusual observations	Outcomes: • Investigate relationships between pressure, volume, and temperature

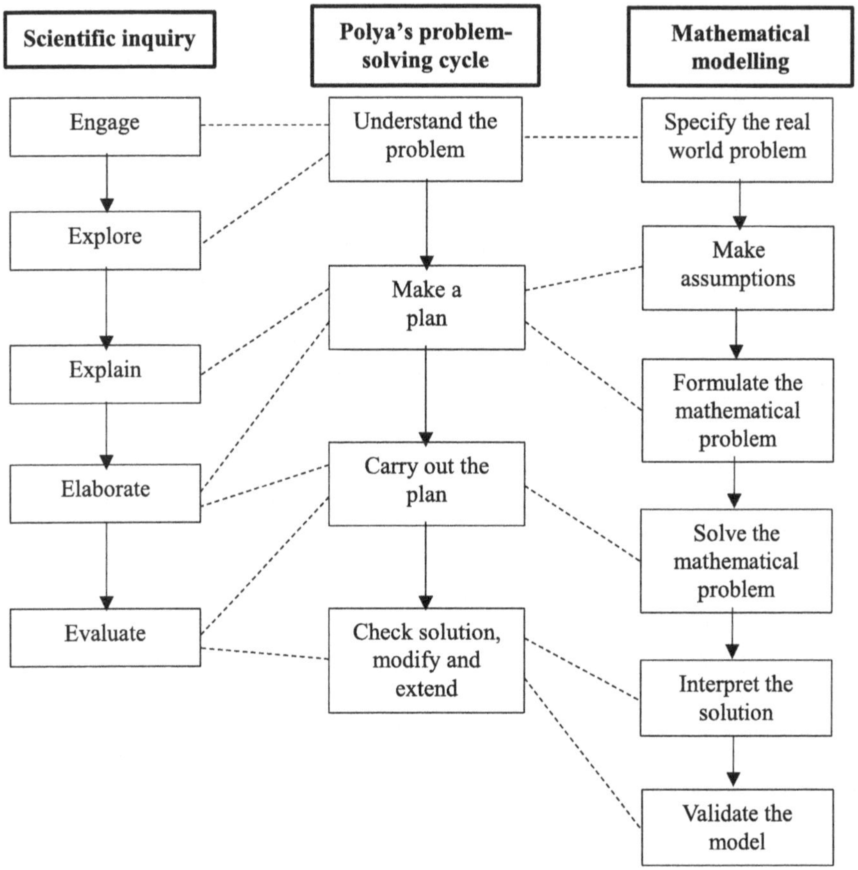

Fig. 19.1 Mapping epistemic knowledge in science and mathematics onto Polya's problem-solving cycle

tool, and mathematics as an *integrative bridge* between the STEM disciplines. When the goal of a STEM activity is to develop new mathematical competencies, the other STEM disciplines provide the context and incentive for learning new mathematics—in other words, mathematics sits in the foreground. In contrast, mathematics that is already familiar to students stays in the background when used as a tool for supporting learning or solving problems in the other STE(M) disciplines. (Note that I think this is still a valid and useful role for mathematics—but it should not be the *only* role for the M in STEM.) The third possibility is for mathematical modelling to serve as a bridge between the STEM disciplines as well as between the mathematical world and the real world—that is, to provide a mechanism for *how* integration is to be achieved. As Kawakami and Saeki point out (this volume), modelling involves back-and-forth movement not only between data and models, but also between the models that are used as idealised representations of reality in all the STEM disciplines. A worthwhile goal for future research and development is to find ways of making these three perspectives on the role of mathematics in integrated STEM education work together—not necessarily in individual STEM tasks, but over a long-term course of study—to add value to students' STEM experiences.

Student Experiences and Outcomes

A consistent finding across reviews of the STEM education literature is that research into student experiences and outcomes is under-developed and has focused mainly on the affective domain, for example, documenting improvements in student engagement, interest, and aspiration for further study of STEM subjects or careers in STEM (e.g., Gao et al., 2020; Goos et al., 2023; Li & Anderson, 2020). Much less is known about the cognitive benefits that students might experience from an integrated STEM approach. These benefits need to be investigated in two directions: how integrated STEM education affects students' learning of mathematics, and how mathematics contributes to students' integrated STEM learning. This research will be challenging, not only because it needs to pinpoint the mathematics-related causes and outcomes of integrated STEM learning, but also because we know so little about how to assess students' interdisciplinary STEM understanding and skills (Gao et al., 2020).

Some of the chapters in this volume provide encouraging evidence of students' mathematics learning in integrated STEM activities, drawing mostly on small scale studies. For example, Torralba and Rouse showed that a small sample of 7th grade students could represent their knowledge of proportional reasoning through maker tasks, using pencil and paper and then digital tools to design and construct a model of their dream home. Kawakami and Saeki presented a detailed case study of one Year 4 student in a teaching experiment involving data driven modelling (the seed dispersal task). Their careful analysis of the student's work identified the data-driven transitions she made between a mathematical model, a statistical model, and models in other STEM subjects, thus documenting evidence of the intricate connections she was making between the "big ideas" of these subjects. The study described by Steffensen gave some glimpses of how responsible citizenship could be developed as an outcome of integrated STEM education in a 7th grade class investigating the experience of being a refugee. The concept of citizenship is becoming more prominent in studies of critical mathematical modelling and critical mathematics education more generally (e.g., Geiger et al., 2023; Maass et al., 2019), which might encourage researchers to expand their conceptions of STEM learning to take account of its ethical and moral implications.

Teacher Education and Professional Development

A common theme in the M in STEM literature suggests that mathematics is the discipline that teachers believe is the most difficult to incorporate into integrated STEM education (Goos et al., 2023). Perhaps this finding is more indicative of teachers' beliefs about mathematics as a computational tool and lack of experience with integrative pedagogies. Although well-prepared teachers are clearly central to effective implementation of integrated STEM education, there is limited research on how to design STEM teacher education and professional development programs that support integrative practices while preserving the disciplinary integrity of mathematics. Nor do we know enough about what kind of mathematical knowledge teachers need for interdisciplinary STEM teaching: there is even some evidence that typical teacher education programs might not provide the specific mathematical content that underpins STEM tasks in areas such as computational thinking, computer graphics and animation, climate change and sustainable development (Preciado-Babb & Friesen, 2021).

In this volume, Fitzpatrick et al. provide a discerning account of the challenges they faced in preparing pre-service primary school teachers to implement a new curriculum framework that introduced STEM as an integrated learning area. One might assume that primary school teachers would find it easier than secondary school teachers to cope with an integrated STEM subject, because primary school teachers are usually trained as generalists who teach all or most subjects to just one class while secondary school teachers tend to specialise in one or two subjects that they teach to multiple classes and year levels. However, in their study, Fitzpatrick et al. realised that they had over-emphasised the (unfamiliar) engineering aspect of the STEM tasks they designed for the pre-service teachers to implement with a primary school class. They had not given enough attention to explicitly identifying mathematics learning outcomes in the task design or to ensuring that the task criteria and materials stimulated mathematical thinking. This study is important because it highlights these key aspects of STEM task design, as well as the need to consider when and where STEM tasks are located in a mathematics learning sequence so that mathematics is positioned as a site of learning and not just a computational tool. These are the types of insights that could usefully inform future research and practice in mathematics (and STEM) teacher education.

Classroom Implementation and Task Design

This volume offers many examples of engaging STEM tasks and thoughtful commentaries on what makes these tasks mathematically rich. In some countries, researchers and practitioners have worked together in large-scale projects to design integrated STEM tasks that have mathematics as a central focus (see Hwang et al.; Larsen et al.).

However, this is still an area that would benefit from further research to explore classroom implementation principles that enable the mathematical affordances of tasks to be fully realised.

English (this volume) argues that design-based mathematical modelling can enhance young children's problem-solving skills in STEM contexts, and she provides examples of primary school students' responses to two design challenges involving creation of a mini-golf course and a sustainable town. Her detailed and vivid account of classroom implementation also illustrates how the students applied their knowledge of mathematics and made design decisions that demonstrated awareness of practical constraints and user/citizen needs. Although not explicitly highlighted in English's account, there are common design elements in how these modelling tasks were implemented:

- Students were engaged in meaningful preparation activities that provided context and activated prior knowledge (e.g., through discussion of student experience or viewing of YouTube clips).
- The tasks were open ended but had specified design constraints.
- Students worked in groups and were provided with the means of recording their work and documenting their design decisions.
- The researchers (task designers) explicitly identified the mathematical knowledge required for each task.
- Critical reflection was encouraged by asking students to identify changes in their designs throughout the design process, the main ideas they had applied, and the mathematics and science knowledge and processes they had used.
- Student groups were required to rate their products, justify their rating, identify changes to improve the design, and evaluate another group's design with recommendations for improvement.

Many of the task design and implementation features listed above are the same as one would expect for good mathematical problems. But the iterative design thinking element seems to be crucial for promoting critical evaluation and improvement of students' designs and also for reducing their fear of making a mistake (see Fitzpatrick et al., this volume). Thus, the process of design-based mathematical modelling creates an integrative bridge between the STEM disciplines while promoting more positive attitudes towards mathematics as a discipline where mistakes can be a source of learning.

Policy, Structures and Leadership

Curriculum reforms in school mathematics are constantly taking place in countries around the world and in a huge range of cultural, geographic, and economic contexts. Many mathematics educators participate in or even lead these initiatives, and yet surprisingly little in-depth research has been conducted to investigate the processes of *mathematics* curriculum reform or the contextual factors that influence its implementation (Shimizu & Vithal, 2023). This "blind spot" also occurs in STEM education research, where little attention has been given to understanding influences on STEM curriculum policy and the impact of this policy on integrated STEM program development—especially the role of mathematics in such programs (Goos et al., 2023).

Research on macro-level curriculum reforms might seem rather distant from the interests of researchers who work "on the ground" with schools, teachers and students to implement these reforms. But some understanding of the educational policy environment and how this interacts with the organisational environments of schools is necessary in order to influence and advocate for the role of mathematics in integrated STEM education. This is because research and practice in education are situated within, and shaped by, a policy context that establishes curricular goals for an education system (national, state, regional, or district) as well as material and social contexts that influence implementation. Remillard and Heck (2014) identified a wide range of influencing factors within their model of the curriculum policy, design, and enactment system: they include societal needs and values; expectations of the school or local community; teacher knowledge and beliefs; teachers' accesses to resources such as textbooks, exemplary tasks, and instructional technologies; assessment and examination requirements; and school structures and leadership practices. These factors interact with each other in complex ways, and their degree of influence on the curriculum system may be either direct or subtle. So, to complement the innovative work on M in STEM task design, pedagogical approaches, and teacher development showcased in this volume, we also need more studies to investigate conditions that support and sustain such innovation over time (e.g., see Anderson & Tully, 2021). In addition, new lines of research are needed to analyse the coherence and relevance of curriculum reforms that promote integrated STEM, with special emphasis on the role of mathematics (Modeste et al., 2023).

Concluding Comments

There are many reasons why the STEM education landscape is challenging for researchers to navigate. First, the rationale for promoting STEM education came from outside the field of education altogether and instead was aligned with government policies for overcoming social and economic challenges to enhance national competitiveness and innovation (Breiner et al., 2012). Educators themselves usually prefer to emphasise alternative rationales, such as the value of integrated STEM education for developing critical and well-informed citizens. Within mathematics education, scholarly discussions could advance knowledge in the STEM field by exploring mathematics-specific rationales for pursuing integrated STEM education.

Second, while approaches to integrated STEM education are becoming more sophisticated and researchers are developing valuable insights into the contribution of mathematics to students' learning in STEM environments, the status of STEM in the official school curriculum is not always clear. Is it an inquiry-based method for enriching cross-curricular links between the separate STEM disciplines? An extra-curricular addition to a discipline-based curriculum? A separate (perhaps elective) school subject? How do each of these alternatives align with the structure and philosophical orientation of the existing curriculum, and what are the implications for the role and status of mathematics as a stand-alone discipline?

Finally, the practical challenges of designing and implementing mathematically rich integrated STEM tasks and learning experiences remain substantial. However, as the contributions to this volume demonstrate, researchers are working innovatively with teachers to position mathematics as a primary learning goal, a computational or representational tool, and a bridge for integrating the inquiry processes of the STEM disciplines. I look forward to seeing further developments in the field of integrated STEM that are similarly research-informed, curriculum-aware, and practice-oriented.

References

Anderson, J., & Tully, D. (2021). Factors which sustain integrated STEM curriculum approaches in secondary school settings. In D. Anderson, M. Milner-Bolotin, R. Santos, & S. Petrina (Eds.), *Proceedings of the 6th International STEM in Education Conference (STEM 2021)* (pp. 20–26). University of British Columbia, Vancouver, Canada, July 5–9. University of British Columbia. https://doi.org/10.14288/1.0402129

Australian Curriculum, Assessment and Reporting Authority [ACARA] (n.d.). *Numeracy*. https://www.australiancurriculum.edu.au/f-10-curriculum/general-capabilities/numeracy/

Australian Curriculum, Assessment and Reporting Authority [ACARA]. (2023). *Understand this learning area: Mathematics*. https://v9.australiancurriculum.edu.au/teacher-resources/understand-this-learning-area/mathematics.

Baldinger, E., Staats, S., Clarkson, L., Gullickson, E., Norman, F., & Akoto, B. (2020). A review of conceptions of secondary mathematics in integrated STEM education: Returning voice to the silent M. In J. Anderson, & Y. Li (Eds.), *Integrated approaches to STEM education* (pp. 67–90). Springer. https://doi.org/10.1007/978-3-030-52229-2_5

Berlin, D. F., & Lee, H. (2005). Integrating science and mathematics education: Historical analysis. *School Science and Mathematics, 105*(1), 15–24. https://doi.org/10.1111/j.1949-8594.2005.tb18032.x

Breiner, J., Jarkness, S., Johnson, C., & Koehler, C. (2012). What is STEM? A discussion about conceptions of STEM in education and partnerships. *School Science and Mathematics, 112*(1), 3–11. https://doi.org/10.1111/j.1949-8594.2011.00109.x

Doig, B., & Williams, J. (2019). Introduction to interdisciplinary mathematics education. In B. Doig, J. Williams, D. Swanson, R. Borromeo Ferri, & P. Drake (Eds.), *Interdisciplinary mathematics education: The state of the art and beyond* (pp. 1–6). Springer. https://doi.org/10.1007/978-3-030-11066-6_1

Doig, B., Williams, J., Swanson, D., Borromeo Ferri, R., & Drake, P. (Eds.) (2019). *Interdisciplinary mathematics education: The state of the art and beyond*. Springer. https://doi.org/10.1007/978-3-030-11066-6

English, L. (2016). STEM education K-12: Perspectives on integration. *International Journal of STEM Education, 3*(3). https://doi.org/10.1186/s40594-016-0036-1

Fitzallen, N. (2015). STEM education: What does mathematics have to offer? In M. Marshman, V. Geiger, & A. Bennison (Eds.), *Mathematics education in the margins. Proceedings of the 38th annual conference of the Mathematics Education Research Group of Australasia* (pp. 237–244). MERGA.

Gao, X., Li, P., Shen, J., & Sun, H. (2020). Reviewing assessment of student learning in interdisciplinary STEM education. *International Journal of STEM Education, 7*(24). https://doi.org/10.1186/s40594-020-00225-4

Geiger, V., Gal, I., & Graven, M. (2023). The connections between citizenship education and mathematics education. *ZDM Mathematics Education, 55*(5), 923–940. https://doi.org/10.1007/s11858-023-01521-3

Goos, M. & Askin, C. (2005). Towards numeracy across the curriculum: Integrating mathematics and science in the middle years. In R. Zevenbergen (Ed.), *Innovations in numeracy teaching in the middle years* (pp. 125–141). Australian Curriculum Studies Association.

Goos, M., Carreira. S., & Namukasa, I. (2023). Mathematics and interdisciplinary STEM education: Recent developments and future directions. *ZDM—Mathematics Education, 55*, 1199–1217. https://doi.org/10.1007/s11858-023-01533-z

Hobbs, L., & Porsch, R. (Eds.) (2022). *Out-of-field teaching across teaching disciplines and contexts.* Springer.https://doi.org/10.1007/978-981-16-9328-1

Honey, M., Pearson, G., & Schweingruber, A. (2014). *STEM integration in K-12 education: Status, prospects, and an agenda for research.* National Academies Press. https://doi.org/10.17226/18612

Just, J., & Siller, H. (2022). The role of mathematics in STEM secondary classrooms: A systematic literature review. *Education Sciences, 12*, 629. https://doi.org/10.3390/educsci12090629

Kaiser, G., & Sririman, B. (2006). A global survey of international perspectives on modelling in mathematics education. *ZDM—Mathematics Education, 38*(3), 302–310. https://doi.org/10.1007/BF02652813

Leung, A. (2020). Boundary crossing pedagogy in STEM education. *International Journal of STEM Education, 7*, 15. https://doi.org/10.1186/s40594-020-00212-9

Li, Y., & Anderson, J. (2020). Focusing on students and their experiences in and through integrated STEM education. In J. Anderson & Y. Li (Eds.), *Integrated approaches to STEM education* (pp. 177–183). Springer. https://doi.org/10.1007/978-3-030-52229-2_10

Li, Y., Wang, K., Xiao, Y., & Froyd, J. (2020). Research and trends in STEM education: A systematic review of journal publications. *International Journal of STEM Education, 7*(11). https://doi.org/10.1186/s40594-020-00207-6

Maass, K., Geiger, V., Romero Ariza, M., & Goos, M. (2019). The role of mathematics in interdisciplinary STEM education. *ZDM—Mathematics Education, 51*(6), 869–884. https://doi.org/10.1007/s11858-019-01100-5

Ministry of Education, New Zealand. (2023). *What is numeracy about?* https://ncea.education.govt.nz/literacy-and-numeracy/numeracy/learning

Modeste, S., Giminéz, J., Lupiáñez Gómez, J. L., Carvalho e Silva, J., & Nguyen, T. N. (2023). Coherence and relevance relating to mathematics and other disciplines. In Y. Shimizu & R. Vithal (Eds.), *Mathematics curriculum reforms around the world: The 24th ICMI study* (pp. 151–172). Springer. https://doi.org/10.1007/978-3-031-13548-4_11

National Council of Teachers of Mathematics [NCTM]. (2000). *Principles and standards for school mathematics.* NCTM.

Norwegian Directorate for Education and Training. (2012). *Framework for basic skills.* https://www.udir.no/in-english/Framework-for-Basic-Skills/

Organisation for Economic Co-operation and Development [OECD] (2018). *PISA 2022 mathematics framework (draft).* https://www.oecd.org/pisa/publications/pisa-2021-assessment-and-analytical-framework.htm

Preciado-Babb, A., & Friesen, S. (2021). Does school mathematics support STEM education? Exploring specialised mathematics knowledge for STEM education. In D. Anderson, M. Milner-Bolotin, R. Santos, & S. Petrina (Eds.), *Proceedings of the 6th International STEM in Education Conference (STEM 2021)* (pp. 331–336). University of British Columbia, Vancouver, Canada, July 5–9. University of British Columbia. https://doi.org/10.14288/1.0402129

Remillard, J., & Heck, D. (2014). Conceptualizing the curriculum enactment process in mathematics education. *ZDM—Mathematics Education, 46*(5), 705–718. https://doi.org/10.1007/s11858-014-0600-4

Sakurai, J., & Goos, M. (2023). Revisiting tools in numeracy learning: The role of authentic digital tools. *Frontiers in Education, 8*, Article 1291407. https://doi.org/10.3389/feduc.2023.1291407

Shimizu, Y., & Vithal, R. (2023). School mathematics curriculum reforms: Widespread practice but under-researched in mathematics education. In Y. Shimizu & R. Vithal (Eds.), *Mathematics curriculum reforms around the world: The 24th ICMI study* (pp. 3–21). Springer. https://doi.org/10.1007/978-3-031-13548-4_1

Stohlmann, M. (2020). STEM integration for high school mathematics teachers. *Journal of Research in STEM Education, 6*(1), 52–63. https://doi.org/10.51355/jstem.2020.71

Vasquez, J., Sneider, C., & Comer, M. (2013). *STEM lesson essentials, grades 3–8: Integrating science, technology, engineering, and mathematics.* Heinemann.

Venville, G., Wallace, J., Rennie, L., & Malone, J. (1998). The integration of science, mathematics, and technology in a discipline-based culture. *School Science and Mathematics, 98*(6), 294–302. https://doi.org/10.1111/j.1949-8594.1998.tb17424.x

Williams, J., Roth, W. -M., Swanson, D., Doig, B., Groves, S., Omuvwie, M., Borromeo Ferri, R., & Mousoulides, N. (Eds.). (2016). *Interdisciplinary mathematics education: A state of the art.* ICME-13 Topical Surveys. Springer. https://doi.org/10.1007/978-3-319-42267-1

Index

J. Anderson and K. Makar (eds.), *The Contribution of Mathematics to School STEM Education*, https://doi.org/10.1007/978-981-97-2728-5

N

P

SPRINGER NATURE

GPSR Compliance

The European Union's (EU) General Product Safety Regulation (GPSR) is a set of rules that requires consumer products to be safe and our obligations to ensure this.

If you have any concerns about our products, you can contact us on ProductSafety@springernature.com

In case Publisher is established outside the EU, the EU authorized representative is:

Springer Nature Customer Service Center GmbH
Europaplatz 3
69115 Heidelberg, Germany

The manufacturer's authorised representative in the EU is Springer
Nature Customer Service Centre GmbH, Europaplatz 3, 69115 Heidelberg,
Germany. If you have any concerns regarding our products, please
contact ProductSafety@springernature.com

Printed and bound by CPI Group (UK) Ltd, Croydon, CR0 4YY
01/10/2026
02225864-0015